ENZYME INHIBITION AND BIOAPPLICATIONS

Edited by **Rakesh Sharma**

Enzyme Inhibition and Bioapplications
Edited by Rakesh Sharma

Published by InTech

Janeza Trdine 9, 51000 Rijeka, Croatia

Publishing Process Manager Anja Filipovic

Technical Editor Teodora Smiljanic

Cover Designer InTech Design Team

Additional hard copies can be obtained from orders@intechopen.com

Enzyme Inhibition and Bioapplications, Edited by Rakesh Sharma
p. cm.
ISBN 978-953-51-0585-5

For EU product safety concerns:
IN TECH d.o.o., Prolaz Marije Krucifikse Kozulić 3, 51000 Rijeka, Croatia,
info@intechopen.com or visit our website at intechopen.com.

Meet the editor

Dr Rakesh Sharma, Ph.D Biochemistry and Ph.D Magnetic Resonance Spectroscopy, is a professor teaching nanotechnology at Amity University. Since 2004 he does research in Biomedical Engineering, National High Magnetic Field Lab. Florida and Center of Nanobiotechnology at Florida State University, Tallahassee. Dr Sharma is affiliated with the department of Medicine, Columbia University, New York since 2000. His research interest is development of enzyme assays.

Contents

Preface

Enzyme is a protein molecule exhibiting specific activity and binding affinity with its substrate molecule to complete enzyme reaction or biocatalytic reaction. Substrate analogues can inhibit the enzyme reaction and act as enzyme inhibitor. Enzyme inhibition (**Enz-ai-m ie-ni-hi-bi-son**) means reducing or blocking an enzyme action on specific location of enzyme active site by specific substrate or analogue so called enzyme inhibitor. In modern times, most of the pharmaceutical as well as nutriceutical compounds are marketed as enzyme inhibitors and such inhibitors exhibit their specific action in enzyme inhibition inside cells, bacteria, virus, animal plants and human body. The action of enzyme inhibitors in drug discovery has become a fundamental approach to pharmacology at any pharmaceutical industry, university research lab or drug research center. The present issue has been compiled from various data sources with aim of incorporating a wide range of basic concepts and applied enzyme inhibition evaluation methods in drug discovery. It is aimed at those who are embarking on drug discovery research projects, immobilized enzyme solid state devices as well as relatively experienced pharmacologists, biochemists and pharmacy scientists who might wish to develop their experiments further to the advanced level. While it is not possible to detail and include every possible technique related with enzyme inhibition evaluation in drug design by using specific inhibitors at specific metabolic mechanism(s), the present issue attempts to provide working tips with examples and analysis relevant to a wide range of more commonly available enzyme inhibition techniques.

The methods and concepts described in this book are aimed at giving the reader a glimpse of some existing enzyme inhibition studies and also methods with context of each enzyme inhibition method applied for, as well as providing some basis of familiarizing oneself with these biochemical methods. While enzyme inhibition has been used as major approach in drug design in the research and industry over last two decades, it was only later part of 20th century that it has become a major part of so many applications in biochemical engineering, biomedical engineering of miniatured clinical chemistry devices in microbiological, bacteriological, immunological, hormonal testing, nanotechnology, physiological monitoring in health science, plant science and environment research work. This is, at least in part, due to the continued development of new solid state polymer platform, pure enzymes and specific inhibitors available, better understanding of enzyme inhibition mechanisms and precise detection methods, new awareness of drug discovery and design, with scanning and monitoring accessories. Thanks to the continued joint efforts of governmental, industrial and academic institutions globally to promote the need of new generations of drugs or enzyme inhibitors and new mechanisms of enzyme inhibition. Regardless of the enzymes or drugs and their brands that are used, one should always be able to understand and justify the use of right inhibitor or drug action on enzyme of right choice to drug design for specific study. With this aim, different approaches of enzyme inhibition methods are presented in separate chapters on the use of different enzyme inhibitors. For learners, basic concepts, mechanistic issues, limitations in drug testing, skepticism in enzyme inhibition approach and drug variability due to non-specific analogues of enzyme inhibitors or substrates is presented with a working enzyme inhibition protocols for drug design and analysis of their inhibitory action.

In chapter 1, the authors have introduced the basic concepts on enzyme, enzyme reaction, inhibitors and types of inhibition with a handful established applications in drug discovery, immobilized enzyme engineering, and biosensing. Major three basic types of enzyme inhibition kinetics is highlighted for beginners. Examples of substrate analogues and their enzyme inhibition behaviour are illustrated with color schemes. Application of immobilized enzyme on chips for environment monitoring and biosensor development is quite intriguing for engineers, scientists and industrialists. In chapter 2, authors overviewed isolates from *Pepermia amplexicaulis* and *Spathelia sorbifolia* plants have examined for CYP inhibitions to exhibit antiprotozoal, chemopreventive and anti-cancer activity. Other plant sources of tea and fruits of *Rhytidophyllum tomentosa, Psidium guajava, Symphytium officinale, Momordica charantia* showed inhibition properties of these teas against a panel of CYP450 enzymes in order to assess the potential for drug interactions with co-medicated pharmaceuticals. The chapter highlights the potential of a few natural products emanating from the Caribbean: chromene amides isolated from *Amyris plumieri*, quassinoids isolated from *Picrasma excelsa*, anhydrosorbifolin isolated from *Spathelia sorbifolia* and 5-Hydroxy-2,7-dimethyl-8-(3-methyl-but-2-enyl)-2-(4-methyl-penta-1,3-dienyl)-chroman-6-carboxylic acid isolated from *Peperomia amplexicaulis.* New information is presented on bioactive screens against CYP enzymes in the presence of five aqueous infusions of popularly used herbs; *Rhytidophyllum tomentosa, Psidium guajava, Symphytium officinale, Momordica charantia* and *Picrasma excelsa. Authors emphasized search of* CYPs 1A1 and 1B1 activity inhibitors as chemoprotectors such as CA1 and quassin for CYP 1A1; and anhydrosorbifolin and 5-Hydroxy-2,7-dimethyl-8-(3-methyl-but-2-enyl)-2-(4-methyl-penta-1,3-dienyl)-chroman-6-carboxylic acid for CYP 1B1 in search of safer herbal remedies in co-adiministration of medicines in cancer prevention. In chapter 3, authors described pharmacomodulation in metalloprotease enzymes by inhibitors. Authors aim to achieve better metallloprotease (MMP) enzyme inhibitors with good MMP-2 selectivities to increase hydrophobicity and rigidity with the dehydro and didehydro analogues and synthesized (analogue 2a-d and 3a-h). Authors displayed sevral examples of inhibitors including pharmacomodulation of galardin®, a powerful broad spectrum MMPI, hydrazide and sulfonylhydrazide-type functions as potential Zinc Binding Compounds for gelatinase enzyme inhibition, double-headed elastase-MMP inhibitor (d-hEMI) able to block elastase and MMP activities, with display of structural-functional models in the last sections. Authors speculate to develop hybrid nanoprobes build from MMPI and fluorescent nanocrystal quantum dots (QDs). Design and chemical synthesis of derivatives of galardin®, selective inhibitors of MMP-2, tagging with QDs. In hope of photo- and chemical stability of QDs, it is possible that apporoach will enable long-term spatiotemporal tracking of the process of inhibition of MMP-2 enzymes to offer better understanding of physiological process of invasion of melanoma. In chapter 4, authors introduce the readers with nonenzymatic glycation of aminotransferases and modulation of these enzymes as model protein. Further authors emphasized the significance of advanced glycation end-products with introduction of structure, targeting with aim to evaluate potential antiglycation activity of two mitochondrial antioxidants, α-phenyl *N-tert*-butyl nitrone (PBN) and *N-tert*-butyl hydroxylamine (NtBHA). In next section, authors established the effect of natural compound on fructose induced AST glycation, basic techniques to determine primary amino groups, enzyme molecular charge of AST modulated by fructose by electrophoresis. Authors attempted to search the compounds with antioxidant and potential antiglycating activities in an illustrated description with a perspective of their use as remedies against diabetic complications. In chapter 5, author introduces the readers with basic mechanism of reduced nitric oxide synthase gene expression and applications of nitric oxide synthase (NOS) inhibition and approaches of nitric oxide (NO) content by using less known multimodal imaging. Nitric oxide imaging techniques utilize mapping NO in tissue using NO specific imaging contrast agents sensitive to fluorescence, magnetic resonance and electron spin resonance. A handful account is presented on NOS expression inhibitors, possibility of dithiacarbamates, paramagnetic complexes for bioimaging of NO. For learners, MRI protocol is given as example. In the light of recent developments in multimodal bioimaging of NO/NOS expression by bioluminescence, fluorescence techniques, a handful information is given in cells, animals, plants and humans body in different diseases including endothelial cell injury, apoptosis, renal, liver, lung, muscle, brain, inflammation, bones, retina with emphasis on multimodal techniques of calcium, ion channels, iron bound complexes. Author speculated the future possibility of NO/NOS bioimaging by combined radioimaging techniques and it remains to see if real-time imaging becomes routine modality. In chapter 6, transcriptional bursting in E.coli tryptophan operon is introduced to explain stochastic dynamics of the tryptophan synthase enzyme feedback inhibition regulatory mechanism at various levels of Trp operon genes. Authors have described in chapter various components of trp operon structure, model development, parameter estimation, and numerical methods with results on stochastic stimulations to evidence the least response time with inhibition-less strain. Authors further investigated that the repressor-operator interaction stimulates transcriptional bursting which shows several dynamic effects on transcriptional bursting possibly by a feedback enzyme-inhibition regulatory mechanism. In chapter 7, a hepatocellular dysfunction criteria with possible mechanism of hepatocellular dysfunction is proposed to evaluate liver degeneration due to amoebic infection and hepatic recovery by nitroimidazole administration. The main focus is the evaluation of regulatory enzyme inhibition effects on major energy metabolism. Liver regeneration is an excitement to reverse the process of enzyme inhibition in defense. Author introduces enzymes in liver abscess, programmed cell death, hepatocellular criteria of hypoxia, loss of metabolic integrity with low NADH/ATP, parenchymal lysosomal enzyme inhibition illustrated by regulatory behavior changes of glucokinase, phiosphofructokinase, pyruvate kinase, phosphodiesterase, respiratory burst, superoxide dismutase, cytochrome oxidase, adenylate cyclase, inhibition of drug metabolizing microsomal enzymes, phagocytosis, and DNA synthesis. A model is proposed on proteolysis in isolated lysosomes to establish the mechanism of degradation and proteolysis inhibition. Further, role of lysosomal enzymes is proposed for liver regeneration and recovery after nitroimidazole treatment. The emerging state of art is presented on role of enzyme inhibition in liver transplantation and tissue engineering. Still, art of liver regeneration is not free from several challenges such as lack of ideal stimulator of hepatic recovery. In attempt to stimulate the liver regeneration by nitroimidazole, it was explored that nitroimidazole is becoming a multi-organ therapeutic drug for tumor, tuberculosis, myocardial infarction, hypoxia and diagnostic tool in imaging, chemosensor, and tissue engineering in addition to hepatic recovery. In chapter 8, authors describe reversal inhibition of tyrosine protein phosphatases and explore it by redox reactions. Investigators describe a structural mechanistic action of ROS in the tyrosine phosphatase enzymatic activity to demonstrate how it interacts with their target molecules; the reversible regulation of this enzyme by oxidants and antioxidants; and the major consequences of this tightly controlled mechanism on cell signalling. Authors describe tyrosine phosphatase family, catalytic domain of tyrosine phosphatase, sources of redox radical, reactive nitrogen, antioxidant mechanisms, redox inhibition, and mechanism of cellular signalling. Authors further emphasize an urgent need of a mechanism(s) available on specific tyrosine phosphatase action to identify the sources of reactive radical species formation. In chapter 9, a novel technique 'response surface methodology' is proposed using lipase inhibition by Novozyme 435-catalyse process to produce fatty acid methyl ester from waste frying oil. The issue of preserving lipase in the reaction is addressed without any loss of enzyme activity due to methanol in the reactor. The chapter describes high yield lipase catalysed process, optimization conditions, properties of feedstock, economic biodiesel production with lipase reutilization, and kinetics of tert-butanol system. Authors highlighted several issues such as short reaction time, flexible enzyme process, and utilization of low cost raw materials as environment-friendly chemical catalysis in economic enzyme process. In chapter 10, author introduces urease inhibition, chemical reactions of urease organic, metal inhibitors, and urease as virulence factor for the urinary tract infections, gastrointestinal infection by inhibition of urease. Author describes urease inhibition mechanism, kinetics, action of inhibition, potential inhibitors, and structure activity relations. Author speculates the potentials of inhibitors in biomedical science with more applications in pharmaceutical, agriculture, environmental research such as sensors, adsorbent and commercial devices.

In last three decades, science of enzyme inhibition has grown by leaps and bounds in the field of biochemistry, medicine and pharmacology. With tremendous developments in technology, now enzymes are valuable research tools as biomarkers, biosensors, detection miniature biodevices, point-of-care pocket monitors in emergency care. Use of enzyme inhibitors in valuable drug testing has rocked the business in pharmaceutical industries and biotechnology industries in search of suitable and effective drug. Present time, science has grown mainly in three directions.

First, precision and accuracy in measurement of enzyme reaction has made possible to measure and localize the site of enzyme inhibition at the level of **picoscale** to down side of 10^{-12} meters. For example, enzyme reaction of caspases and metalloproteases can be detected and visualized by **vision®** at the site of tumor under powerful microscope. In future, it is speculated that **picotechnology** will replace nanotechnology in terms of more detectable applications at picoscale instead of nanoscale with better understanding of enzyme-substrate or inhibitor complexes. Further, solid phase enzyme devices or immobilized enzyme fixed on polymeric base have emerged as detection tools in agriculture, environment, healthcare, and biotechnology industries.

Second, enzyme inhibitors do behave very precise at the exact location of enzyme active site in the 'lock-key' configuration. The 3D structure and conformation of active site permits very powerful enzyme inhibitors or substrate analogues to bind with enzyme molecule to make desirable product and accomplish the goal as pharmaceutical drug. Many drug classes and variants have been explored and developed for almost each and every disease, infection and immunity. Better sources of structural-functional relationship enzyme-inhibitor and database available have changed the scenario. Now it is a fashion of new generation medication available based on new drug discovery almost every year. It is all possible due to new developments in drug testing or enzyme inhibitor development in pre-clinical and clinical use.

Third, enzyme inhibitors play a major role in enzyme engineering and many other related fields. Design of portable devices using enzyme reaction as detection mechanism need a polymer base with layer of enzyme molecules fixed on its surface and placed in device at the point-of-detection. Such miniature detection devices are becoming popular in detection of microbial, bacterial contamination, pollutes, organic, chemical toxicity, allergens, harmful gases, paints, vapors, bioconversion recovery, molecular biology products, biopharmaceutical products, hormones, agroproducts, nutraceutical products, and list is growing in use at clinical chemistry, microbiology, toxicology, pathology labs and quick diagnostic tools in hospitals.

I acknowledge Anja Filipovic for the guidance in book production, the continuous efforts made by Miss Nisha Keval to assist me in whole editorial work, in shaping the chapters at various points of time in presentable form, and Professor P.V.Pannirselvam from Brazil, Federal University Riogrande Norte, Natal, Brazil in chapter 1.

Finally, I thank all authors and co-authors for their chapter contributions and for selecting the right topics suitable for this book.

I hope that readers will enjoy reading the book and that it will serve the purpose of basic concepts on enzyme inhibition with advanced applications in the field of applied enzyme inhibition in pharmaceutical, biotech industries and academic research.

Rakesh Sharma, Ph.D
MS-Ph.D, ABR II
Professor (Nanotechnology)
Amity University,
India

Research Professor, Center of Nano-Biotechnology,
Florida State University, Tallahassee, FL,
USA

Section 1

Basic Concepts

1

Enzyme Inhibition: Mechanisms and Scope

Rakesh Sharma[1,2,3]
[1]Center of Nanomagnetics Biotechnology, Florida State University, Tallahassee, FL
[2]Innovations and Solutions Inc. USA, Tallahassee, FL
[3]Amity University, NOIDA, UP
[1,2]USA
[3]India

1. Introduction

Enzyme is a protein molecule acting as catalyst in enzyme reaction. Enzyme inhibition is a science of enzyme-substrate reaction influenced by the presence of any organic chemical or inorganic metal or biosynthetic compound due to their covalent or non-covalent interactions with enzyme active site. It is well known that all these inhibitors follow same rule to interplay in enzyme reaction. Present chapter introduces beginners with basic tenets of classic presumptions of enzyme inhibition, types of enzyme inhibitors, different models of enzyme inhibition with established examples cited in literature, and scientific basis of emerging immobilized enzyme technology in different applications. In the end, limitations of using classic presumptions and variants of enzyme inhibition are highlighted with new challenges to achieve best results. Present time, best approach is 'customize new technology with detailed analysis to make it highly efficient' in both drug discovery and enzyme biosensor industry. However, other applications are described in following chapters on pesticides, herbicides.

2. What are enzyme inhibitors?

The enzyme inhibitors are low molecular weight chemical compounds. They can reduce or completely inhibit the enzyme catalytic activity either reversibly or permanently (irreversibly). Inhibitor can modify one amino acid, or several side chain(s) required in enzyme catalytic activity. To protect enzyme catalytic site from any change, ligand binds with critical side chain in enzyme. Safely, chemical modification can be done to test inhibitor for any drug value.

In drug discovery, several drug analogues are chosen and/or designed to inhibit specific enzymes. However, detoxification or reduced toxic effect of many antitoxins is also accomplished mainly due to their enzyme inhibitory action. Therefore, studying the aforementioned enzyme kinetics and structure-function relationship is vital to understand the kinetics of enzyme inhibition that in turn is fundamental to the modern design of pharmaceuticals in industries [Sami et al. 2011]. Enzyme inhibition kinetics behavior and inhibitor structure-function relationship with enzyme active site clarify the mechanisms of enzyme inhibition action and physiological regulation of metabolic enzymes as evidenced in following chapters in this book. Some notable classic examples are: drug and toxin action and/or drug design for therapeutic uses e.g., iodoacetamide deactivates cys amino acid in enzyme side chain; methotrexate in cancer chemotherapy through semi-selectively inhibit DNA synthesis of malignant cells; aspirin inhibits the synthesis of the proinflammatory prostaglandins; sulfa drugs inhibit the folic acid synthesis essential for growth of pathogenic bacteria and so many other drugs. Many life-threatening poisons, e.g., cyanide, carbon monoxide and polychlorinated biphenols are all enzyme inhibitors.

Conceptually, enzyme inhibitors are classified into two types: non-specific inhibitors and specific inhibitors.

The enzyme inhibition reactions follow a set of rules as mentioned in following rules. Presently, computer based enzyme kinetics data analysis softwares are developed using following basic presumptions.

1. Enzyme interacts with substrate in 1:1 ratio at active site to catalyze the reaction.
2. Enzyme binds with substrate at active site in the form of a lock-key 3D arrangement for induced fit.
3. Inhibitor active groups compete with substrate active groups and/or active groups at enzyme allosteric catalytic site in a synergistic manner or first cum first preference (competition) to make enzyme-inhibitor-substrate/enzyme-substrate/enzyme-inhibitor complexes.
4. Enzyme-inhibitor-substrate complex formation depends on active free energy loss and thermodynamic principles.
5. Enzyme and substrate or inhibitors react with each other as active masses and reaction progresses in kinetic manner of forward or backward reaction.
6. Kinetic nature of inhibitor or substrate binding with enzyme is expressed as kinetic constants of a catalytic reaction.
7. Enzyme reaction(s) are highly depend on physiological conditions such as pH, temperature, concentration of reactants, reaction period to determine the rate of reaction.
8. Substrate and inhibitor molecules arrange over enzyme active site on specific sub unit(s) in 3D manner. As a result enzyme-substrate-inhibitor exhibit binding rates depend on allosteric sites or subunit-subunit homotropic or heterotropic interactions.
9. Intermolecular forces between enzyme subunits, substrate or inhibitor active group interactions, physical properties of binding nature: electrophilic, hydrophilic, nucleophilic and metalloprotein nature; hydrogen bonding affect the overall enzyme reaction rates and mode of inhibition (3D orientation of inhibitor molecule on enzyme active site).

Other factors are also significant in determining enzyme inhibition reaction as described in each individual inhibitor in following sections. For basic principles of enzyme units (apoenzyme, holoenzyme, co-factor, co-enzyme) in enzyme catalysis, active energy loss, Michaelis-Menton Equations, LeChatelier's principle, Lineweaber-Burk and semi-log plots, apparent and actual plots, readers are requested to read text books [Schnell et al. 2003, Nelson, et al. 2008, Jakobowski 2010a, Strayer et al. 2011]. Our focus is enzyme inhibition mechanisms with examples in following description. For multisubstrate enzymes, ping-pong mechanism, allosteric mechanisms, and diffusion kinetics, readers are requested to read original papers [Pryciak 2008, Bashor 2008, Jakobowski 2010b]

These inhibitors may act in reversible or irreversible manner. Non-specific irreversible non-competitive inhibitors include all protein denaturating factors (physical and chemical denaturation factors). The specific inhibitors attack a specific component of the holoenzyme system. The action depends on increased amount of substrate or by other means of physiological conditions, toxins. Specific inhibitors can be described in several forms including; *1) coenzyme inhibitors*: e.g., cyanide, hydrazine and hydroxylamine that inhibit pyridoxal phosphate, and, dicumarol that is a competitive antagonist for vitamin K; *2) inhibitors of specific ion cofactor*: e.g., fluoride that chelates Mg^{2+} of enolase enzyme; *3) prosthetic group inhibitors*: e.g., cyanide that inhibits the heme prosthetic group of cytochrome oxidase; and, *4) apoenzyme inhibitors* that attack the apoenzyme component of the holoenzyme; 5) *physiological modulators* of reaction pH and temperature that denature the enzyme catalytic site.

The apoenzyme inhibitors are of two types; i) *Reversible inhibitors;* their inhibitory action is reversible because they make reversible association with the enzyme, and, ii) *Irreversible inhibitors;* because they make inactivating irreversible covalent modification of an essential residue of the enzyme. Apoenzyme inhibitors show effect on K_m and V_{max}. The reversible apoenzyme inhibitors are also called metabolic antagonists. They are of three subtypes; *a) competitive, b) uncompetitive* and *c) non-competitive* or mixed type. For example: enzyme inhibitors are used in drug design.

Discovery of useful new enzyme inhibitors used to be done by trial and error through screening a huge library of compounds against a target enzyme at allosteric catalytic site. This approach is still in use for compounds with combinatorial chemistry and high-throughput screening technology as described in following description based on recent concepts [El-Metwally et al. 2010]. However, rational drug design as an alternative approach uses the three-dimensional structure of an enzyme's active site or transition-state conformation to predict which molecules might be ideal inhibitors as given an example of urease in chapter 11 in this book. 3D-structure shortens the long screening list towards a right set of novel inhibitor which kinetically characterizes and allows specific structural changes in amino acids of catalytic site chain to optimize inhibitor-enzyme binding. Alternatively, molecular docking and molecular mechanics are computer-based methods that predict the affinity of an inhibitor for an enzyme. In following description, a glimpse of these mechanisms is given on different types of inhibitors based on recent classic book [El-Metwally et al. 2010]. Readers are requested to read other classic details from advanced text books [Dixon and Webb, 1979].

3. Irreversible inhibition

The irreversible apoenzyme inhibitors have no structural relationship to the substrate and bind covalently. They also bind stable non-covalently with the active site of the enzyme or destroy an essential functional group of active site. So, irreversible inhibitors are used to identify functional groups of the enzyme active sites at which location they bind. Although inhibitors have limited therapeutic applications because they are usually act as poisons. A subset of irreversible inhibitors called *suicide irreversible inhibitors,* are relatively inactive compounds. They get activated upon binding with the active site of a specific enzyme. After such binding, the suicide irreversible inhibitor is activated by the first few intermediary steps of the biochemical reaction - like the normal substrate. However, it does not release any product because of its irreversible binding at the enzyme active site. Inhibitors make use of the normal enzyme reaction mechanism to get activated and subsequently inactivate the enzyme. Due to this very nature, suicide irreversible inhibitors are also called *mechanism-based inactivators* or *transition state analog inhibitors.* Thus, inhibitor exploits the transition state stabilizing effect of the enzyme, resulting in a better binding affinity (lower K_i) than substrate-based designs. An example of such a transition state inhibitor is active form of the antiviral drug oseltamivir (Tamiflu; see Figure 1); this drug mimics the planar nature of the ring oxonium ion in the reaction of the viral enzyme neuraminidase [El-Metwally et al. 2010]. After drug activation in the liver, the drug replaces sialic acid as the normal substrate found on the surface proteins of normal host cells. It prevents the release of new viral particles from infected cells. It has been used to treat and prevent Influenza virus A and Influenza virus B infections. Most of such inhibitors are classified as tight-binding competitive inhibitors in other references of enzymes. However, their reaction kinetics is essentially irreversible.

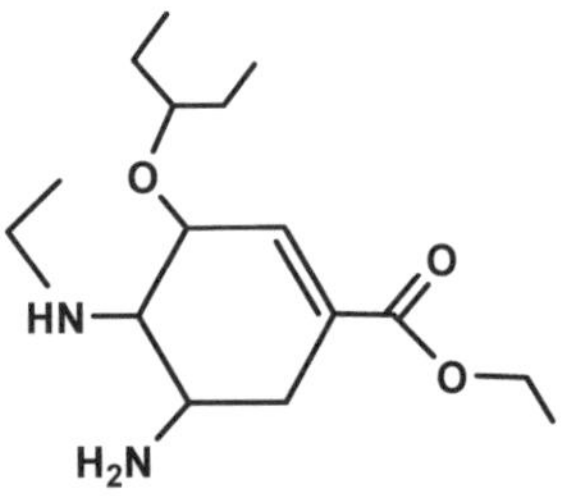

Fig. 1. The transition state analog oseltamivir - the viral neuraminidase inhibitor.

The present art of drug discovery and design of new drugs is based on suicidal irreversible inhibitors. Chemicals are synthesized based on knowledge of 3D conformation of substrate-active site binding at specific binding rates in presence of co-factors, co-enzyme (enzyme reaction mechanisms) to inhibit at specific enzyme active site with minimal side-effects due to its non-specific binding nature. Transition state analogs are extremely potent and specific inhibitors of enzymes because they have higher affinity and stronger binding to the active site of the target enzyme than the natural substrates or products. However, exact design of drugs that precisely mimic the transition state is a challenge because of unstable structure of transition state in the free-state. Prodrugs undergo initial reaction(s) to form an overall electrostatic and three-dimensional intermediate transition state complex form with close similarity to that of the substrate. These prodrugs serve as guideline for drug development to form transition state suitable for stable modification; or, using the transition state analog to design a complementary catalytic antibody; called *Abzyme.* Example: Abzymes are used in catalytic antibodies and ribozymes in catalytic ribosomes [El-Metwally et al. 2010].

- Abzymes are antibodies generated against analogs of the transition state complex of a specific chemical. The arrangement of amino acid side chains at the abzyme variable regions is similar to the active site of the enzyme in the transition state and work as artificial enzymes. For example, an abzyme was developed against analogs of the transition state complex of cocaine esterase, the enzyme that degrades cocaine in the body [El-Metwally et al. 2010]. Thus, this abzyme has similar esterase activity that is used as injection drug to rapidly destroy cocaine in the blood of addicted individuals to decreasing their dependence on it.

- Thrombin inhibition is common in saliva of leeches and other blood-sucking organisms. They contain the anticoagulant hirudin that irreversibly inhibits thrombin, and, to regain thrombin action synthesis of new thrombin molecules is required. This made it unsafe as an anticoagulation drug. However, based on hirudin structure, rational drug design synthesized 20-amino acids peptide known as bivalirudin that is safe for long-term use because of its reversible effects on thrombin; despite its high binding affinity and specificity for thrombin.
- Ornithine decarboxylase by difluoromethylornithine is used to treat African trypanosomiasis (sleeping sickness). The enzyme initially decarboxylates difluoromethylornithine instead of ornithine and releases a fluorine atom, leaving the rest of the molecule as a highly electrophilic conjugated imine. The later reacts with either a cysteine or lysine residue in the active site to irreversibly inactivate the enzyme.
- Inhibition of thymidylate synthase by fluoro-dUMP. Imidazole antimycotic drugs are examples of such group that inhibit several subtypes of cytochrome P450 [Sharma, 1990]. The mechanisms of toxicities and antidotes of irreversible inhibitors are of medical pathological importance. Because of the irreversible inactivation of the enzyme, irreversible inhibition is of long duration in the biological system because reversal of their action requires synthesis of new enzyme molecules at the enzyme gene-transcription-translation level.
- Inhibition of acetylcholine esterase (ACE) by diisopropylfluorophosphate (DPFP), the ancestor of current organophosphorus nerve gases (e.g., Sarin and Tabun) and other organophosphorus toxins (e.g., the insecticides Malathion and Parathion and chlorpyrifos). ACE hydrolyzes the acetylcholine into acetate and choline to terminate the transmission of the neural signal form the neuromuscular excitatory acetylcholine presynaptic cell to somatic neuromuscular junction (see Figure 2). DPFP as a potent neurotoxin inhibits ACE and acetylcholine hydrolysis. Failure of hydrolysis leads to persistent acetylcholine excitatory state and improper vital function particularly respiratory muscles that may lead to suffocation; with a lethal dose of less than 100 mg. DPFP inhibits other enzymes with the reactive serine residue at the active site, e.g., serine proteases such as trypsin and chymotrypsin, but the inhibition is not as lethal as that of acetylcholine esterase. Similar to DPFP, malaoxon the toxic reactive derivative from Malathion (after its metabolism by the liver) binds initially reversibly and then irreversibly (after dealkylation of the inhibitor) to the active site serine and inactivates ACE and other enzymes. Lethal doses of oral Malathion are estimated at 1 g/kg of body weight for humans.
- Inhibition of ACE by these poisons leads to accumulation of acetylcholine that over-stimulates the autonomic nervous system (including heart, blood vessels, and glands), thereby accounting for the poisoning symptoms of vomiting, abdominal cramps, nausea, salivation, and sweating. Acetylcholine is also a neurotransmitter for the somatic motor nervous system, where its accumulation resulted in poisoning symptom of involuntary muscle twitching (muscle fasciculation), convulsions, respiratory failure and coma. Intoxication of Malathion is treated by the antidote drug Oxime that reactivates the acetylcholine esterase and by intravenous injection of the anticholinergic (antimuscarinic) drug atropine to antagonize the action of the excessive amounts of acetylcholine [El-Metwally et al. 2010].

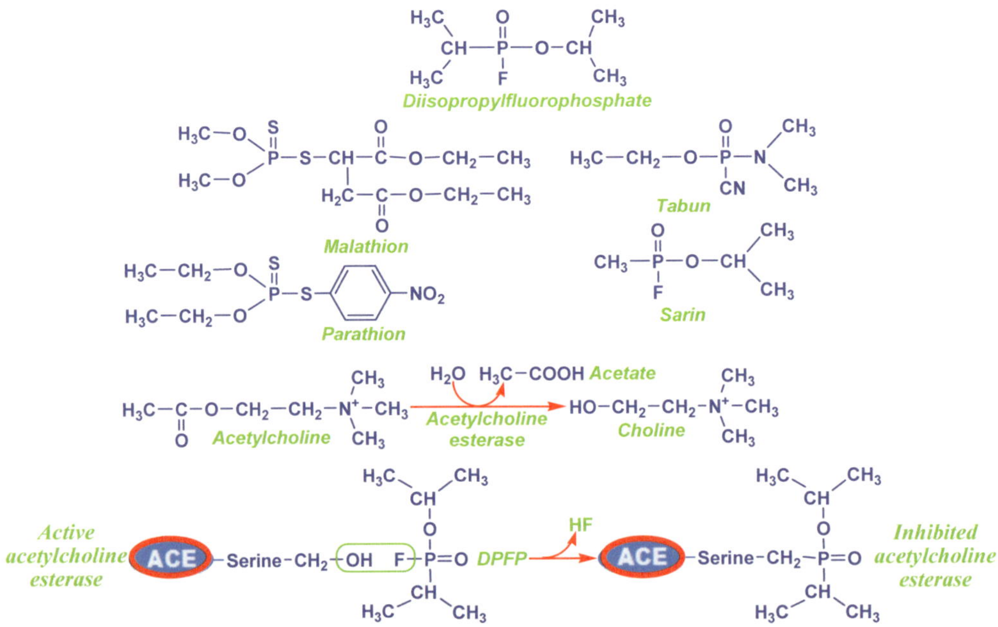

Fig. 2. Organophosphorus compounds and the suicidal irreversible mechanism-based inhibition of the enzyme acetylcholine esterase by diisopropylfluorophosphate. Malathion and parathion are organophosphorus insecticides. The nerve gases Tabun and Sarin are other organophosphorus compounds.

Another example of irreversible inhibition is iodoacetate inhibition of the glycolytic glyceraldehyde-3-phosphate dehydrogenase (GPD). Iodoacetate is a sulfhydryl compound that covalently alkylates and blocks the sulfhydryl group at the active site of the enzyme. Iodoacetate also inhibits other enzymes with -SH at the active site (Figure 3).

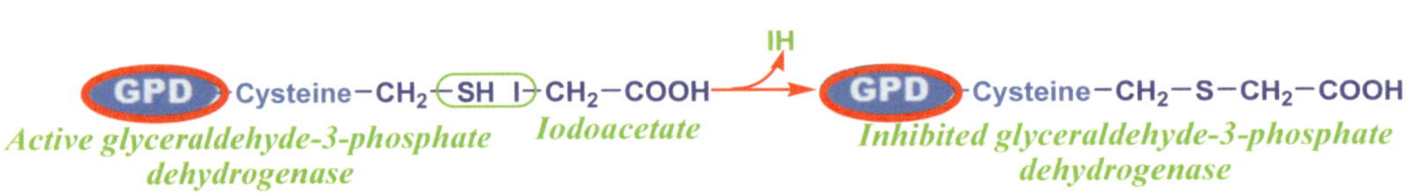

Fig. 3. The suicidal irreversible mechanism-based inhibition of the enzyme glyceraldehyde-3-phosphate dehydrogenase by iodoacetate.

- Allopurinol - the anti-gout drug - is a suicidal irreversible mechanism-based inhibitor of the enzyme xanthine oxidase that works as oxidase or dehydrogenase. The enzyme commits suicide by initial activating allopurinol into a transition state analog - oxypurinol - that bind very tightly to molybdenum-sulfide (Mo-S) complex at the active site (Figure 4). This enzyme accounts for the human dietary requirement for the trace mineral molybdenum. The molybdenum-sulfide (Mo-S) complex binds the substrates and transfers the electrons required for the oxidation reactions.

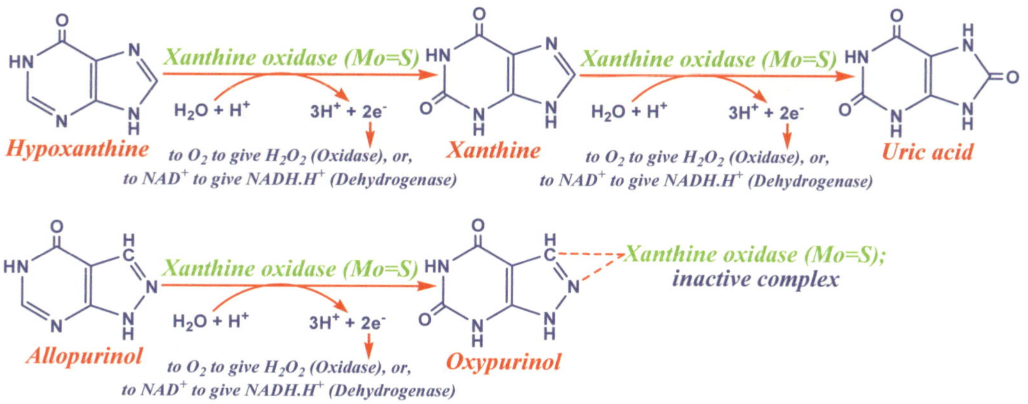

Fig. 4. The suicidal irreversible mechanism-based inhibition of the enzyme xanthine oxidase by allopurinol.

- Guanosine analogue antiviral drug aciclovir - acycloguanosine (2-amino-9-((2-hydroxyethoxy)methyl)-1H-purin-6(9H)-one), as one of the most commonly-used antiviral drugs, it is primarily used for the treatment of herpes simplex and herpes zoster (shingles) viral infections. Aciclovir (see Figure 5) started a new era in antiviral therapy, as it is extremely selective and low in cytotoxicity. Aciclovir as a prodrug differs from previous nucleoside analogues in that it contains only a partial nucleoside structure: the sugar ring is replaced by an open-chain structure. It is selectively converted into acyclo-guanosine monophosphate (acyclo-GMP) by viral thymidine kinase, which is far more effective (3000 times) in phosphorylation than cellular thymidine kinase. Subsequently, the monophosphate form is further phosphorylated into the active triphosphate form, acyclo-guanosine triphosphate (acyclo-GTP), by cellular kinases. Acyclo-GTP is a very potent inhibitor of viral DNA polymerase; it has approximately 100 times greater affinity for viral than cellular polymerase. As a substrate, acyclo-GTP is incorporated into viral DNA, resulting in chain termination. Acyclo-GTP is fairly rapidly metabolized within the cell, possibly by cellular phosphatases.

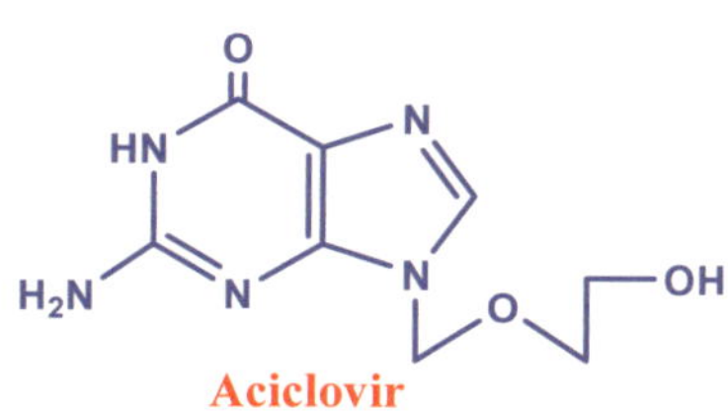

Fig. 5. Aciclovir; the prodrug for the suicidal irreversible inhibition of the viral DNA polymerase.

- *The antibiotic penicillin* is another transition state analog suicidal inhibitor that binds irreversibly covalently to serine at the active site of the bacterial enzyme glycopeptide transpeptidase. The enzyme is a serine protease required for synthesis of the bacterial cell wall and is essential for bacterial growth and survival. It normally cleaves the peptide bond between two D-alanine residues in a polypeptide. Penicillin structure contains a strained peptide bond within the β-lactam ring that resembles the transition state of the normal cleavage reaction, and thus penicillin binds very readily to the enzyme active site. The partial reaction to cleave the imitating penicillin peptide bond activates penicillin to bind irreversibly covalently to the active site serine (Figure 6).

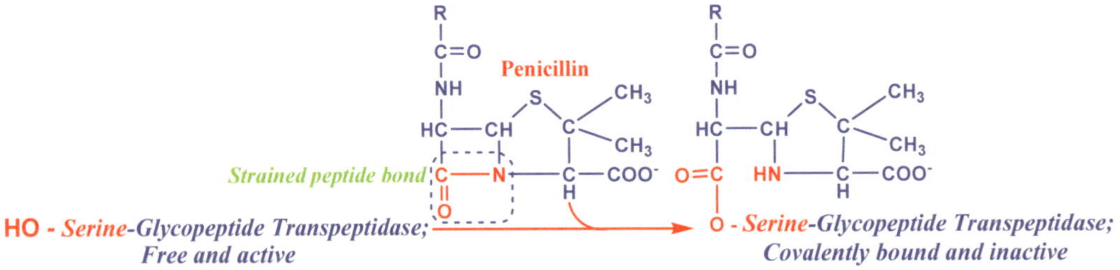

Fig. 6. The suicidal irreversible mechanism-based inhibition of the bacterial enzyme glycopeptide transpeptidase by the antibiotic penicillin.

- *Aspirin (acetylsalicylic acid)* provides an example of a pharmacologic drug that exerts its effect through the covalent acetylation of an active site serine in the enzyme cyclooxygenase (prostaglandin endoperoxide synthase). Aspirin resembles a portion of the prostaglandin precursor that is a physiologic substrate for the enzyme.
- *Heavy metal toxicity* is caused by tight binding of a metal such as mercury, lead, aluminum, or iron, to a functional group at the active site of an enzyme. At high concentration of the toxin, heavy metals are relatively nonspecific for the enzymes they inhibit and inhibit a large number of enzymes. For example, it is impossible to specify which particular enzyme is implicated in mercury toxicity that binds reactive -SH groups at the active sites. Lead developmental and neurologic toxicity is caused by its ability to replace the normal functional metal in target enzymes; particularly Ca^{2+} in important enzymes, e.g., Ca^{2+}-calmodulin and protein kinase C. Because of their irreversible effect, heavy metals are routinely use as fixatives in histological preparations.

Kinetically, the irreversible inhibitors decrease the concentration of active enzyme and in turn decrease the maximum possible concentration of ES complex with ultimate reduction in the reaction rate of the inactivated individual enzyme molecules. The remaining unmodified enzyme molecules are normally functional considering their turnover number and K_m. For example: Natural poisons act as Enzyme inhibitors and Inhibitory enzymes

In nature, animals and plants are rich in poisons as secondary metabolites, peptides and proteins that can act as enzyme inhibitors. Natural toxins are small organic molecules and act as natural inhibitors for enzymes in metabolic pathways and non-catalytic proteins.

- Neurotoxins are natural inhibitors, toxic but valuable for therapeutic uses at lower doses. For example, glycoalkaloids from Solanaceae family plants (potato, tomato and eggplant) act as acetylcholinesterase inhibitors to increase the acetylcholine neurotransmitter, muscular paralysis and then death. Many natural toxins are secondary metabolites. These neurotoxins also include peptides and proteins. An example of a toxic peptide is alpha-amanitin, found in death cap mushroom and acts potent enzyme inhibitor, in this case preventing the RNA polymerase II enzyme from transcribing DNA. The algal toxin microcystin is also a peptide and is an inhibitor of protein phosphatases. This toxin can contaminate water supplies after algal blooms and is a known carcinogen that can also cause acute liver hemorrhage and death at higher doses. Proteins can also be natural poisons or antinutrients, such as the trypsin inhibitors that are found in some legumes, potato, and tomato. Several invertebrate and vertebrate venoms contain protein and peptide enzyme inhibitors for, e.g., plasmin, renin and angiotensin converting enzymes. Inhibitory enzymes are enzymes that irreversibly inhibit other enzymes by chemically modifying them. In the broad sense, they include all proteases and lysosomal enzymes. Some of them are toxic plant products, e.g., ricin, a glycosidase that is an extremely potent protein toxin found in castor oil beans. It inactivates ribosomes by cleavage the eukaryotic 28S rRNA and reduces protein synthesis and a single molecule of ricin is enough to kill a cell.

4. Reversible inhibition

Reversible inhibitors may be competitive, noncompetitive, or uncompetitive inhibitors relative to a particular substrate. Products of enzymatic reactions are reversible inhibitors of the enzymes. A decrease in the rate of an enzyme caused by the accumulation of its own product plays an important role in the balance and most economic usage of metabolic pathways. It prevents one enzyme in a sequence of reactions from generating a new product more than the capacity of the next enzyme in that sequence, e.g., inhibition of hexokinase by accumulating glucose 6-phosphate.

With the reduction in the inhibitor concentration, the enzyme activity is regenerated due to the non-covalent association and the reversible equilibrium with the enzyme. The equilibrium constant for the dissociation of enzyme inhibitor complexes is known as K_i that equals [E][I]/[EI] [Cheng et al. 1973]. The inhibition efffect of K_i on the reaction kinetics is reflected on the normal K_m and or V_{max} observed in Lineweaver-Burk plots; in a pattern dependent on the type of the inhibitor [Nelson et al. 2008]. The inhibitor is removable by several ways. *The three common types of reversible inhibitions are:*

- Competitive reversible inhibition.
- Uncompetitive reversible inhibition.
- Mixed reversible inhibition (or non-competitive inhibition).

4.1 Competitive reversible inhibition

The competitive inhibitor is structurally related to the substrate and binds reversibly at the active site of enzyme and occupies it in a mutually exclusive manner with the substrate. Therefore, the competitive inhibitor competes with the substrate for the active site. The binding is mutually exclusive because of their free competition. According to the law of mass action, relatively higher inhibitor concentration prevents the substrate binding. Since the reaction rate is directly proportional to [ES], reduction in ES formation for EI formation lowers the rate. Increasing substrate towards a saturating concentration alleviates competitive inhibition. In the time enzyme-substrate complex releases the free enzyme and a product, the enzyme-inhibitor complex does release neither free enzyme nor a product.

Reversible inhibition is of short duration in the biological system because it depends on substrate availability and/or rate of the catabolic clearance of the inhibitor (Figure 7).

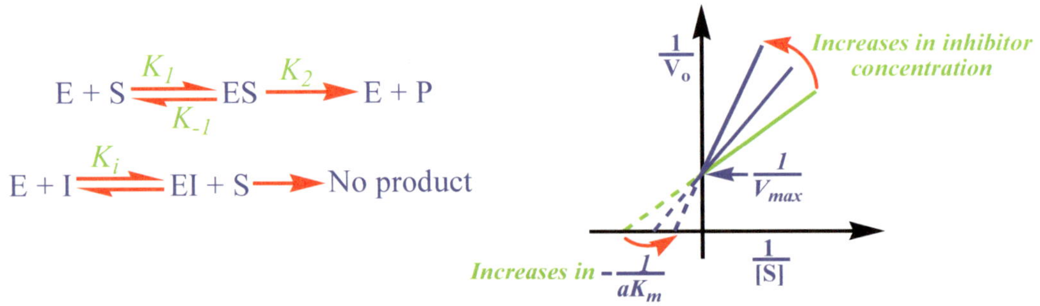

Fig. 7. The equation and the effect of the competitive inhibitor on the double reciprocal plot of the substrate-reaction rate relationship.

Kinetically, the inhibitor (I) binds the free enzyme reversibly to form enzyme inhibitor complex (EI) that is catalytically inactive and cannot bind the substrate. The competitive inhibitor reduces the availability of free enzyme for the substrate binding. Thus, the K_m of the normal reaction is increased to a new K_m (aK_m) as a function of the inhibitor concentration (expressed in the "a" factor - *a*pparent K_m in presence of the inhibitors), where the substrate concentration at $V_o = ½\, V_{max}$ is equal to aK_m. The "a" can be calculated from the change in the slope of the line at a given inhibitor concentration;

$$a = 1 + \frac{[I]}{K_I},\ where,\ K_I = \frac{[E][I]}{[EI]} \tag{1}$$

Therefore, competitive inhibitors do not affect the turnover number (active site catalysis per unit time) or the efficiency of the enzyme because once enzyme is free, enzyme behaves normally. The Michaelis-Menten equation for competitive inhibitors becomes

$$V_o = \frac{V_{max}[S]}{aK_m + [S]} \tag{2}$$

Consequently, the double reciprocal form of the equation is also modified so as the line slope becomes $\frac{aK_m}{V_{max}}$ and the intercept with y-Axis stays at $\frac{1}{Vmax}$ but the intercept with the x-axis at $-\frac{1}{aK_m}$ will differ according to the concentration of the competitive inhibitor. The later property is characteristic for competitive inhibitors.

Examples include the classical competitive inhibitory effect of *malonic acid* on succinate dehydrogenase (SD) of the Krebs' cycle that reversibly dehydrogenates succinate into fumarate. Other less potent competitive inhibitors of succinate dehydrogenase include; oxalate, glutamate and oxaloacetate. The common molecular geometric feature of these compounds is the presence of two negatively charged -COOH groups suggesting that the active site of the flavoprotein SD has specifically positioned two positively charged binding groups (Figure 8).

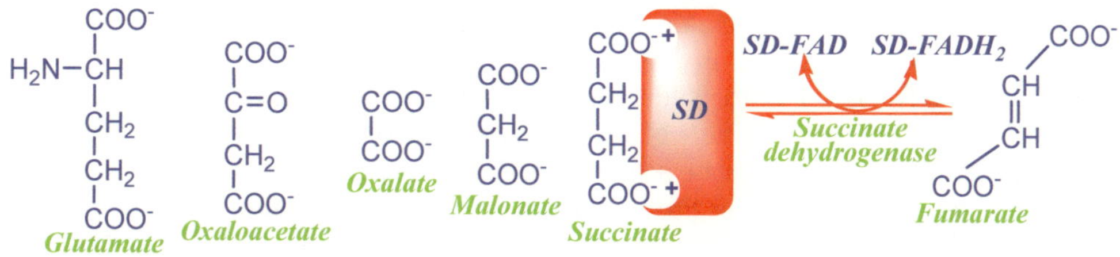

Fig. 8. The substrate and different competitive inhibitors of succinate dehydrogenase (SD).

Methotrexate - competitive inhibitor of dihydrofolate reductase (DHFR) is another example. The drug is used as anticancer antimetabolite chemotherapy particularly for pediatric leukemia. It hinders the availability of tetrahydrofolate as a carrier for one-carbon moieties important for anabolic pathways -particularly synthesis of purine nucleotides for DNA replication (Figure 9).

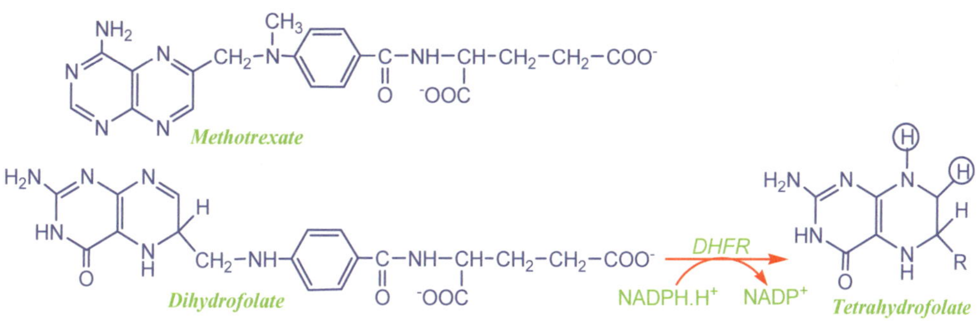

Fig. 9. The substrate and methotrexate as a competitive inhibitor for dihydrofolate reductase.

Sulfanilamides - the simplest form of Sulfa drugs - were among earliest antibacterial chemotherapeutic drugs classified as enzyme inhibitors. They are competitive inhibitors of the bacterial folic acid synthesizing enzyme system from *p*-aminobenzoic acid. Bacterial cannot absorb pre-made folate that is necessary to be synthesized *de novo*. Structural similarity of sulfanilamide (and other sulfas derived from it) to *p*-aminobenzoic acid made them competitive inhibitors to the enzyme (Figure 10).

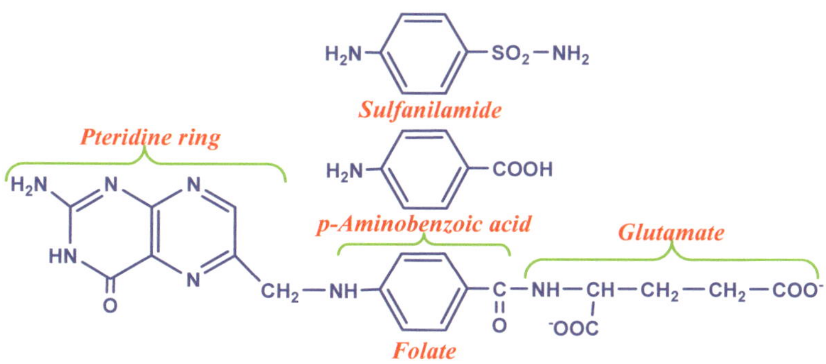

Fig. 10. The *p*-aminobenzoic acid substrate and sulfanilamide as a competitive inhibitor during the bacterial folate synthesis.

Male erectile impotence was a major medical problem. Now a group of chemicals with molecular structural similarity to cGMP is promising that competitively inhibit the cGMP-phosphodiesterase-5. They include sildenafil citrate (Viagra; Figure 11), vardenafil (Levitra) and tadalafil (Cialis). The inhibition of this enzyme that has a limited tissue distribution including the penile cavernous tissue spares cGMP. Accumulation of cGMP leads to smooth muscle relaxation (vasodilation) of the intimal cushions of the helicine arteries, resulting in increased inflow of blood and an erection.

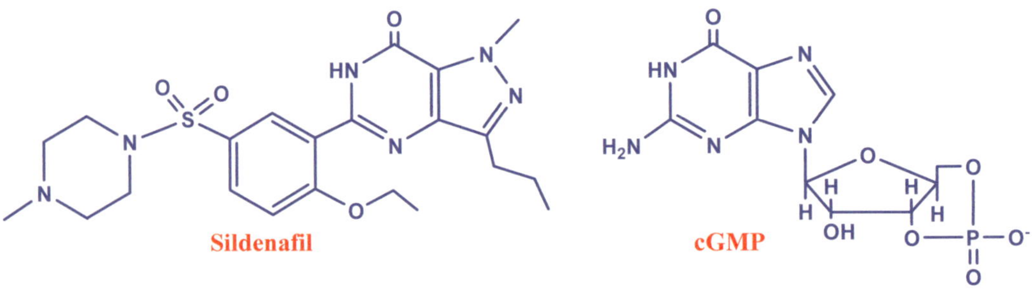

Fig. 11. The cGMP substrate and sildenafil a competitive inhibitor of the cGMP-phosphodiesterase-5.

Another example of these substrate mimics competitive inhibitors are the peptide-based protease inhibitors, a very successful class of antiretroviral drugs used to treat HIV, e.g., ritonavir that contains three peptide bonds (see Figure 12).

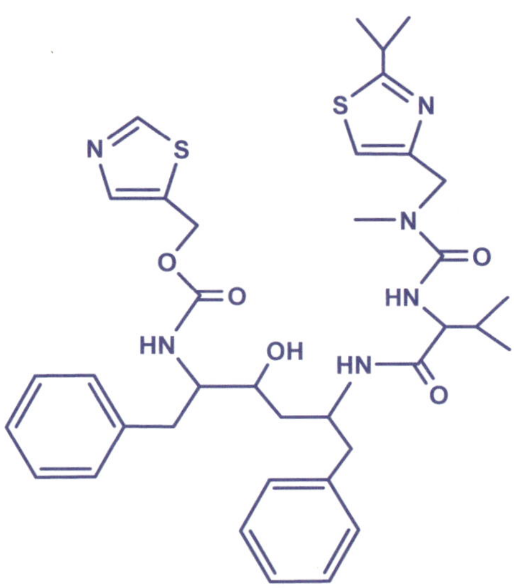

Fig. 12. The peptide-based competitive protease inhibitor ritonavir.

Reversible competitive inhibitors of acetylcholinesterase, such as edrophonium, physostigmine, and neostigmine, are used in the treatment of myasthenia gravis and in anesthesia. The carbamate pesticides are also examples of reversible acetylcholinesterase inhibitors.

4.2 Uncompetitive reversible inhibition

Uncompetitive inhibitor has no structural similarity to the substrate. It may bind the free enzyme or enzyme substrate complex that exposes the inhibitor binding site (ESI). Its binding, although away from the active site, causes structural distortion of the active and allosteric sites of the complexed enzyme that inactivates the catalysis. This leads to a decrease in both K_m and V_{max}. Increasing substrate towards a saturating concentration does not reverse this type of inhibition and reversal requires special treatment, e.g., dialysis. This type of inhibition is also encountered in multi-substrate enzymes, where the inhibitor competes with one substrate (S2) to which it has some structural similarity and is uncompetitive for the other (S1). The reaction without the inhibitor would be; $\mathbf{E + S_1 \Leftrightarrow ES_1 + S2 \Leftrightarrow ES_1S_2 \Rightarrow E + Ps}$ and with uncompetitive inhibitor becomes; $\mathbf{E + S_1 \Leftrightarrow ES_1 + I \Rightarrow ES_1I}$ (prevents S2 binding) $\Rightarrow$ **no product**. It is a rare type and the inhibitor may be the reaction product or a product analog.

Kinetically, uncompetitive inhibition modifies the Michaelis-Menten equation by (a') factor that proportionates with the inhibitor concentration to be:

$$V_o = \frac{V_{max}[S]}{K_m + a'[S]} \tag{3}$$

and in the double-reciprocal equation to be:

$$\frac{1}{V_o} = \frac{a'}{V_{max}} + \frac{K_m}{V_{max}} X \frac{1}{[S]} \tag{4}$$

while y-intercept is at $\frac{a'}{V_{max}}$ and x-intercept is at $-\frac{a'}{K_m}$, whereas, the line slope stays $\frac{K_m}{V_{max}}$. This gives a number of lines in the Lineweaver-Burk plot that are parallel to the normal line with decreased $1/V_{max}$ and $-a'/K_m$ proportional to concentrations of the uncompetitive inhibitor. The later is characteristic to uncompetitive inhibition (Figure 13).

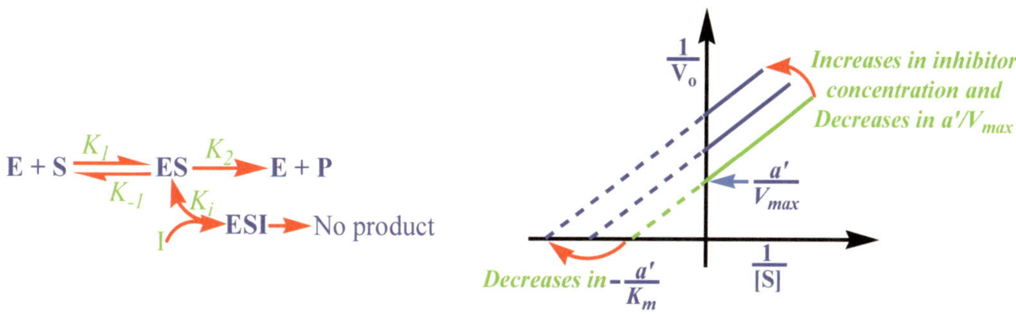

Fig. 13. The equation and the effect of the uncompetitive inhibitor on the double reciprocal plot of the substrate-reaction rate relationship.

Uncompetitive reversible inhibition is rare, but may occur in multimeric enzymes. Examples of uncompetitive reversible inhibitors include; inhibition of lactate dehydrogenase by oxalate; inhibition of alkaline phosphatase (EC 3.1.3.1) by L-phenylalanine, and, inhibition of the key regulatory heme synthetic enzyme; δ-aminolevulinate synthase and dehydratase and heme synthetase by heavy metal ion, e.g., lead. Heavy metals, e.g., lead, form mercaptides with -SH at the active site of the enzyme (2 R-SH + Pb $\Rightarrow$ R-S-Pb-S-R + 2H).

Oxidizing agents, e.g., ferricyanide also oxidizes -SH into a disulfide linkage (2 R-SH ⇒ R-S-S-R). Reversion here requires treatment with reducing agents and/or dialysis.

4.3 Mixed (noncompetitive) inhibition

The mixed type inhibitor does not have structural similarity to the substrate but it binds both of the free enzyme and the enzyme-substrate complex. Thus, its binding manner is not mutually exclusive with the substrate and the presence of a substrate has no influence on the ability of a non-competitive inhibitor to bind an enzyme and *vice versa*. However, its binding - although away from the active site - alters the conformation of the enzyme and reduces its catalytic activity due to changes in the nature of the catalytic groups at the active site. EI and ESI complexes are nonproductive and increasing substrate to a saturating concentration does not reverse the inhibition leading to unaltered K_m but reduced V_{max}. Reversal of the inhibition requires a special treatment, e.g., dialysis or pH adjustment. Some classifications differentiate between non-competitive inhibition as defined above and *mixed inhibition* in that the EIS-complex has residual enzymatic activity in the mixed inhibition.

Kinetically, mixed type inhibition causes changes in the Michaelis-Menten equation so as

$$V_o = \frac{V_{max}[S]}{aK_m + a'[S]} \tag{5}$$

Mixed type inhibition - as the name imply - has a change in the denominator with K_m modified by factor (a) as in competitive inhibition, and [S] modified by factor (a') as in uncompetitive inhibition. In the double reciprocal equation 6,

$$\frac{1}{V_o} = \frac{a'}{V_{max}} + \frac{aK_m}{V_{max}} X \frac{1}{[S]} \tag{6}$$

A line slope is $\frac{aK_m}{V_{max}}$, and the intercept with y-axis is at $\frac{a'}{V_{max}}$ and with x-axis is at $\frac{a'}{aK_m}$. This results in progressive decreases in V_{max} and progressive increases in K_m proportional to the increase in the mixed inhibitor concentration. The double reciprocal plot shows a number of lines reflecting decreases in V_{max}/increases in K_m but their intercept is to the left of the y-axis. Mixed type inhibitor would be called non-competitive only if [a = a'], where, it will only lower V_{max} without affecting the K_m (Figure 14).

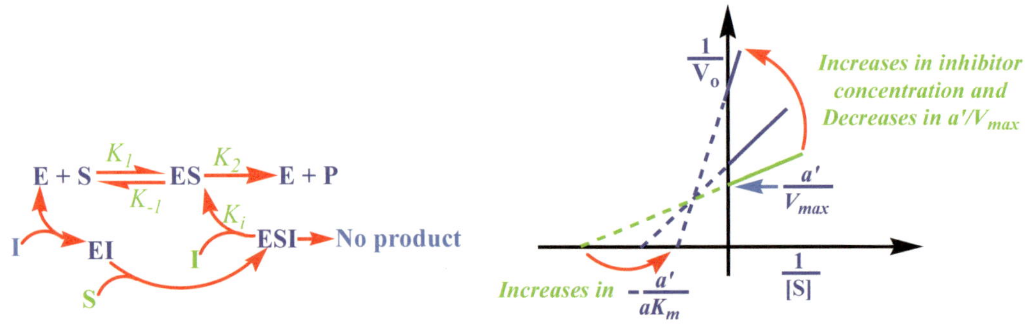

Fig. 14. The equation and the effect of the mixed type (noncompetitive) inhibitor on the double reciprocal plot of substrate-reaction rate relationship.

Examples of noncompetitive inhibitors are mostly poisons because of the crucial role of the targeted enzymes. Cyanide and azide inhibits enzymes with iron or copper as a component of the active site or the prosthetic group, e.g., cytochrome c oxidase (EC 1.9.3.1). They include the inhibition of an enzyme by hydrogen ion at the acidic side and by the hydroxyl ion at the alkaline side of its optimum pH. They also include inhibition of; carbonic anhydrase by acetazolamide; cyclooxygenase by aspirin; and, fructose-1,6-diphosphatase by AMP. Cyanide binds to the Fe^{3+} in the heme of the cytochrome aa_3 component of cytochrome c oxidase and prevents electron transport to O_2. Mitochondrial respiration and energy production cease, and cell death rapidly occurs. The central nervous system is the primary target for cyanide toxicity. Acute inhalation of high concentrations of cyanide (e.g., smoke inhalation during a fire and automobile exhaust) provokes a brief central nervous system stimulation rapidly followed by convulsion, coma, and death. Acute exposure to lower amounts can cause lightheadedness, breathlessness, dizziness, numbness, and headaches. Cyanide is present in the air as hydrogen cyanide (HCN), in soil and water as cyanide salts (e.g., NaCN), and in foods as cyanoglycosides. Comparison of the three types of the reversible enzyme inhibitors is presented in Table 1.

In a special case, the mechanism of *partially competitive inhibition* is similar to that of non-competitive, except that the EIS complex has catalytic activity, which may be lower or even higher (partially competitive activation) than that of the enzyme-substrate (ES) complex. This inhibition typically displays a lower V_{max}, but an unaffected K_m value. We compare three main types of inhibitors in terms of reaction properties as shown in Table 1 and Figure 15.

Competitive inhibitor	Uncompetitive inhibitor	Mixed (noncompetitive inhibitor)
• The inhibitor binds the catalytic/substrate binding site. • It competes with substrate for binding. • Inhibition is reversible by increasing substrate concentration. • V_{max} constant, the substrate concentration has to be increased as reflected on increased K_m.	• Substrate binding exposes the inhibitor binding site away from the catalytic/substrate binding site. • Increasing substrate concentration does not reverse the inhibition. • The inhibited reaction rate parallel the normal one as reflected on decreased both V_{max} and K_m.	• The inhibitor binds each of the free enzyme and the substrate-enzyme complex away from the catalytic/substrate binding site. • Increasing substrate concentration does not reverse the inhibition. • Only V_{max} is decreased proportionately to inhibitor concentration, • K_m is unchanged since increasing substrate concentration is ineffective.

Table 1. Comparison of the different types of reversible inhibition is shown in Table with a quick view of mechanism in sketches as below.

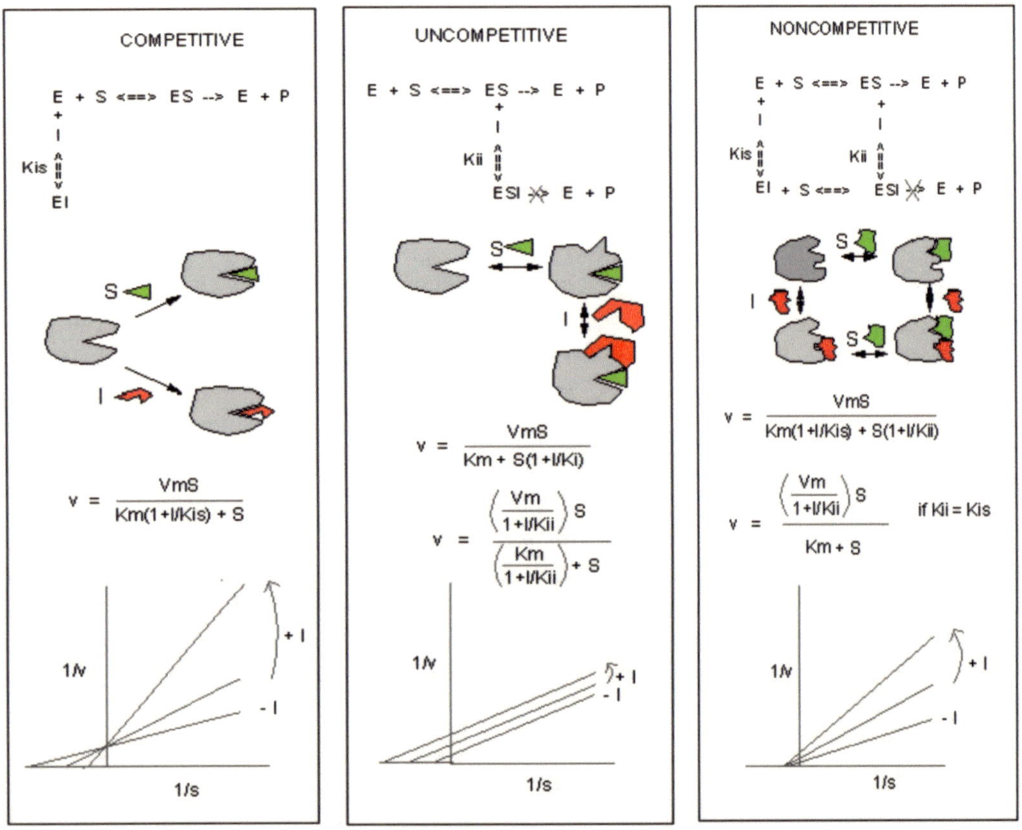

Fig. 15. Sketch of three different enzyme inhibition by competitive, uncompetitive and noncompetitive types are shown with illustration of enzyme-substrate or inhibitor binding, kinetics and graphs.

In last decade, role of membrane receptors was explored in relation with enzyme inhibition. Membrane receptors or transmembrane proteins bind with natural ligands such as hormones, neurotransmitters in tissue membranes. Receptor-ligand binding modulates the binding of drugs with enzyme. Such ligand binding behavior also influences the analysis of competitive, uncompetitive and noncompetitive inhibition by biological effect of prodrugs on enzymes. It usually involves a shape change in the receptor, a transmembrane protein, which activates intracellular activities. The bound receptor usually does not directly express biological activity, but initiates a cascade of events which leads to expression of intracellular activity. However, occupied receptor actually expresses biological activity itself. For example, the bound receptor can acquire enzymatic activity, or become an active ion channel with similar competitive, noncompetitive behavior. Drugs targeted to membrane receptors can have biological effects similar to the natural ligands, they are called **agonists,** or conversely they may inhibit the biological activity of the receptor, they are called **antagonists** [Jakobowski 2010a].

4.4 Agonist

An **agonist** or test drug or substrate is similar to natural ligand and binds with receptor to produce a similar biological effect as the natural ligand. Agonist binds at the same binding site in competition with natural ligand to show full or partial response. So, it is called **partial agonist**. If receptor has a basal (or constitutive) activity in the absence of a bound ligand, it is called **inverse agonist**. If either the natural ligand or an agonist binds to the receptor site, the basal activity is increased. If an inverse agonist binds, the activity is decreased. Ro15-4513 and benzodiazepines (Valium) bind with the GABA receptor. As a result, GABA receptor is "activated" to become a ion channel allowing the inward flow of Cl^- into a neural cell, inhibiting neuron activation. Ro15-4513 binds to the benzodiazepine site, which leads to the opposite effect of valium, the inhibition of the receptor bound activity - a chloride channel as shown in Figure 16.

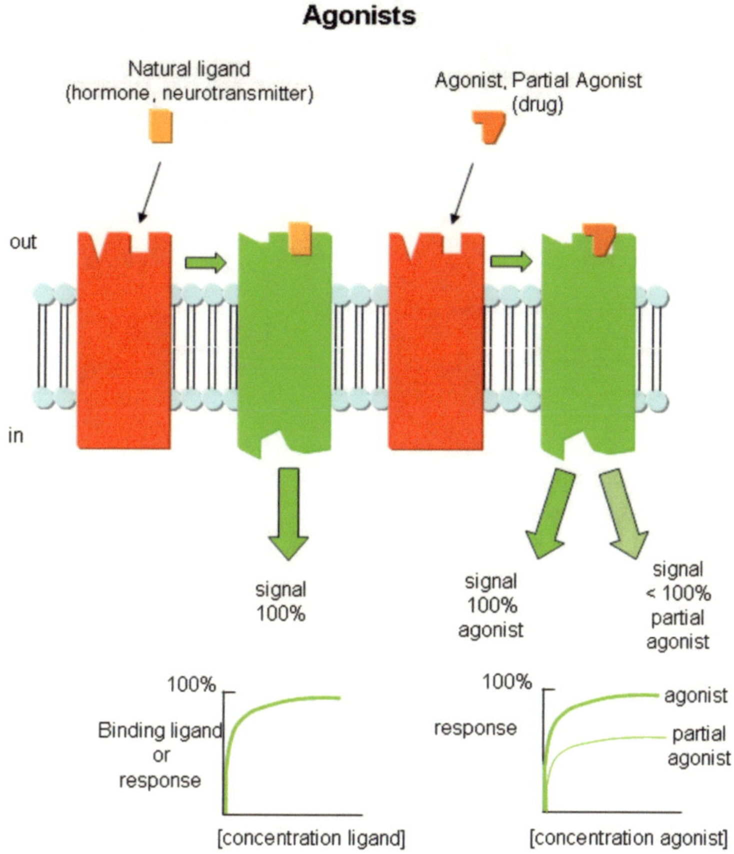

Fig. 16. A sketch is shown for membrane receptor binding with ligand (agonist) acting like as enzyme. Reproduced with permission [Jakobowski 2010a].

4.5 Antagonist

Antagonist or test inhibitor can inhibit the effects of the natural ligand (hormone, neurotransmitter), agonist, partial agonist, and inverse agonists. We can think of them as inhibitors of receptor activity behaving as competitive, noncompetitive and irreversible antagonists as shown in Figure 17. For further details, readers are requested to read advanced text book [Nelson et al. 2008, Dixon and Webb 1979]

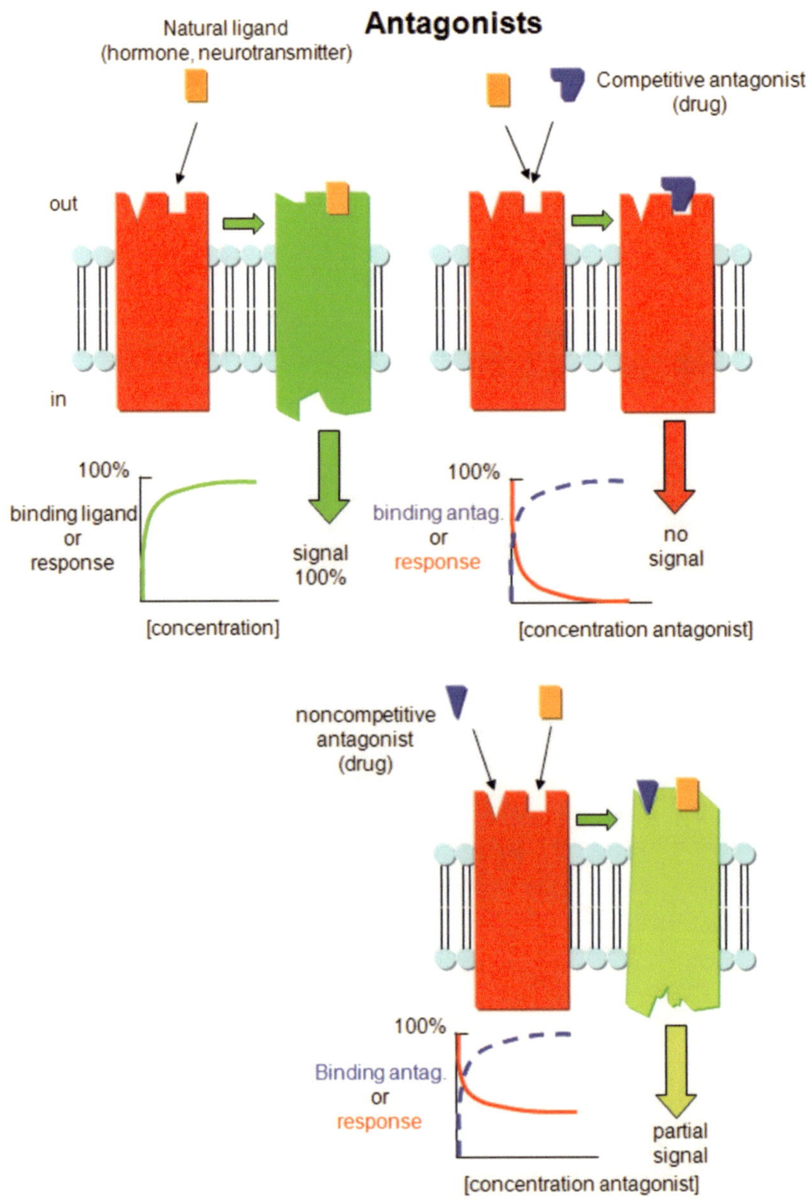

Fig. 17. Sketch is shown for membrane receptor binding with ligand (acting as agonist) and antagonist (acting as inhibitor) in competition with agonist to bind with enzyme. Reproduced with permission [Jakobowski 2010a]

5. Inhibition by physiological modulators

5.1 Temperature of reaction

Some endothermic or exothermic chemical compounds change the temperature of reaction. Enzyme reaction experiences inhibition at higher or lower than optimal physiological temperature. For example, human body optimal temperature of human body is 37 °C. For most of the enzyme reactions, enzyme activity usually increases at 0 to about 40-50 °C in the absence of catalysts. As a general rule of thumb, reaction velocities double for each increment of 10°C rise. At higher temperatures, the activity decreases dramatically as the enzyme denatures as shown in Figure 18.

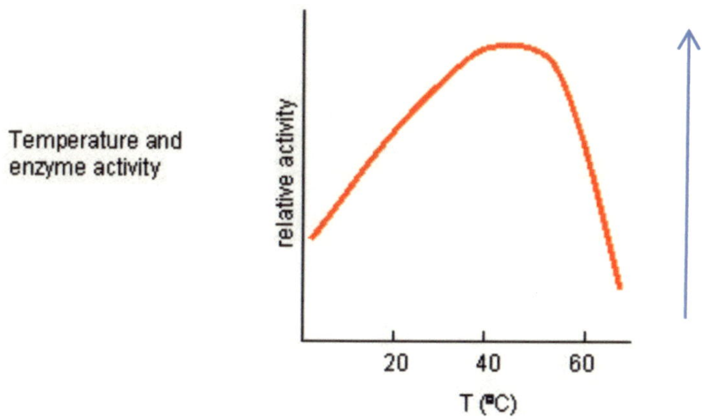

Fig. 18. Figure shows the effect of temperature change on the rate of enzyme reaction. Notice the initial rise of rate of reaction and sudden fall near to optimal temperature 37-42 °C.

5.2 Hydrogen ion concentration or pH of reaction

Think of all the things that pH changes might affect. Many chemicals such as acids or alkaline chemical compounds if mixed in enzyme reaction medium can change the pH. As a result, reaction rate changes. It might

- affect E in ways to alter the binding of S to E, which would affect K_m
- affect E in ways to alter the actual catalysis of bound S, which would affect k_{cat}
- affect E by globally changing the conformation of the protein
- affect S by altering the protonation state of the substrate

The easiest assumption is that certain side chains necessary for catalysis must be in the correct protonation state. Thus, some side chain, with an apparent pKa of around 6, must be deprotonated for optimal activity of trypsin which shows an increase in enzyme activity with the increase in range centered at pH 6. Which amino acid side chain would be a likely candidate to participate in enzyme inhibition? It all depends on net charge on active group of each amino acid in the active site chain. The pH of reaction thus depends on net pK_a value of amino acids and presence of acid or alkaline nature of substrate effects on enzyme kinetics by formation of EH, ESH as shown in Figure 19. It can be modeled at the chemical and mathematical level to calculate velocity(v), V_m(apparent) and K_m(apparent) as shown in Equations 7-9. Different enzymes show different behavior of enzyme catalyzed reactions such as chymotrypsin, cholinesterase, papain, and papsin show distinct graphs (see Figure 20). For further details, readers are requested to read text books [Nelson et al. 2008, Berg et al. 2011]

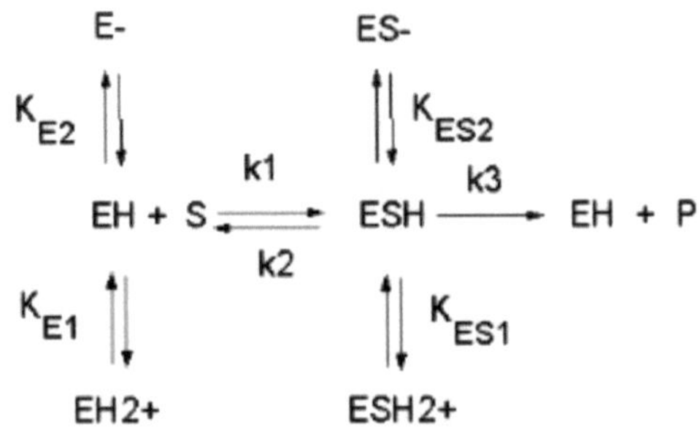

$$V = \frac{\text{Vm app S}}{\text{Km app} + S} \tag{7}$$

$$\text{Vm app} = \frac{\text{Vm}}{1 + H+/\text{Kes1} + \text{Kes2}/H+} \tag{8}$$

$$\text{Km app} = \frac{\text{Km}(1 + H+/\text{Ke1} + \text{Ke2}/H+)}{1 + H+/\text{Kes1} + \text{Kes2}/H+} \tag{9}$$

Fig. 19. Chemical equations showing the mechanism of pH effects on enzyme catalyzed reactions. Different mathematical equations 7-9 illustrate the modeling pH effects on enzyme catalyzed reactions.

5.2.1 Three dimensional nature of enzyme-inhibitor complex at enzyme active site

The role of non-covalent interactions such as hydrogen bonding, hydrophobic interaction and orientation of inhibitor and enzyme in an organized fashion was well described in classic paper [Amtul et al., 2002]. 3D nature of enzyme reaction can be understood as following. There are two sites on enzyme molecule: 1. at allosteric site, inhibitor binds with enzyme, and 2. at active site, substrate binds with enzyme. However, substrate and inhibitor interact with each other by non-covalent interactions of their chemical groups. Inhibitors interact at allosteric site and known as 'pharmacohores'. Presently, structure-based design and testing, mechanistic biological approach is a state-of-art to develop new pharmacohores. The non-covalent interactions determine the chemoselectivity of the substrate and enzymes during formation of the ESI complex. In other words, ESI complex provides enzyme as a platform to perform catalysis. 3D geometrical shape and topology of active site match with orientation of chemical groups in substrate molecule that fit together in a 'lock and key' arrangement. Several possibilities happen to make enzyme-inhibitor complexes such as bidentate, tri-, tetra- and polydentate, trigonal, pyramidal, tetrahedral, polyhedral charge transfer complexes due to co-ordinate interactions between metallic co-factor with hydrophilic groups on inhibitor(s). In this process, geometry of amino acid side chains at allosteric site changes due to hydrogen bonding between amino acid residues. Suboptimal

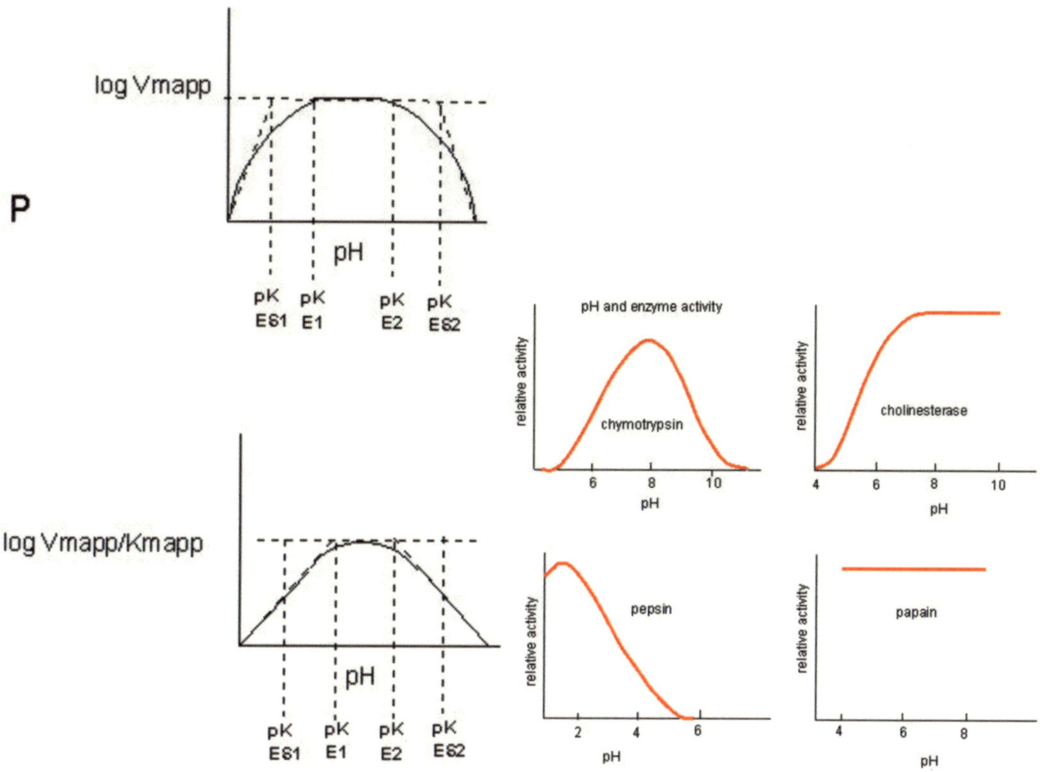

Fig. 20. Graphs of different pH effects on enzyme catalyzed reactions as log V_m(app) and V_m/K_m(app) are shown on left. Different enzymes such as chymotrypsin, cholinesterase, pepsin and papain are illustrated with different rates of enzyme reaction. Reproduced with permission [Jakobowski 2010a]

interactions of metal-solvent, oxygen-water molecular bridge, free energy content loss, subunit-subunit biophysical interactions as a result play a significant role in inhibitor-enzyme complex formation and completion of enzyme catalysis.

For more details, readers are requested to read recent reference papers on 3D mechanistic studies on enzymes. Specific example on urease is cited in chapter 11 in this book. Now science is shifting to develop crystallized enzyme molecules, better structural-functional relationship in enzyme catalysis and immobilized enzyme chips.

In following description, factors are discussed on different practical considerations that influence the enzyme reaction rates, enzyme inhibition kinetics, % binding efficiency on enzyme solid support with a glimpse of known theories and concepts on real-time, cheaper, economic, user-friendly immobilized enzyme technology.

When actual and practical considerations are analyzed to work in enzyme reactor, the scenario becomes complicated. Several factors such as inhibitor chemical state, substrate structure, enzyme 3D conformation or peptide subunit interactions, physiological reaction conditions in reactor and enzyme carrier supports also contribute in inhibition kinetics and rates of reaction to form ES,ESI and P. Every year list of new factors grows in new enzyme systems.

Author believes that more and more contributory factors introduced, will influence enzyme reaction rate kinetics and more and more additive kinetic constants are introduced with new variants to define the action of inhibitors on enzyme catalysis.

Other factors to keep in mind for new possibilities are:

1. enzyme autoinhibition and enzyme molecular structural-functional factors affecting 3D conformation of active site compatible with active groups of substrate or inhibitor
2. porosity and diffusion across the enzyme support material and availability of exposed active sites to react
3. real-time recording the instant formation of ESI or ES or EP or EI on solid phase enzyme support organic chip
4. sustrate-inhibitor interactions, % binding of active site with each additive
5. computer based semi-corrected or averaged calculations of kinetic constants of inhibition kinetics
6. thermodynamic states of the enzyme reaction in reactor and fluctuating physiological and physical states of substrate, inhibitor, enzyme complexes in reactor.
7. synergy of inhibitors, substrate, subunits in enzyme on active site

For all these factors and details, readers are expected to read advanced text books on enzyme inhibition and enzyme engineering. Readers will experience a wide variation in the scientific analysis of enzyme inhibition data in different enzyme reactors used in different studies. High efficiency with desired results of enzyme inhibitors is the new challenges to optimize reaction, scale-up, and phase out unwanted physiological factors from reaction. In following section, these issues are addressed. Author believes that above mentioned description is just iceberg from a large hidden treasure or unknown factors contributing enzyme inhibition to give desired outcome.

6. Immobilized enzyme systems

In search of economic, efficient and practical enzyme platforms to test enzyme inhibitors, new user-friendly immobilized enzyme technology is available now. It is based on principle that an enzyme molecule is contained within confined space for the purpose of retaining and re-using enzyme on solid medium in processing system or equipment. There are many advantages of immobilized enzymes and methods of immobilization such as low cost, suitability of reusable model system in membrane-bound enzymes in cell. However, some disadvantages are expansive methods of adsorption or covalent bound or matrix trapping or membrane trapping immobilization methods, low measurement of enzyme activity with mass transfer limitations. For knowledge sake, the entrapment of enzyme molecules on matrix, diffusion phenomenon and kinetics are important to understand. A brief description is given for interested readers on classic concepts and scientific basis of porous or non-porous enzyme supports, theory of enzyme immobilization and efficiency of reaction outcome. For more details of each aspect, readers are requested to read individual research papers.

Matrix entrapment is done by mixing enzyme solution with polymer fluid in matrices such as Ca-alginate, agar, polyacrylamide, collagen. **Membrane entrapment** is done by confining enzyme solutions between semi-permeable membrane hollow fibers made of nylon, cellulose, polysulfone, polyacrylate etc. Surface immobilization by **adsorption** is done by attaching enzymes on stationary solids such as alumina, porous glass, cellulose, ion-exchange resin, silica, ceramic, clay, starch etc. by physical forces keeping active sites intact. **Covalent bonding** is done by enzyme retention on support surfaces by covalent binding between functional groups such as amino, carboxylic, sulfhydryl, hydroxyl groups on the enzyme and those on the support surface keeping enzyme active site(s) free (see Figure 21) [Laider et al. 1980].

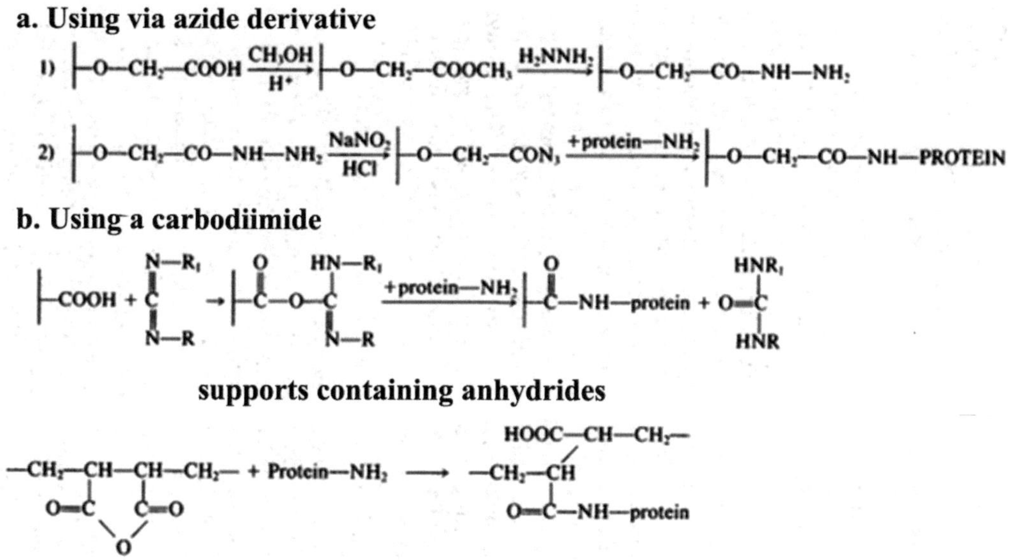

Fig. 21. Scheme of immobilization of enzyme is shown with chemical groups involved in binding of enzyme on solid surface. Reproduced with permission from reference Lieder et al.1980.

Diffusional limitations are observed to various degrees in all immobilized enzyme systems. This occurs because substrate must diffuse from the bulk solution up to the surface of the immobilized enzyme prior to reaction. The rate of diffusion relative to enzyme reaction rate determines whether limitations on intrinsic enzyme kinetics is observed or not as shown in Figures 22 [Laider et al.1980]. However, rate of diffusion across and within matrix is determinant of immobilized enzyme reaction as shown in Figure 22 and 23.

In immobilized enzyme reaction, two major effects due to diffusion and product inhibition are first observed by Lineweaber-Burk plots in classic study [Rees, 1984]. The diffusional effects and product inhibition both influenced the shape of Lineweaver-Burk plot (see Figure 22). In case of substrate inhibition effects binding of more than one substrate molecule(s) lead to inhibition showing same type of curved Lineweaver-Burk plot as those observed for diffusional limitation and product inhibition in immobilized enzymes. Combination of these two effects lead to intermediate behavior, such as normal Michaelis-Menten kinetics as shown in Figure 24, 25 by curves [Rees, 1984]. However, immobilized enzyme system also suffers from both diffusion and product inhibition effects. As a consequence, it is important to consider diffusion effects and product inhibition effects while extracting catalytic parameters from kinetic data for immobilized enzyme systems. Use of non-porous support in enzyme immobilization minimizes the diffusion effects to some extent.

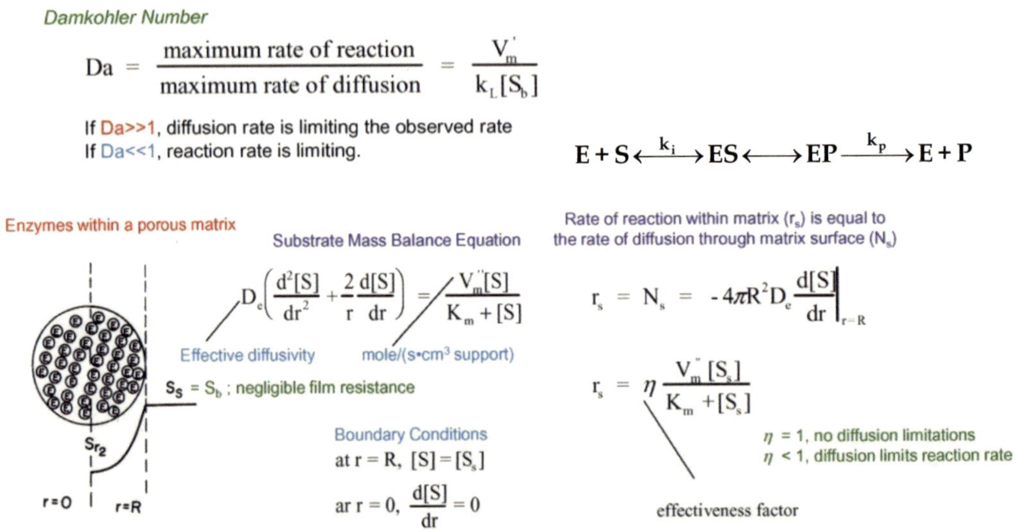

Fig. 22. A sketch of porous matrix is shown (on left) and a scheme of substrate mass balance Equation to calculate rate of immobilized enzyme reaction r_s is shown (on right)

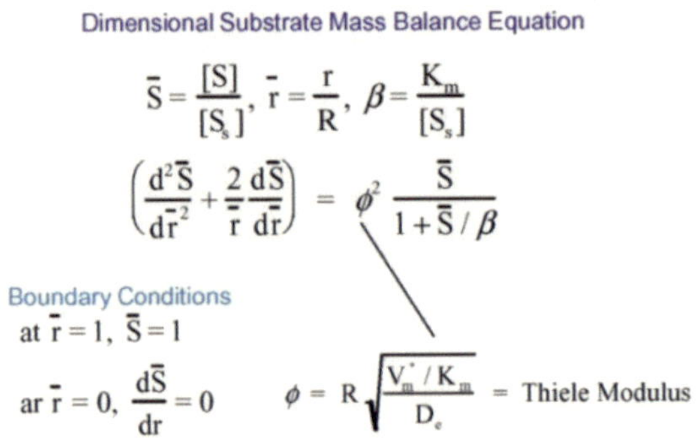

Fig. 23. A scheme of substrate mass balance is shown to calculate S with boundary conditions.

Enzyme kinetics predicts the efficiency of reaction. Kinetics of immobilized enzymes depends on conformational alterations within the enzyme due to the immobilization procedure, or the presence and nature of the immobilization support. Immobilization can greatly affect the stability of an enzyme such as any strain into the enzyme will inactivate the enzymes under denaturing conditions (e.g. higher temperatures or extremes of pH). An example of unstrained multipoint binding between the enzyme and the support to cause substantial stabilization is illustrated in Figure 20. From mechanistic standpoint, a lesser conformational change within the protein structure will initiate enzyme inactivation. As a result, covalent immobilization processes involve an initial freely-reversible stage. Covalent links may form, break and re-form till an unstrained covalently-linked structure is created. However, additional stabilization is derived from maximum enzyme-support compatibility, least enzyme molecule interactions, least proteolytic and microbiological attacks.

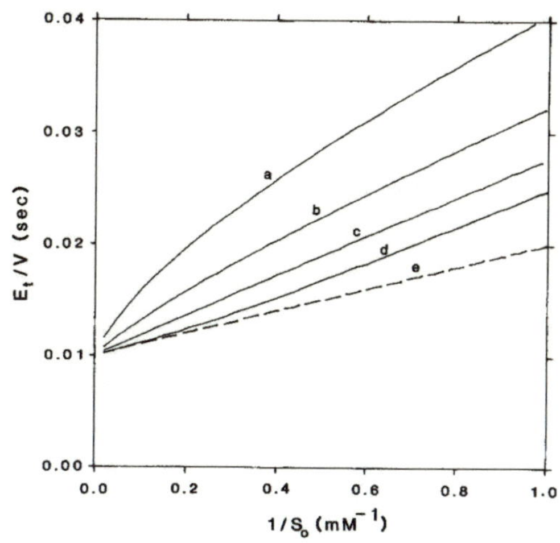

Fig. 24. Effect of one or more inhibitor molecules on enzyme kinetics and their inhibition effect dependent on 1/So. Reproduced with permission from Rees et al. 1984.

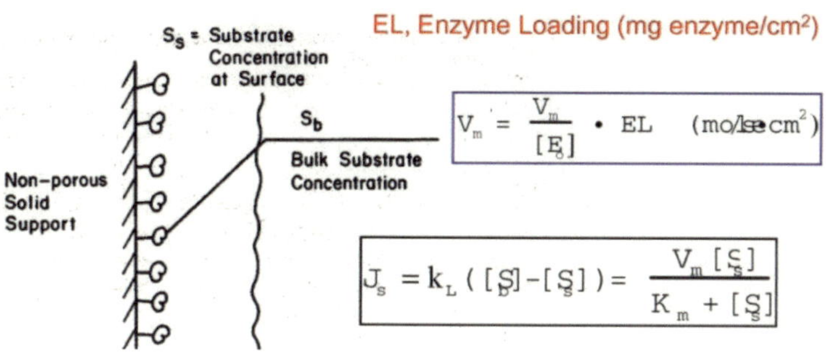

Fig. 25. A scheme of immobilized enzyme action is shown on non-porous solid support. Notice the dependence of V_m on available immobilized enzyme active sites (E_L).

The kinetic constants (e.g. K_m, V_{max}) of immobilized enzymes may be altered by the process of immobilization due to internal structural changes and restricted access to the active site. Thus, the intrinsic specificity (k_c/K_m) of such enzymes may well be changed relative to the soluble enzyme. An example of trypsin is illustrated in Figure 21, where the freely soluble enzyme hydrolyses fifteen peptide bonds in the protein pepsinogen but the immobilized enzyme hydrolyses only ten. The apparent value of these kinetic parameters, when determined experimentally, may differ from the intrinsic values. This fact may be due to changes in the properties of the solution in the immediate vicinity of the immobilized enzyme, or the effects of molecular diffusion within the local environment. The relationship between these intrinsic and apparent parameters is shown below in Figure 26. Typically, nonporous microenvironment consists of the internal solution plus part of the surrounding solution which is influenced by the surface characteristics of the immobilized enzyme. Partitioning of substances occurs between these two environments. Substrate molecule (S)

Intrinsic parameters of the soluble enzyme

↓

Intrinsic parameters of the immobilized enzyme

↓

Apparent parameters due to partition and diffusion

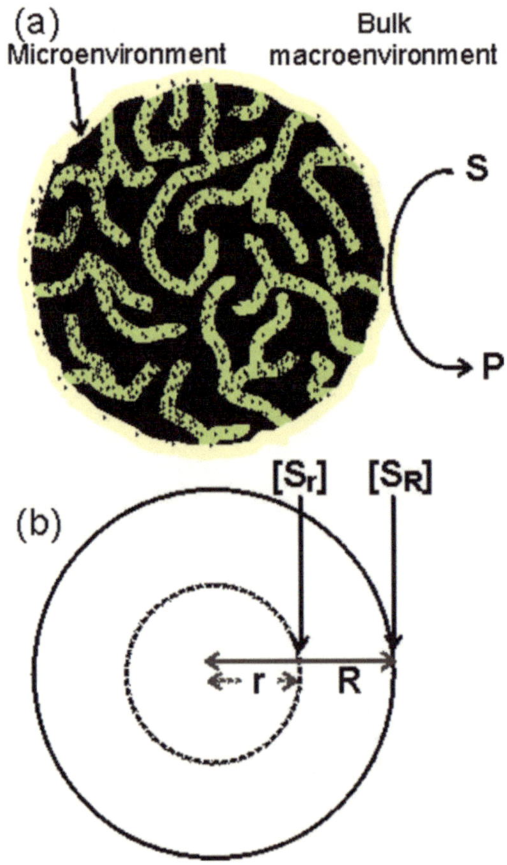

Fig. 26. A schematic cross-section of an immobilized enzyme particle (a) shows the macroenvironment and microenvironment. Triangular dots represent the enzyme molecules. Courtesy: Pangandai V. Pennirselvam, Ph.D UFRN, Lagoa Nova-Natal/RN Campus Universitário. North East, Brazil.

diffuses through the surrounding layer (external transport) in order to reach the catalytic surface and gets converted to product (P). In order for all immobilized enzyme to be utilized, substrate must diffuse within the pores in the surface of the immobilized enzyme particle (internal transport) [Pryciak 2008]. The degree of stabilization is determined by strength of the gel, and hence the number of non-covalent interactions. As a result, intrinsic parameters of enzyme result with specific apparent parameters dependent on partition and diffusion as shown in Figure 27.

- The porosity (e) of the particle can be expressed as ratio of the volume of solution contained within the particle to the total volume of the particle. The tortuosity (t) is the average ratio of the path length, via the pores, between any points within the particle to their absolute distance apart.
- The tortuosity, which is always greater than or equal to unity, depends on the pore geometry. The diagram exaggerates dimensions for the purpose of clarity.
- The concentration of the substrate at the surface of the particles [S_r] depends on radius R or internal concentration [S_i] at any smaller radius (r) is the lower value.

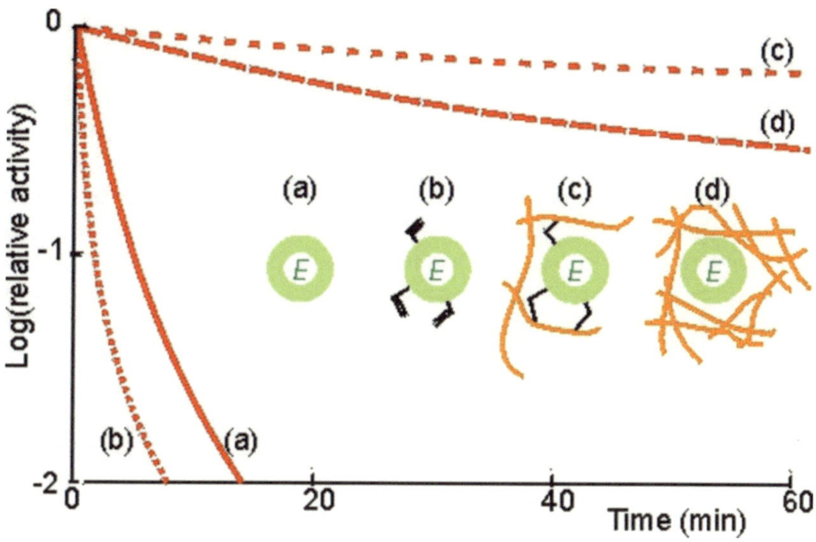

Fig. 27. Illustration of the use of multipoint interactions for the stabilization of enzymes. (a) -------- activity of free un-derivatized chymotrypsin. (b) ..... activity of chymotrypsin derivatized with acryloyl chloride. (c) -- -- -- activity of acryloyl chymotrypsin copolymerized within a polymethacrylate gel. Up to 12 residues are covalently bound per enzyme molecule. Lower derivatization leads to lower stabilization. (d) ----- activity of chymotrypsin non-covalently entrapped within a polymethacrylate gel. All reactions were performed at 60°C using low molecular weight artificial substrates. The immobilized chymotrypsin preparations showed stabilization of up to 100,000 fold, most of which is due to their multipoint nature although the consequent prevention of autolytic loss of enzyme activity must be a significant contributory factor. Reproduced with permission from Martinek et al, 1977a,b.

In general, the use of immobilized enzyme can be divided into two major categories of applications: in biosensors and bioreactors. However, list is growing in the other fields of ecological, environmental, agriculture, health, oceanic, space and earth sciences.

7. New developments in art of enzyme inhibition

Now a day, immobilized enzymes are used in industries and have value as medicinal and industrial enzyme products. Good examples of industrial enzymes are amylase, glucoamylase, trypsin, pepsin, rennet, glucose isomerase, penicillinase, glucose oxidase, lipase, invertase, pectinase, cellulase in medicinal use. With emergence of new inhibitors in the quest of drug discovery, several new inhibition mechanisms are expected in case of new substrate analogues. New substrate–enzyme active site interactions are envisaged due to different binding intricacies. Some examples of emerging concepts are outlined in following description and readers are expected to read advanced literature on these applications.

- ***Slow-tight inhibition:*** Slow-tight inhibition occurs when the initial enzyme-inhibitor complex EI undergoes isomerizing conformational change to a more tightly binding complex. However, the overall inhibition process is reversible. This manifests itself as slowly increasing enzyme inhibition. Under these conditions, traditional Michaelis-Menten kinetics gives a false value of a time-dependent K_i. The true value of K_i can be obtained through more complex analysis of the on (k_{on}) and off (k_{off}) rate constants for inhibitor association.
- ***Substrate and product inhibition:*** Substrate and product inhibition is where either the substrate or product of an enzyme reaction inhibits the enzyme's activity. This inhibition may follow the competitive, uncompetitive or mixed patterns. In substrate inhibition there is a progressive decrease in activity at high substrate concentrations. This may indicate the existence of two substrate-binding sites in the enzyme. At low substrate, the high-affinity site is occupied and normal kinetics is followed. However, at higher concentrations, the second inhibitory site becomes occupied, inhibiting the enzyme. Product inhibition is often a regulatory feature in metabolism and can also be a form of negative feedback; see allosteric regulation [Pryciak 2008, Bashor 2008].
- ***Antimetabolites:*** They are chemicals that interfere with the normal metabolism of normal biochemical metabolite(s). This in most of case is due to their structural similarity to such physiological substrates and therefore works as competitive enzyme inhibitors. They include antifolates such as methotrexate, hydroxyurea and purine and pyrimidine analogues. They are mainly used as cytotoxic anticancer drugs through inhibiting DNA and RNA synthesis and cell division. An example of nitroimidazole is described in detail on its metabolic effects at cellular level in this book [Sharma 2012a].
- ***Antienzyme:*** Intestinal parasites, e.g., Ascaris, protect themselves from digestion by expressing on their surface substances that are protein in nature which inhibit the action of digestive enzymes, e.g., pepsin and trypsin. The blood plasma and extracellular fluids are containing several types of protease inhibitors particularly important in controlling the blood clot formation and dissolution and matrix and cytokine homeostasis. Most of these inhibitors are peptides and several of them are also isolated from raw egg white, potatoes, tomatoes and Soya bean and other plant sources. Most of the natural peptide protease inhibitors are similar in structure to the amino acid sequence of the peptide substrates of the enzyme. Designed peptide protease inhibitors are important drugs, e.g., captopril that is a metalloprotease angiotensin-converting enzyme peptide inhibitor. Inhibiting this enzyme prevent activation of angiotensin and therefore prevent vasoconstriction to lower blood pressure. Crixivan is an anti-retroviral aspartyl protease peptide inhibitor used in the treatment of Human

Immunodeficiency Virus (HIV)-induce acquired immunodeficiency syndrome (AIDS). It inhibits the HIV protease that cleaves the large multidomain viral protein into active enzyme subunits. Because these peptide inhibitors may not be specific, they have several side-effects as drugs.

- ***Antibodies*** against several nonfunctional plasma enzymes have clinical diagnostic importance since they are longer living than the enzyme itself and hence reflect the disease history better. In this respect, autoimmune antibodies are clinically important in diagnosis of autoimmune diseases, e.g., anti-glutamic acid decarboxylase antibodies in type 1 diabetes mellitus.
- ***Biosensors:*** Light inhibits most enzyme activity although some enzymes, e.g., amylase are activated by red or green light and also specific DNA repairing enzymes (e.g., UV-specific endonuclease) are activated by the blue and UV light. Ultraviolet rays and ionizing radiations cause denaturation of most enzymes. Most enzymes contain sulfhydryl (-SH) groups at their active sites which upon oxidation by oxidants and free radicals by oxidants and free radicals inactivate the enzyme. Examples: Effect of radiations, light and oxidants on the rate of the enzyme catalyzed reaction.
- Other application of membrane bound redox enzymes constitutes them as a scaffolding enzyme arrangement into systems for multi-step catalytic processes. The reconstruction of portions of this redox catalytic machinery, interfaced to an electrical circuit leads to novel biosensing devices or biosensors. An example of nitric oxide synthase enzyme is cited in this book [Sharma, 2012b].
- In **EzNET®** water purifying system, nitrate pollution is eliminated. Enzyme is immobilized on "beads" with an electron-carrying dye as shown in Figure 28. Reduction of nitrate to environmentally safe nitrogen gas is driven by a low voltage direct current.

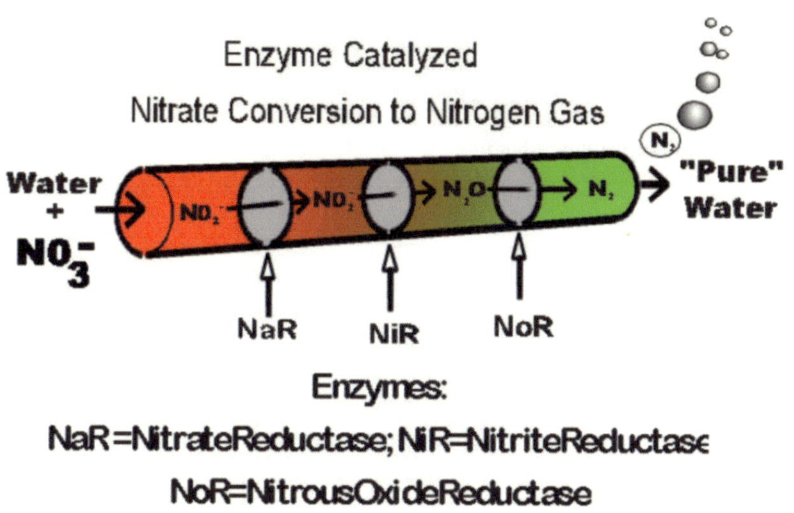

Fig. 28. EzNET® system shows immobilized enzyme on "beads" with an electron-carrying dye. In this system, reduction of nitrate generates environmentally safe nitrogen gas driven by a low voltage direct current. Source: The Nitrate Elimination Co., Inc. 2000.

- In biolumescence detection for toxicity of HPV chemicals or drug development, 62 kDa MW oxygenase (yellow green light emitted at 560 nm) enzyme gives 88 photon/cycle light output proportional to [ATP] according to:

Luciferin + luciferase + ATP → luciferyl adenylate-luciferase + pyrophosphate

Luciferyl adenylate-luciferase + O2 → Oxyluciferin + luciferase + AMP + light

Strong inhibition of luciferase by chloroform or HPV chemicals indicates the efficiency of immobilized recombinant luciferase enzyme system as shown in Figure 20. Inhibition by chloroform is much reduced in the mutant Luciferase compared to the wild type Luciferase as shown in Figures 29, 30.

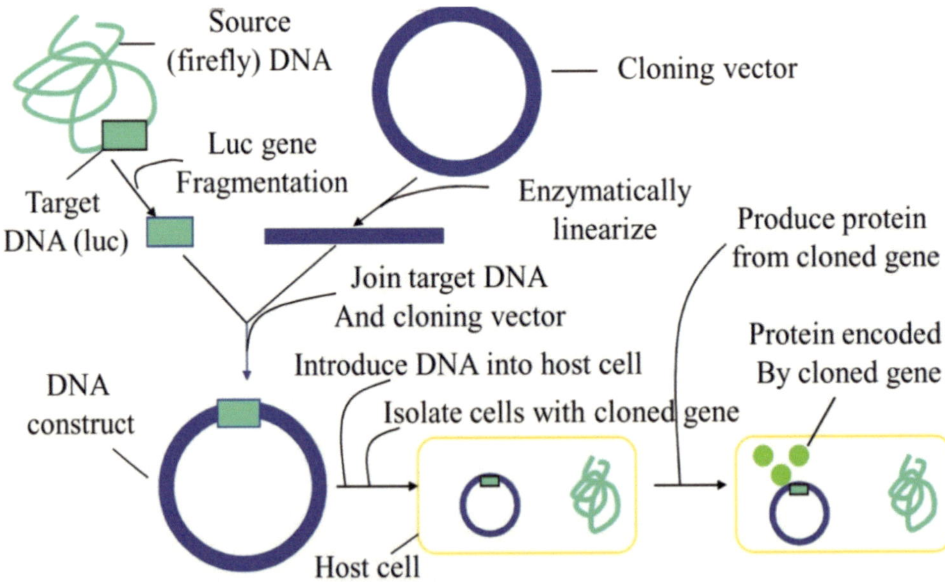

Source: Kim et al. AIChEngg Annual Meeting 2003, San Francisco, CA.

Fig. 29. A sketch of recombinant luciferase is shown illustrating the gene clone.

- In the search for new therapeutics, the high throughput screening (HTS) of ligands for key target proteins, enzymes represent the principal hit identification tool for early drug discovery [Bartolini et al. 2009]. However, output depends on cost-based or amount-based limitation of target availability, need of speed, automation and easy coupling of the enzyme assay with separation systems (affinity chromatography of immobilized proteins) and appropriate detectors. Good example is targeting in drug discovery represented by enzyme inhibition mechanism in monolithic immobilized enzyme reactors (IMERs) to represent different phases of the drug discovery pathway-starting with active compounds (hit) identification, through drug development and lead optimization, early ADMET (absorption, distribution, metabolism, excretion, toxicity) studies and quality control of protein drugs. Some details are described in chapters in this book [Bartolini et al. 2005, 2007]. Interested readers are requested to read advanced text books on these aspects. Different IMER have own requirements for optimal performances to show an increased data output, reliability and stability to translate into cost reduction for potential applications in pharmacy industry [Bartolini et al. 2005, 2007].

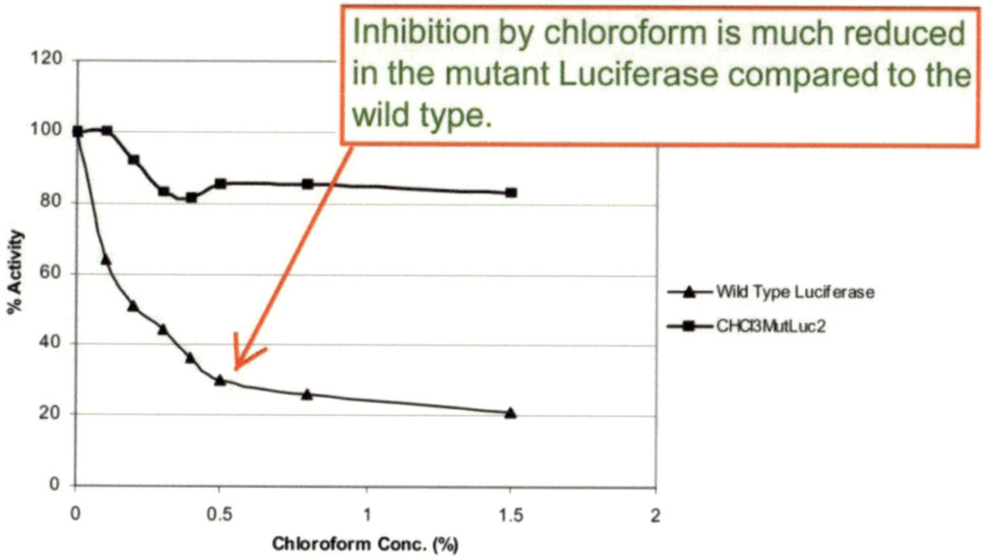

Source: Kim et al. AIChEngg Annual Meeting 2003, San Francisco, CA.

Fig. 30. Inhibition of luciferase activity by increasing the concentration of chloroform.

8. Softwares and computerization in enzyme inhibition kinetics

Recently softwares have popped up to visualize custom visual interface to see curve fits in real-time, graph transforms, equations using kinetic data entry in terms of substrate, inhibitor, activator, velocity, and standard deviation of the velocity. Data tables are directly generated linked to the Fitting Panel of software. The data and results analysis is transferred in user-friendly lay-out, ANOVA window, % inhibition using Monte-Carlos fits, and receptor or ligand binding calculator. For interested readers, VISUALENZYMICS 2010® is available for statistical analysis for enzyme kinetics.[http://www.softzymics.com/visualenzymics.htm].

9. Limitations and challenges

Above mentioned description on mechanism and applications shows a clear issue on need of careful analysis for enzyme inhibition factors, presumptions of enzyme reaction, use of new immobilized enzyme support and enzyme recording/monitoring methods. Challenge is that most of times, basic presumptions do not hold true in enzyme reactors and addition of new factors further complicate the calculation of reactor outcome. Most of the times, computer based kinetic calculations average out outcome as less realistic with more chances of variants. Other major challenge is that each time enzyme reactor outcome depends on individual inhibitor and individual enzyme reactor at different times. It is less reproducible and unpredictable because of synergy, interplay of known and unknown physical, physiological, biological, molecular factors affecting reaction kinetics.

10. Impact of enzyme inhibition science in business

The major current and emerging therapeutic markets for enzyme inhibitors used in human therapeutics are very high. New information is available on biochemistry for enzyme inhibitors and classes of enzyme inhibiting products with broad current or potential therapeutic applications in large markets. However, more than 100 enzyme inhibitors are currently marketed and double than those are under development. A better understanding of the emerging enzyme inhibitors on enzyme mechanism is main key. These include selected indications for asthma and chronic obstructive pulmonary disease (COPD), cardiovascular diseases, erectile dysfunction, gastrointestinal disorders, hepatitis B virus infection, hepatitis C virus infection, herpesvirus infections, human immunodeficiency virus (HIV)/acquired immune deficiency syndrome (AIDS) and rheumatoid arthritis and related inflammatory diseases. Key information from the business literature and thorough enzyme inhibition research is the basis of expert opinion on commercial potential and market sizes from enzyme industry professionals. Since initial reports on chemical immobilization of proteins and enzymes first appeared ~30 years ago, immobilized proteins are now widely used for the processing of products in industries from food business to environmental control. In recent years, use of chemical immobilization was extended to immobilized antibodies or antigens in bioaffinity chromatography. In coming years, it is speculated that immobilization techniques of proteins and enzymes will have greater impact on point-of-care medical and health business.

11. Conclusion

Enzyme inhibition is significant biological process to characterize the enzyme reaction, extraction of catalysis parameters in bio-industry and bioengineering. Conceptual models of inhibition define the interactions of substrate-enzyme or inhibitor-enzyme or both substrate-enzyme-inhibitor in the moiety of active site. In recent years, application of enzymes and enzyme inhibition science have gone in healthcare, pharmaceutical, bio-industries, environment, and biochemical enzyme chip industries with great impact on healthcare and medical business. Last decade has shown the measurement and accuracy of enzyme detection up to the scale of picometer and enzyme industry is entering in the area of picotechnology. Immobilized enzyme technology has given a new way of economic tools in drug discovery and biosensor industry. Every year new enzyme inhibitors are discovered useful as drugs but success still needs to minimize challenges.

12. Acknowledgements

Author acknowledges the suggestions of Dr Pagandai V. Pannirselvam, MTech, Ph.D at Centro de Technologia, UFRN, Lagoa Nova-Natal/RN Campus Universitário. North East, Brazil. Author contributed to explain intriguing issues on enzyme inhibition and highlighted the need of better understanding on mechanism of inhibitors before applying them in industries.

13. References

Amtul, Z. Atta, Ur. R., Siddiqui, R.A., Choudhary, M. I. (2002). Chemistry and mechanism of urease inhibition. *Current Medicinal Chemistry,* Vol 9, pp 1323-1348.

Bartolini M., Cavrini V., Andrisano V. (2005) *J. Chromatogr A,* Choosing the right chromatographic support in making a new acetylcholinesterase microimmobilized enzyme reactor for drug discovery. Vol 1065, pp 135-144.

Bartolini M, Greig NH, Yu QS, Andrisano V. (2009) Immobilized butyrylcholinesterase in the characterization of new inhibitors that could ease Alzheimer's disease. *J Chromatogr A.* Vol 1216(13), pp 2730-38.

Bartolini M., Cavrini V., Andrisano V. Characterization of reversible and irreversible acetylcholinesterase inhibitors by means of an immobilized enzyme reactor. *J. Chromatogr. A* (2007) Vol 1144, pp 102 –10.

Bartolini M, Andrisano V. (2009) Immobilized enzyme reactors into the drug discovery process: The Alzheimer's Disease case. Web Source: http://www.farm.unipi.it/npcf3/pdf/BartoliniManuela.pdf

Bashor, C.J., Helman,N.C., Yan, S., Lim, W.A. Using Engineered Scaffold Interactions to Reshape MAP Kinase Pathway Signaling Dynamics. *Science.*Vol 319 (5869), pp1539-1543

Berg, J.M., Tymoczko, J.L., Stryer, L. (2011) Biochemistry ISBN-13: 978-1429231152, Freeman WH and Company.

Cleland, W.W.(1979) Substrate inhibition, *Methods Enzymol.* Vol 63, pp 500-513.

Dixon,M., Webb,E.C. (1979) Enzymes, 3rd ed., Academic Press, New York.

El-Metwally, T.H., El-Senosi, Y. (2010) Enzyme Inhibition. Medical Enzymology: Simplified Approach.Chapter 6, Nova Publishers, NY. pp 57-77.

Jakbowski H. (2010a) Personal communication. Online study. Chapter 6- Transport and Kinetics. C. Models of Enzyme Inhibition and D. More complicated Enzymes. Internet source. http://employees.csbsju.edu/hjakubowski/classes/ch331/transkinetics/olcomplicatedenzyme.html

--ibid- (2010b) http://employees.csbsju.edu/hjakubowski/classes/ch331/transkinetics/olinhibition.html

Laider, K., Bunting, P. (1980) The kinetics of immonbilized enzyme systems. *Methods Enzymol.* Vol 64, pp 227-248.

Martinek, K., Klibanov, A.M., Goldmacher, V.S. & Berezin, I.V. (1977a) The principles of enzyme stabilization 1. Increase in thermostability of enzymes covalently bound to a complementary surface of a polymer support in a multipoint fashion. *Biochimica et Biophysica Acta,* Vol 485, pp 1-12.

Martinek, K., Klibanov, A.M., Goldmacher, V.S., Tchernysheva, A.V., Mozhaev, V.V., Berezin, I.V. & Glotov, B.O. (1977b) The principles of enzyme stabilization 2. Increase in the thermostability of enzymes as a result of multipoint noncovalent interaction with a polymeric support. *Biochimica et Biophysica Acta* Vol 485, pp 13-28.

Nelson, D.L., Cox, M.M. (2008) Lehninger Principles of Biochemistry. 5th Edition ISBN-13: 978071677108, Freeman W.H. and Company.

Pryciak, P. (2008) Customized Signaling Circuits. Science 319, pg 1489.

Rees, D.C. (1984) A general solution for the steady state kinetics of immobilized enzyme systems. *Bulletin of Mathematical Biology*, Vol 46, 2,pp 229-234.

Sami, A.J., Shakoor, A.R.. (2011) Cellulase activity inhibition and growth retardation of associated bacterial strains of Aulacophora foviecollis by two glycosylated flavonoids isolated from Mangifera indica leaves. *Journal of Medicinal Plants Research* (2011) Vol. 5(2), pp. 184-190.

Sharma,R. (1990) The effect of nitroimidazoles on isolated liver cell metabolism during development of amoebic liver abscess. Dissertation submitted to Indian institute of Technology, Delhi and CCS University.

Sharma, R. (2012a) Mechanisms of Hepatocellular Dysfunction and Regeneration: Enzyme Inhibition by Nitroimidazole and Human Liver Regeneration. In: *Enzyme Inhibition: Concepts and Bioapplications*. Chapter 7, InTech Web Publishers, Croatia. ISBN 979-953-307-301-8.

Sharma, R. (2012b) Inhibition of Nitric Oxide Synthase Gene Expression: *In Vivo* Imaging Approaches of Nitric Oxide with Multimodal Imaging. In: *Enzyme Inhibition: Concepts and Bioapplications*. Chapter 8, InTech Web Publishers, Croatia. ISBN 979-953-307-301-8.

Section 2

Applications of Enzyme Inhibition

2

Cytochrome P450 Enzyme Inhibitors from Nature

Simone Badal, Mario Shields and Rupika Delgoda
University of the West Indies/
Natural Products Institute
Jamaica

1. Introduction

1.1 Cytochrome P450

Cytochrome P450 (CYP) is a heme containing enzyme superfamily that catalyzes the oxidative biotransformation of lipophilic substrates to hydrophilic metabolites facilitating their removal from cells. The CYPs were first recognized by Martin Klingenberg (Klingenberg, 1958) while studying the spectrophotometric properties of pigments in a microsomal fraction prepared from rat livers. When a diluted microsomal preparation was reduced by sodium dithionite and exposed to carbon monoxide gas, a unique spectral absorbance band with a maximum at 450nm appeared. The ferric ion in the resting heme, binds easily with CO following reduction, and the complex's maximal absorbance band, unique amongst hemeproteins, serves as the signature of CYP enzymes.

CYPs are mostly located in the endoplasmic reticulum, and to some extent in mitochondrial fractions of hepatic and extra-hepatic tissues. Even though these enzymes are ubiquitous in the body (Table 1), of the 18 families in mammals identified, 11 are expressed in a typical human liver (CYP1A2, CYP2A6, CYP2B6, CYP2C8/9/18/19, CYP2D6, CYP2E1, and CYP3A4/5). In addition, five of these enzymes (CYPs 1A2, 2C9, 2C19, 2D6 and 3A4) expressed at high levels in the liver demonstrate a broad substrate selectivity which accounts for about 95% of drug metabolism (Nelson, 2009; Treasure, 2000).

The metabolism of a drug can be altered by another drug or foreign chemical and such interactions can often be clinically significant. As a result, the FDA (Food and Drug Administration) and other regulatory agencies such as the Department of Health and Human Services (DHHS), Centers for Disease Control and Prevention (CDS) and Hazard Analysis Critical Control Point (HACCP) among others expect information on the relationship between each new drug to CYP enzymes (substrate, inhibitor and or inducer) making these enzymes vital in the process of drug discovery. One of the major concerns is avoiding drug interactions, an issue whose importance increases with the aging of population (Guengerich, 2003) along with the increase in the practice of polypharmacy.

Organ	CYPs detected
Nasal mucosa	2A6, 2A13, 2B6, 2C, 2J2, 3A
Trachea	2A6, 2A13, 2B6, 2S1
Lung	1A1, 1A2, 1B1, 2A6, 2A13, 2B6, 2C8, 2D6, 2E1, 2F1, 2J2, 2S1, 3A4, 3A5, 4B1
Oesophagus	1A1, 1A2, 2A, 2E1, 2J2, 3A5
Stomach	1A1, 1A2, 2C, 2J2, 2S1, 3A4
Small Intestine	1A1, 1B1, 2C9, 2C19, 2D6, 2E1,2J2, 2S1, 3A4, 3A5
Colon	1A1, 1A2, 1B1, 2J2, 3A4, 3A5

Table 1. Human cytochrome P450 genes expressed in different parts of the respiratory and gastrointestinal tracts (adopted from Ding and Kaminsky, 2003).

1.2 Classification of CYP enzymes

All eukaryotic CYPs except fungal CYP55s are membrane bound; 18 mammalian CYP enzyme structures are known and 15 of these are of human origin; [1A2, 2A6, 2A13, 2B4 rabbit, 2B6, 2C5 rabbit, 2C8, 2C9, 2D6, 2E1, 2R1, 3A4, 7A1, 8A1, 19A1, 24A1 rat, 46A1, 51A1, (Nelson and Nebert, 2011)]. CYPs sharing >40% sequence identity are categorised within the same family while those with >55% sequence identity are placed within the same subfamily. The CYP superfamily members are named according to a nomenclature system that was established in the mid-1980s (Nebert *et al.*, 1987), however, the last comprehensive revision was published in 1996 (Nelson *et al.*, 1996).

CYP2 is the largest CYP450 family in mammals with 13 subfamilies and 16 genes in humans. CYPs2C8, 2C9, 2C18 and 2C19 jointly metabolise more than 50 drugs whilst CYP2D6 metabolises more than 70 drugs (Meyer and Zanger, 1997). CYP3A is the most abundantly expressed CYP450 gene in the human liver and gastrointestinal tract (Nelson, 1999) and is known to metabolise more than 120 commonly prescribed pharmaceutical agents.

CYPs1Al and 1B1 are predominately expressed in extra-hepatic tissues (Guengerich and Shimada, 1991; Shimada *et al.*, 1992) while CYP1A2 is expressed primarily in the liver. As a result, constitutive levels of CYP1A2 are much greater than those of CYPs1A1 and 1B1 (Shimada *et al.*, 1992; Shimada *et al.*, 1994b) whose levels are usually induced by PAHs. All 3 members of the CYP1 family are upregulated by halogenated and polycyclic aromatic hydrocarbons such as those found in cigarette smoke and charred food.

1.3 Importance of CYP enzyme inhibition

1.3.1 Involvement in drug interactions

The metabolism of a drug can be altered by another drug or foreign chemical and such interactions can often be clinically significant. The observed induction and inhibition of CYP enzymes by various traditional remedies have led to the general acceptance that natural therapies can have adverse effects. This is contrary to the popular beliefs in countries where there is an active practice of ethnomedicine. Drug-herb interactions may involve competitive, noncompetitive, or uncompetitive inhibition of drug metabolizing enzymes or enzyme induction by the phytopharmaceutical (Delgoda and Westlake, 2004).

Several epidemiological surveys including ones conducted by our laboratory (Delgoda *et al.*, 2004; Delgoda *et al.*, 2010; Picking *et al.*, 2011) have indicated high usage of herbal medicines along with prescription medicines with low physician awareness. With over 80% of the prescription medicine users also seeking some form of herbal remedy in Jamaica, the chances of drug interactions rises and this prompted investigations into likely pharamacokinetic, metabolism based interactions between the two types of medicines.

The CYP enzymes, responsible for the metabolism of over 90% of drugs in the market is unsurprisingly associated with numerous metabolism related drug interactions (Guengerich, 1997), including those of drugs and herbs (Ioannides, 2002; Delgoda and Westlake, 2004). The inhibition of CYP3A4 by fucocoumarins found in grapefruit juice leading to clinically observable toxicities with drugs and the induction of the same CYP3A4 enzyme by ingredients found in St. John's wort leading to subtherapeutic interferences with cycloporin provide suitable examples for the involvement of CYP enzymes in drug herb interactions. While clinical studies provide the ultimate proof for relevant drug interactions, *in-vitro* laboratory evaluations with CYP enzymes, has provided a convenient, economical and useful starting point for screening those herbs that may ultimately cause clinically observable drug interactions. Human liver microsomes, heterologously expressed enzymes and hepatocytes although with limitations, have provided convenient means for such initial assessements.

In this chapter, we describe for the first time, the initial inhibitory impact of four commonly consumed infusions on six major CYP enzymes. Our findings support that the teas are moderate to weak CYP inhibitors and so we postulate that they would unlikely result in drug interactions.

1.3.2 CYP inhibition and its relation to chemoprevention

Approximately five decades of systematic drug discovery and development have established a reliable collection of chemotherapeutic agents (Yarbro, 1992; Chabner, 1991). These chemotherapeutic agents have assisted with numerous successes in the treatment and management of human cancers (Chabner *et al.*, 1991).

Chemoprevention is the ability of compounds to protect healthy tissues via the prevention, inhibition or reversal of caricnogenesis. The inhibition of CYP1 enzymes is one such route among others that include the induction of cell cycle arrest, the induction of phase II enzymes and the inhibition of inflammatory. The CYP1 family has been linked with the activation of pro-carcinogens which is facilitated by the regulation of the aryl hydrocarbon receptor. As such research has shown that inhibiting CYP1 enzymes plays a key role in protecting healthy cells from the harmful effects of activated carcinogens.

Among the polycyclic hydrocarbons that are activated into reactive metabolites by CYPs 1A1 and 1B1 is benzo-a-pyrene [BaP]. Metabolites from BaP include phenols, polyphenols, quinines, epoxides and dihydrodiols. Among these dihydrodiols; (-)-benzo[a]pyrene-trans-7,8-dihydrodiol (BPD) and (+)-anti-benzo[a]pyrene-trans-7,8-dihydrodiol-9,10-epoxide (anti-

BPDE) are carcinogenic, however the latter is the ultimate carcinogen as it has been shown to bind DNA predominantly at the N^2-position of guanine to produce primarily N^2-guanine lesions, benzo-a-pyrene 7,8-diol-9,10-epoxide-N^2-deoxyguanosine (BPDE-N^2-dG) adduct (Osborn *et al.*, 1976). It is proposed that BPDE-N^2-dG is linked to the high frequency of p53 G→T transversions observed in lung cancer of smokers (Hainaut and Pfeifer, 2001; Pfeifer *et al.*, 2002). Further mutations in the p53 gene have also been found and these include transversions, G→A and G→C (Shukla *et al.*, 1997; Schiltz *et al.*, 1999). Similar to the role of CYP1A1 in the activation of BaP is that of the aromatic amines; amino-3-methylimidazo[4,5-*f*]quinoline (IQ), 2-amino-1-methyl-6-phenylimidazo[4,5-b]pyridine (PhIP) and 2-amino-3,8-dimethylimidazo-[4,5-flquinoxaline (MeIQx). CYP1A2 plays an important role in the N-oxidation of these aromatic amines which have been linked to colon and urothelium cancers (Landi *et al.*, 1999), thus highlighting the role of CYP1 enzymes in carcinogenic activation and thus their potential as preventative targets. Fig.1 is a schematic representation of the process of carcinogenesis at the cellular level.

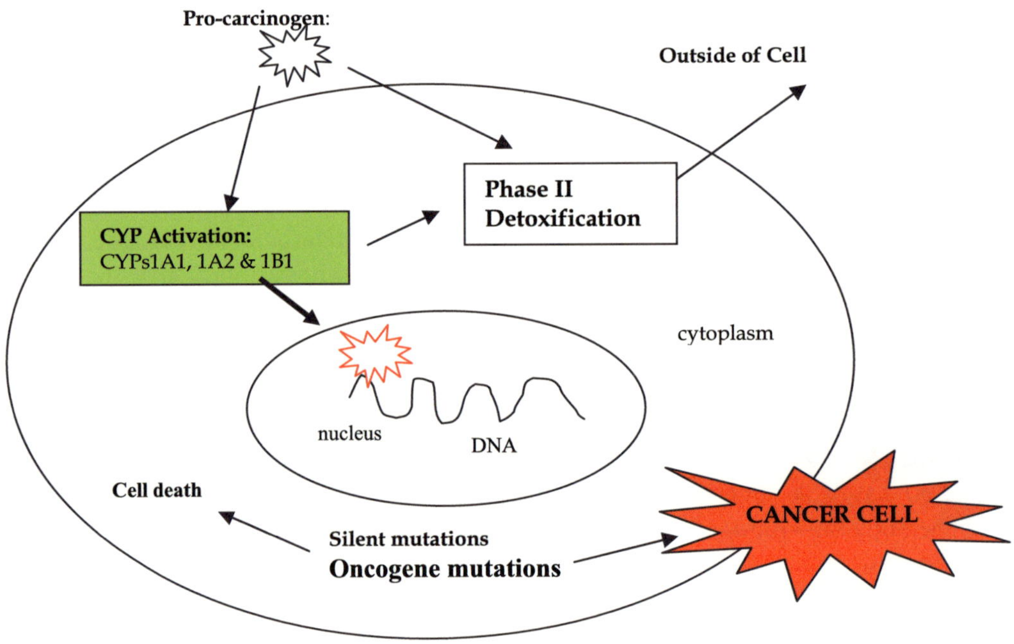

Fig. 1. A schematic representation of carcinogenesis via the activation of CYP1 enzymes. Upon the activation of the pro-carcinogens by the CYP1 enzymes, they have the ability to bind to DNA, which can lead to mutations and then the formation of cancer cells.

One of the first reported chemoprotectants was disulfiram (Stoner *et al.*, 1997) which inhibited the action of dimethylhydrazine via the inhibition of CYP1 enzymes. Other chemopreventive agents are discussed by Chang and others (Chang *et al.*, 2002) who report that Ginseng decreases the incidence of 7,12 dimethyldenz(a)anthracence (DMBA)-initiated tumorigenesis in mice via the inhibition of CYPs1A1, 1A2 and 1B1. Also, the flavanoid, galangin was found to be an agonist of the aryl hydrocarbon receptor and consequently was responsible for an increased level of CYP1A1 expression, however this effect was counteracted by its ability to inhibit the enzyme directly and so is deemed an effective chemo-preventive agent (Ciolino and Yeh, 1999). Resveratrol was also found to exhibit chemo-preventive properties via the inhibition of CYP1A1 expression *in vivo* by preventing the binding of the AhR to promoter sequences that regulate the CYP1A1 transcription and also by the direct potent inhibition of CYPs1A1 and 1B1 (Ciolino *et al.*, 1998; Chen *et al.*, 2004).

1.4 CYP inhibition and its relation to chemoprevention

Bioactivity of isolates from the Jamaica plants, *Amyris plumieri*, *Peperomia amplexicaulis*, *Spathelia sorbifolia* and *Picrasma excelsa* are reported in this chapter. *Amyris plumieri* is found in the Caribbean, Central America and Venezuela and plants of this genus have been used in folk medicine against skin irritation while isolates have been found to exhibit anticancer and antimycobacterial properties (Fuente *et al.*, 1991, Hartwell, 1968). Even though both *Peperomia amplexicaulis* and *Spathelia sorbifolia* are not commonly consumed in Jamaica, isolates from these plants have been shown to exhibit antiprotozoal, chemopreventive and anti-cancer activity (Mota *et al.*, 2009; Cassady *et al.*, 1990) and previously examined for CYP inhibitions (Badal *et al.*, 2011; Shields *et al.*, 2009) and overviewed in this chapter. Infusions of the plant *Picrasma excelsa*, known as Jamaican bitterwood tea, are commonly consumed to lower blood sugar levels in diabetics who are already on prescription medicines. All other plants investigated in this chapter; *Rhytidophyllum tomentosa*, *Psidium guajava*, *Symphytium officinale*, *Momordica charantia* are frequently consumed in the form of teas or the fruits of the appropriate plants. We therefore investigated the inhibition properties of these teas against a panel of CYP450 enzymes in order to assess the potential for drug interactions with co-medicated pharmaceuticals.

2. Materials and methods

2.1 Chemicals

All CYP substrates and metabolites were purchased from Gentest Corporation (Woburn, MA, U.S.A.). All other chemicals were purchased from Sigma-Aldrich (MO, U.S.A.).

2.2 CYP microsomes

Escherichia coli membranes expressing human CYP2D6, CYP3A4, CYP1A1, CYP1A2 and each containing P450 reductase, were a gift from Dr. Mark Paine and Prof. Roland Wolfe (University of Dundee, UK). CYP2C19 expressed in baculovirus-insect cells (supersomes) were purchased from Gentest Corporation, Woburn, MA

2.3 Preparation of infusions from medicinal plants

The selection of the plants for screening and method of preparation were based on the survey conducted by Delgoda *et al* (Delgoda *et al.*, 2010). The teas were prepared by infusing 100ml of boiling deionized water per 1g of dried, finely ground material (leaf, bark or wood chips), for 10 minutes. The resulting liquor was suctioned filtered through type 1 Watman filter paper. A portion of the filtrate was then centrifuged at 13000 × g for 5 minutes to remove suspended solids.

2.4 Separation of active ingredients from medicinal plants

Infusions were freeze dried and re-dissolved in water just prior to use, unless otherwise stated. 25μl infusions were loaded onto a microsorb C18 column (ID 4.6mm, 25cm, 5m) and separated using the appropriate solvent systems using Varian Prostar HPLC system (Varian Inc. USA).

2.5 CYP inhibition assays

Routinely, appropriate volumes of potassium phosphate buffer (KPB), test inhibitor, CYP, and the substrates were added to a NADPH regenerating mixture and made up to 400 μL, and monitored fluorometrically on a continuous basis for 10mins as described elsewhere (Shields, 2009), using CYP450 substrates,3-[2-(N,N-Diethyl-N-methylamino)ethyl]-7-methoxy-4methylcoumarin (AMMC), 7-Benzyloxy-4-trifluoromethylcoumarin (BFC), as substrates for CYP3A4 and CYP2D6 respectively and 7-ethoxy-3-cyanocoumarin (CEC) as substrate for CYPs 1A1, 1A2, 2C19 and 2C9. In other instances (as specified in each case), a 96-well plate assay was employed as detailed in (Badal *et al.*, 2011). Fluoroscence was monitored using a Varian Cary Eclipse Fluorescence spectrophotometer.

Positive control experiments were conducted with varying concentrations of furafaylline (≥98%) (0.5-10μM), quinidine (≥90%, 1-50nM) and ketoconazole (≥98%) (2-100nM) with CYP1A2, CYP2D6 and CYP3A4 respectively.

2.6 Data analysis

IC_{50} and K_i values were determined by fitting the data in Sigma Plot (version 10.0) and enzyme kinetics module, using non linear regression analysis. The data listed represent the average values from three different determinations.

3. Results

3.1 Optimising experimental conditions

To verify the accuracy of experimental techniques employed to detect CYP inhibition, assays with known inhibitors were carried out with furafylline (against CYP1A2), ketoconazole (against CYP1A1, CYP1B1 and CYP3A4), (−)-N-3-benzyl-phenobarbital (NBPB, against CYP2C19) and quinidine (against CYP2D6) and the obtained IC_{50} values (0.8±0.2, 0.04±0.01, 6.3±1.7, 0.06±0.01, 0.3±191 0.01, and 0.03±0.01μM respectively) compared well with published values (0.99, b10, b10, 0.06, 0.25 and 0.01μM respectively; Shields, 2009; Badal *et al.*, 2011; Powrie, 2010; Stresser *et al.*, 2004; Cali, 2003 and McLaughlin *et al.*, 2008).

3.2 Natural products as CYP inhibitors

Several classes of natural products were examined in our laboratory for their inhibitory properties towards CYP450 enzymes. Chromene amides (CAs) isolated from *Amyris plumieri*, quassinoids isolated from *Picrasma excelsa*, anhydrosorbifolin isolated from

Spathelia sorbifolia and chroman 6 isolated from *Peperomia amplexicaulis.* Structures for these can be seen in Figs. 2.1, 2.2 and 2.3 and in addition obtained IC_{50}s can be seen in Table 2. Both CA1 and quassin exhibited the most potency against CYP1A1. Both Anhydrosorbifolin and chroman 6 and CAs, 1, 2 and 3moderately (IC_{50} between 1 and 10µM) inhibited the activities of CYP1 family.

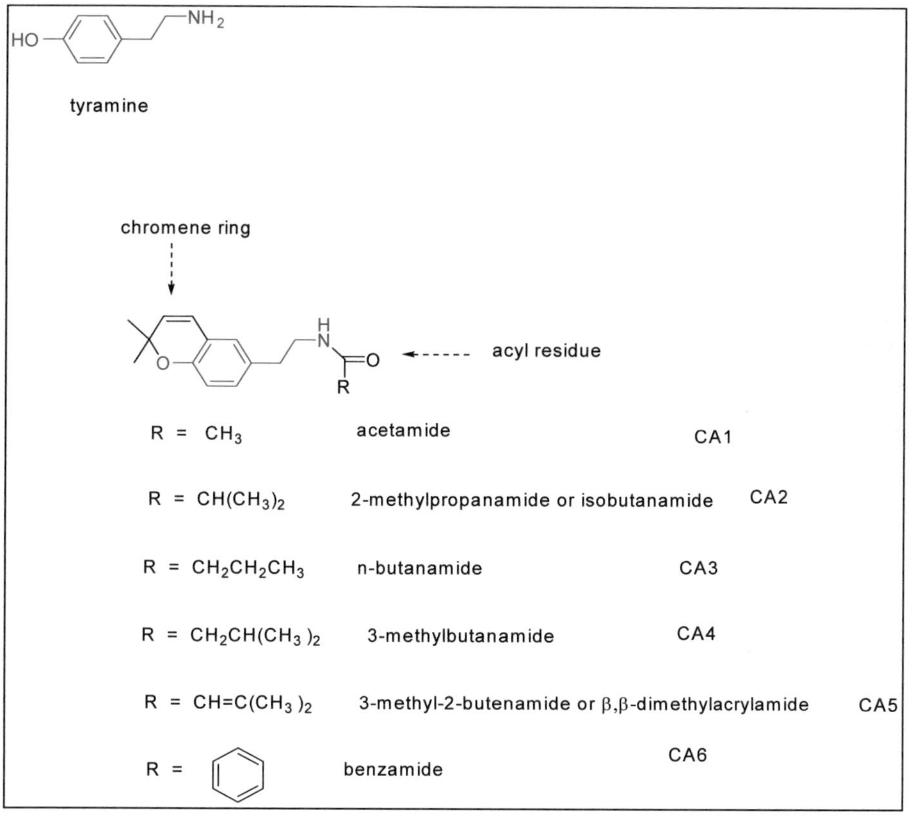

Fig. 2.1. Chromene amides

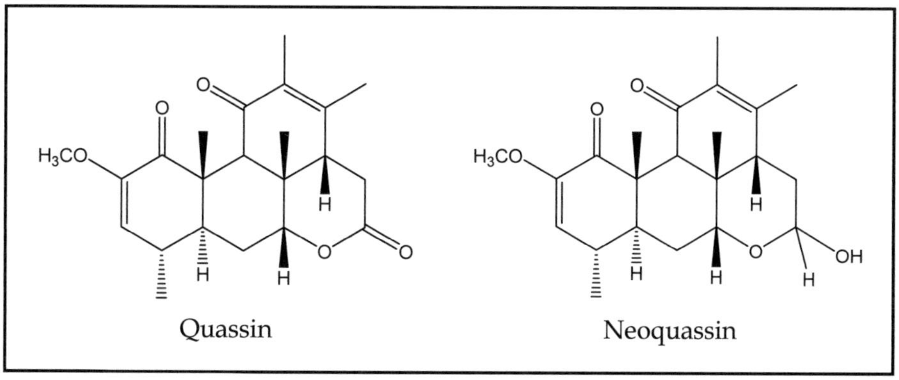

Fig. 2.2. Quassinoids

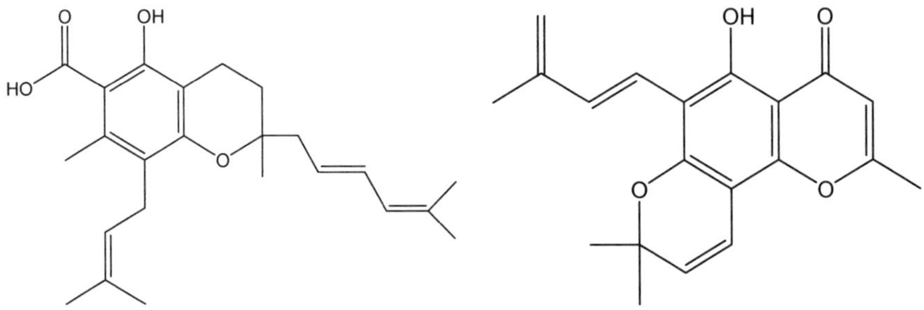

5-Hydroxy-2,7-dimethyl-8-(3-methyl-but-2-enyl)-2-(4-methyl-penta-1,3-dienyl)-chroman-6-carboxylic acid

Anhydrosorbifolin

Fig. 2.3. Others

Compounds	CYP isoforms						
	1A1	1A2	1B1	2C9	2C19	2D6	3A4
CA1	1.31 ± 0.42 K_i =0.37	32.80 ± 4.45	15.36 ± 0.42	nd	0.77 ± 0.39	2.22 ± 0.69	1.14±0.4 8
CA2	1.63 ± 0.53 K_i=2.40	6.25 ± 1.85	37.04 ± 1.51	nd	1.09 ± 0.52	359.88 ± 144.55	15.48±0. 45
CA3	2.43 ± 0.62 K_i=1.39	189.84 ± 7.60	179.30 ± 20.5	nd	2.43 ± 0.28	11.70 ± 5.40	122.93±5 .95
CA4	14.39 ± 7.40	18.59 ± 0.67	18.14 ± 1.02	nd	2.55 ± 1.85	84.40 ± 3.5	7.63±1.2 6
Quassin	9.2 K_i=10.8	57.6	ND	92.5	262.5	217.8	47.0
Neoquassin	11.9 K_i=11.3	85.3	ND	80.6	113.4	184.1	24.5
Anhydro-sorbifolin	4.9	1.9	1.4	nd	nd	nd	nd
Chroman 6	2.1	5.8	5.6	nd	nd	nd	nd

Table 2. Summary of IC_{50} and K_i values (µM) obtained from the interaction of isomers of chromene amides, quassinoids along with anhydrosorbifolin and chroman 6 using heterologously expressed CYP microsomes. ND: Not determined due to intrinsic fluorescence and quenching/enhancement of the metabolite nd: not done

3.3 Herbal infusions with CYP inhibitors

Hot water infusions of five popular herbs; *Rhytidophyllum tomentosa, Psidium guajava, Symphytium officinale, Momordica charantia* and *Picrasma excelsa* were characterized for impact as shown in Fig.3 and calculated IC_{50} values on the activities of CYP enzymes are shown in Table 3.

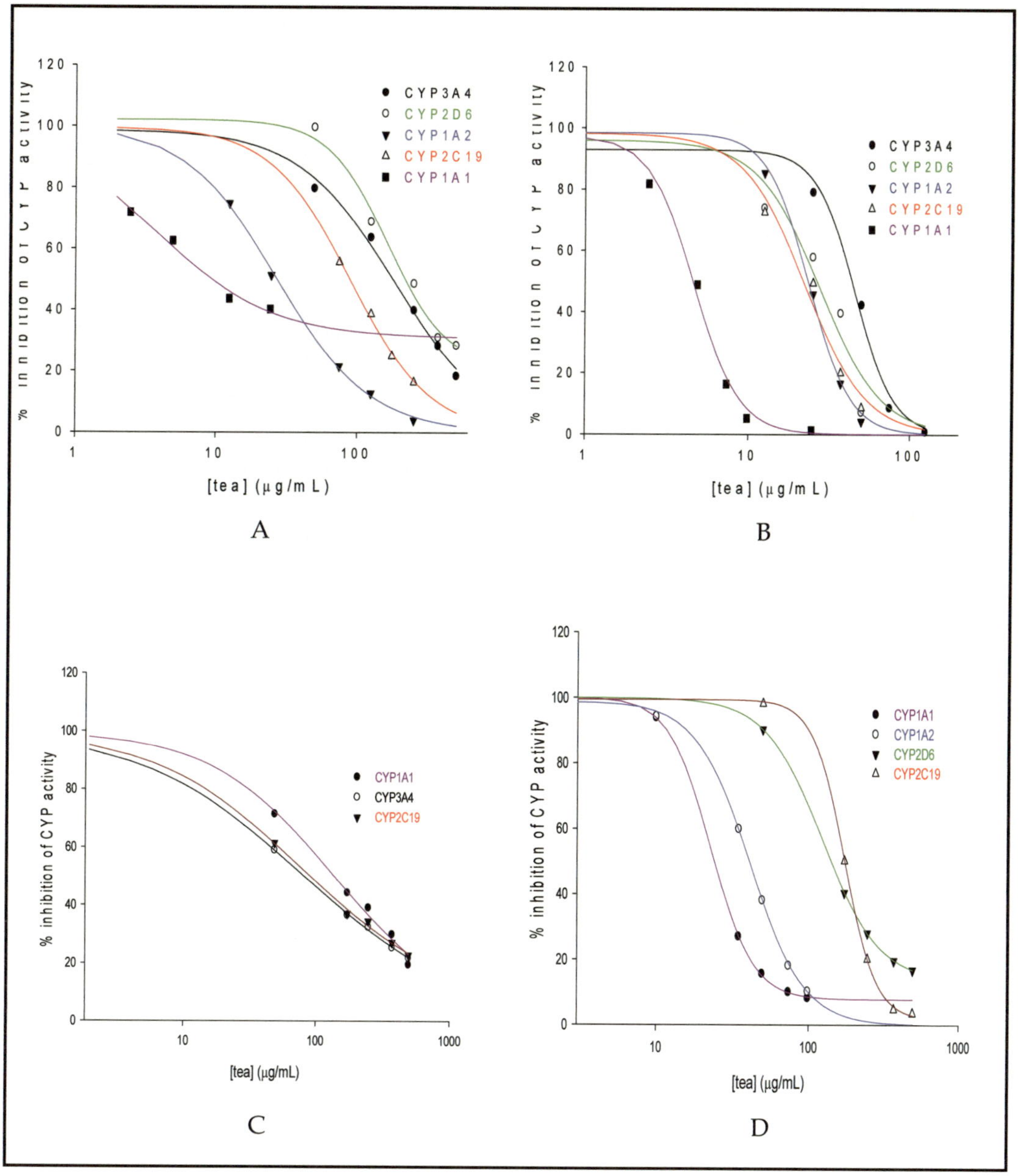

Fig. 3. Inhibition of CYP activity by medicinal plant infusions.

3.4 Identification of active ingredients

Due to the potency displayed against the activities of CYP450 enzymes, *Psidium guajava* was selected for further characterization. Preliminary separation of the freeze-dried extract of *Psidium guajava* by reverse phase HPLC (see Fig.4) revealed several resolved peaks and LC-MS analysis at the same time and the results are summarized in Table 3. Two peaks were identified as quercetin and hyperin whose structures are shown in Fig.5 and these displayed

50% inhibition against the activity of CYP2D6 enzymes as shown in Fig.6. Previously modelled active site of CYP1A1 with bound quassin is displayed in Fig.7 where key residues in the enzyme are identified; Asp313, Thr11, Ser124, Phe123, Ile386 and Leu496 in the interaction between quassin and neoquassin.

Inhibition of CYP activity by *Rhytidophyllum tomentosa infusion* (A); *Psidium guajava* (B); *Momordica charantia* (C); *and Symphytium officinale* (D). Different volumes of reconstituted freeze-dried infusion were added to the incubation mixture, along with the CYP isoform, substrate, and 6GPDH, and monitored fluorometrically over time, as described in Materials and Methods. Control enzyme activity (mean ± SEM) for CYP3A4, CYP1A1, CYP2D6, CYP1A2, CYP2C19, and CYP2C9 was 0.147 ± 0.037, 0.907 ± 0.095, 0.005 ± 0.000, 1.45 ± 0.04, 0.054 ± 0.016, and 0.057 ± 0.004μM/min/pmol of CYP, respectively. Curves for CYP2D6 and CYP1A2 in for *Memordica charantia* (C) and for CYP3A4 in by *Symphytium officinale* (D) were not included because their IC_{50} values exceeded 200μg/mL.

Isoform	IC_{50} (μg/mL)				
	Rhytidophyllum tomentosa	*Psidium guajava*	*Momordica charantia*	*Symphytium officinale*	*Picrasma excelsa**
CYP1A1	10.2	4.9	137.1	24.0	15.0
CYP1A2	28.3	24.0	>200.0	40.3	19.1
CYP2D6	158.0	26.3	>200.0	127.9	>200
CYP2C19	93.8	23.3	91.0	172.7	199.9
CYP3A4	178.1	48.7	82.3	>200.0	122.8

*Values for *Picrasma excelsa* were obtained from Shields *et al*, 2008.

Table 3. Summary of IC_{50} values obtained for the extracts from Fig. 2

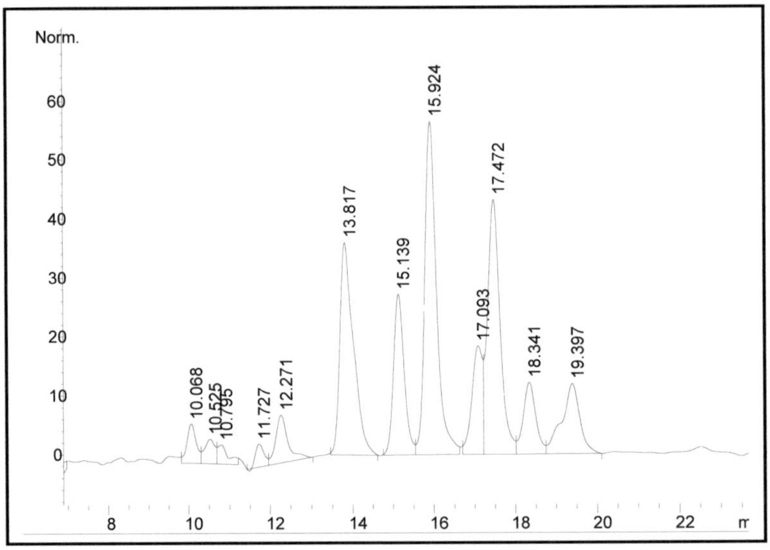

Fig. 4. HPLC profile of *Psidium guajava* extract

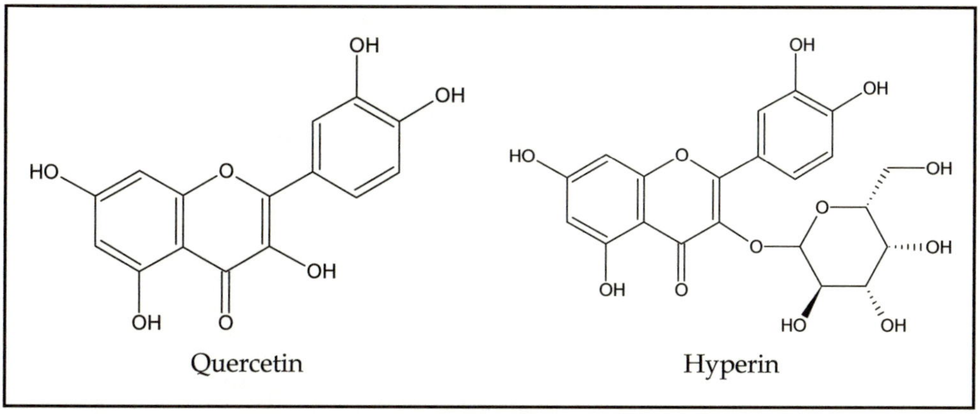

Fig. 5. Structures of quercetin and hyperin (quercetin-3-D-galactoside).

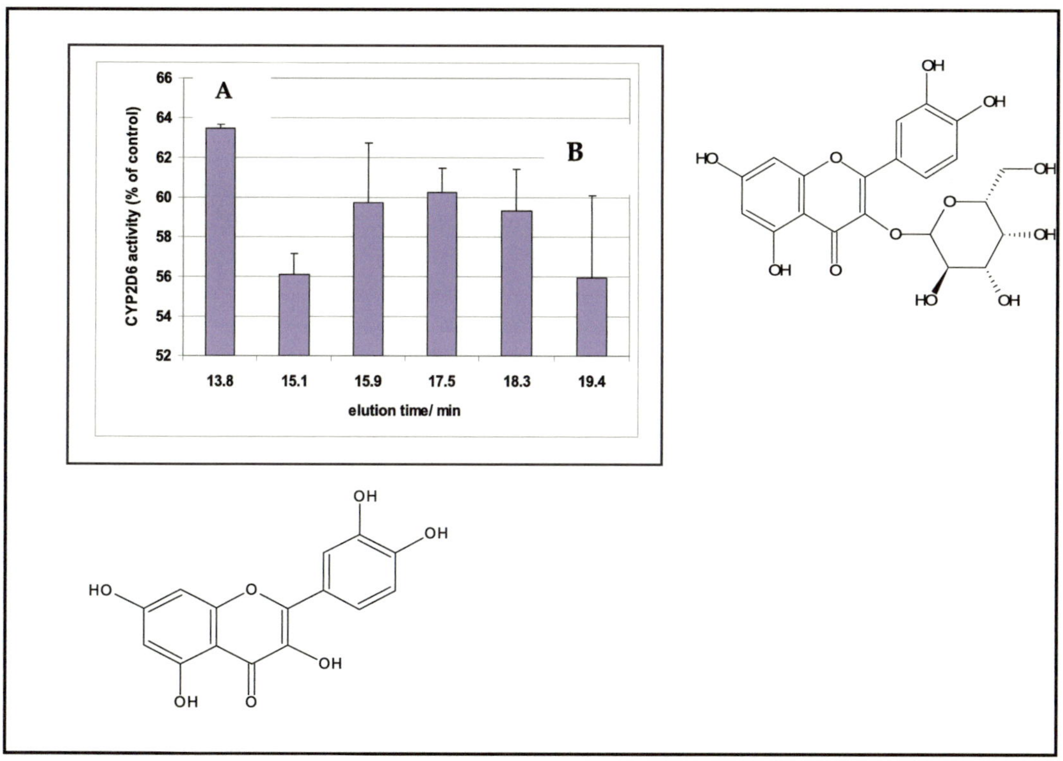

Fig. 6. HPLC profile of *Psidium guajava* (adapted from Shields *et al.*, 2009).

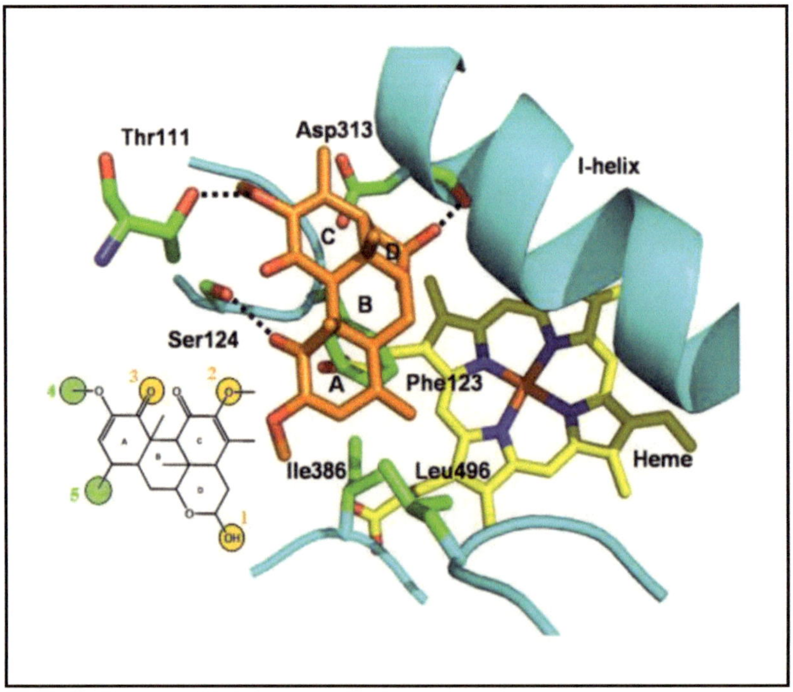

Fig. 7. Interaction of quassinoids with CYP1A1

4. Discussion

Cytochrome P450 enzymes have been of particular interest in the field of drug discovery for numerous reasons including the involvement of these enzymes in the metabolism of over 95% of the drugs on the market and the potential of drug-drug interaction through metabolism. CYP1 family which is under the regulation of the aryl hydrocarbon receptor have been extensively researched and implicated in drug resistance as well as carcinogenesis. CYP1B1 in particular, found in elevated levels in cancer tissues such as those in colon is thought to provide a novel pathway for drug discovery and optimisation for cancer treatment. Inhibitors of the activities of CYP1 enzymes are now accepted as potential chemoprotectants by preventing the activation of polycyclic aromatic hydrocarbons such as benzo-a-pyrene. Catalysed by CYP1A1 and CYP1B1, metabolites of this pro-carcinogen, (+)-anti-benzo[a]pyrene-trans-7,8-dihydrodiol-9,10-epoxide (anti-BPDE) and that 3-hydroxybenzo [a]pyrene (3HBaP) have been shown to bind to DNA predominantly at the N^2-position of guanine to produce N^2-guanine lesions, benzo-a-pyrene 7,8-diol-9,10-epoxide-N^2-deoxyguanosine (BPDE-N^2-dG) adduct (King *et al.*, 1976). Thus inhibitors of CYP1 enzymes hold the potential to prevent the formation of such damaging precursors that initiate malignant cancers of the breast, colon, lung and urothelium among others.

In this chapter we highlight the potential of a few natural products abundant in the Caribbean: chromene amides isolated from *Amyris plumieri*, quassinoids isolated from *Picrasma excelsa*, anhydrosorbifolin isolated from *Spathelia sorbifolia* and chroman 6 isolated from *Peperomia amplexicaulis.* We also report for the first time bioactive screening of CYP enzymes in the presence of five aqueous infusions of popularly used herbs; *Rhytidophyllum tomentosa, Psidium guajava, Symphytium officinale, Momordica charantia* and *Picrasma excelsa.*

Potent and selective inhibition of the CYP1 enzymes were found amongst the investigated natural compounds in particular chromene amides and quassinoids. CA1 displayed potent inhibition against the activity of CYP1A1, with a K_i of 0.37μM and an IC_{50} value of 1.31μM while quassin inhibited this enzyme with an IC_{50} value of 9.2μM and K_i of 10.8μM with selectivity extended throughout all CYP enzymes investigated except CYP2C19 for CA1. The degree of potency and selectivity with which both compounds inhibited this enzyme warrants further research as possible chemoprotectants. Previously known and studied natural compounds deemed to possess chemoprotective properties due to their ability to inhibit CYP1A1 include; quercetin (IC_{50}=1.36μM, Leung *et al.*, 2007), curcumin (IC_{50}=20μM), demethoxycurcumin (IC_{50}=21μM), ε-viniferin (IC_{50}=1μM), resveratrol (IC_{50}=30μM), and sanguinarine (K_i =2μM). Both test compounds compare well with these known chemoprotectants and thus warrant further research.

Both CYPs 1A1 and 1A2 share approximately 70% similarity in amino acid and the specificity with which inhibition targeted towards CYP1A1 activity is noticeable in the compounds CA1, CA3 and quassin. As such we unlocked the interaction between quassin and CYP1A1 in previous publication (Shields *et al.*, 2009). One of the first active site models for CYP1A1 was demonstrated with quasin and important residues were highlighted; Asp313, Thr11, Ser124, Phe123, Ile386 and Leu496 as shown in Fig.7 as being critical for binding quassinoids.

CYP1B1 has been drawing keen interest for novel and anticancer therapeutics. Findings of the over-expression of CYP1B1 in many tumour tissues compared with normal surrounding cells, have led to the search for pro-drugs reliant on CYP1B1 metabolism for the conversion into cytotoxic therapeutics. Although the role of such over- expression is yet to be fully understood, it has been linked with drug resistance and in the promotion of cell survival (Martinez *et al.*, 2008). The modification in the expression levels of CYP1B1 has been shown to modulate tumour progression (Castro *et al.*, 2008) and thus specific inhibitors are expected to be of therapeutic/preventive benefit. Although the potency and the specificity of the chromene amides examined this chapter against CYP1B1 is not particularly high, structure-activity relations may guide towards chromene amides with putative improvement. Anhydrosorbifolin and chroman 6 displayed the most potency against this enzyme deeming further investigations worthwhile.

Little impact towards the CYP1 family was observed in the presence of CA2 which could be due to the isopropyl group on this chromene amide, compared with CA1. Even though the K_i against CYP1A1 was increased (to 2.63μM), the IC_{50} value remained more or less the same as CA1 (1.63μM), displayed moderate to low potency against CYPs 1A2 and 1B1. The structural change made to CA2 was more significant in binding CYP2D6 as the inhibition dropped over a 100 times (IC_{50}=360μM for CA2 vs 2μM for CA1). Hence, CA2 displayed characteristics of a useful molecular probe where all significant drug metabolizing enzymes can be inhibited except for the activity of CYP2D6. Chain elongation and the loss of branching in the n-propyl end unit to form CA3, have a dramatic impact on the affinity to CYP1A2 and CYP1B1. The inhibition potency dropped 6 folds against CYP1A2 (from 32.8µM for CA1 to 189.8µM for CA3) and 10 folds against CYP1B1 (from15.4µM to 179.3µM). Thus, CA3 appears to show increased selectivity in its inhibition against CYP1A1. Exploring other side groups as well as shifting the position of the existing side groups can be explored in hopes of increasing the potency of CAs towards the activity of CYP1B1

There is a large consumption of natural medicines alone and concurrently with prescription medicines in the Caribbean, as in many parts of the world. This is confirmed in a recent pilot study done in which 80% of prescription medicine consumers also take natural remedies (Delgoda *et al.*, 2004; Picking *et al.*, 2011). Also, adverse drug reactions (ADRs) accumulates to over 2 million per year in the United States alone (Gurwitz *et al.*, 2000), therefore, the ability to predict drug interactions involving the CYP enzymes has become a key component of the drug discovery process (Forti and Wahlstrom, 2008). Providing the FDA with the metabolic profile of a new drug entity with CYP enzymes is the first step towards avoiding adverse reactions (Delgoda and Westlake, 2004). Known drug -herb interactions with clinical impact include grapefruit juice with felopidium, tricyclic anti depressants which is medicated through CYP3A4 inhibition.

We report for the first time hot water infusions of *Rhytidophyllum tomentosa, Psidium guajava, Symphytium officinale, Momordica charantia* in comparison with reported *Picrasma excelsa* being tested against CYP enzymes activity. The greatest potency was observed in the presence of *Psidium guajava* that inhibited CYP1A1 with an IC_{50} of 4.9µg/mL. All of the infusions are commonly consumed, be it, the fruit or as teas. With the high levels of polypharmacy practise that exist among Caribbean people and the world, having a metabolic profile on teas or commonly consumed plants become of grave importance. Results displayed in Table 3 point to is minimal risk through CYP mediated drug interactions as the teas weakly inhibited the main drug metabolising enzymes. Because, the activity of *Psidium guajava* towards CYP1A1 was the most potent we further evaluated the identification of the active ingredients that could be responsible for the observed bioactivity towards the CYP1A1 activity. The active ingredients were found to be quercetin and hyperin (see Figs. 4 & 5), compounds known as inhibitors of CYP enzymes, where quercetin isolated from St. John's Wort has been previoulsy shown to inhibit activities of CYPs 1A2, 2C19 and CYP2D6 with IC_{50}s of 3.87 µM, 6.23 µM and 20.99 µM respectively; while hyperforin isolated from *Ginko biloba* shown to inhibit the activity of CYP3A4 with an IC_{50} of 4.30µM (Moltke *et al.*, 2004; Zou *et al.*, 2002). Inhibitions against the CYP enzymes appear to be between moderate and weak which confirms data in our lab, thus evoking moderate concern for potential interactions with co-medicated pharmaceuticals through CYP mediated metabolism.

5. Conclusion

Review of compounds that have potent and selective inhibitory properties against the activities of CYP1 family in particular CYPs 1A1 and 1B1 aid in identification of useful chemoprotectors. CA1 and quassin warrant further research because they were both potent against the activity of CYP1A1 while anhydrosorbifolin and chroma 6 targeted

CYP1B1. *In vitro* and *in silico* models as demonstrated for the first time in this chapter are useful tools in the process of drug development to approximate the risk of drug interactions and in the process of target improvement of key enzymes in chemoprevention. In particular, for herbal remedies they confer useful models for evaluating the risks of adverse effects arising from interactions with co-administered prescription medicines, for which drug-interaction information is not mandated by the regulatory agencies. Once the initial risk is estimated, clinical drug-interaction studies can be launched, thus providing a cost-effective sieving process prior to embarking on rigorous and expensive investigations.

6. Acknowledgments

We are grateful to the International Foundation for Science (IFS), Sweden, the University of the West Indies post graduate fund, the Forestry Conservation fund and the Luther Speare Scholarship for financial support. We are also grateful to Professor Helen Jacobs for provision of select natural products.

7. References

Badal, S.; Williams, S. G. Huang, G. Francis, S. Vedantam, P. Dunbar, O. Jacobs, H. Tzeng, J. Gangemi J. and Delgoda, R. (2011). Cytochrome P450 1 enzyme inhibition and anticancer potential of chromene amides from *Amyris plumieri*, *Fitoterapia*. Vol.82 pp. 230-236.

Cassady, J.M.; Baird, W.M. Chang, C.J. (1990). Natural Products as a Source of Potential Cancer Chemotherapeutic and Chemopreventive Agents. *Journal of natural products*. Vol.53 pp 23-41.

Castro, D.J.; Baird, W.M. Pereira, C.B. Giovanni, J. Löhr, C. Fischer, K. Yu, Z. Gonzalez, F.J. Krueger, S.K. Williams D.E. (2008). Fetal mouse cyp1b1 and transplacental carcinogenesis from maternal exposure to Dibenzo[a,l]pyrene. *Cancer Prev Res*, Vol.1 pp. 128-34.

Chabner, B.; (1991). Anti-cancer drugs. Principles and Practice, 4th Edition. Philadelphia, Lippincott pp. 325-417

Chang, T.; Chen, J.S. Benetton, S. (2002). In Vitro Effect of Standardized Ginseng Extracts and Individual Ginsenosides on the Catalytic Activity of Human CYP1A1, CYP1A2, and CYP1B1. *Drug Metab. Dispo,s* Vol.30 pp. 378-384.

Chang, S.; Johnston Jr, P. Frokjaer-Jensen, C. Lockery, S. and Hobert, O. (2004). Hobert, MicroRNAs act sequentially and asymmetrically to control chemosensory laterality in the nematode. *Nature*, Vol.430 pp.785-789.

Chen, Z.H.; Hurh, Y.J. Na, H.K. Kim, J.H., Chun, Y.J. Kim, D.H. Kang, K.S. Cho, M.H. Surh, Y.J. (2004). Resveratrol inhibits TCDD-induced expression of CYP1A1 and CYP1B1 and catechol estrogen-mediated oxidative DNA damage in cultured human mammary epithelial cells. *Carcinogenesis*, Vol. 25 pp. 2005-2013.

Ciolino, H.; Daschner, P. and Yeh, G. (1998). Resveratrol inhibits transcription of CYP1A1 in vitro by preventing activation of the aryl hydrocarbon receptor. *Cancer Res* Vol.58 pp. 5707-5712.

Delgoda, R.; and Westlake, A. (2004). Herbal Interactions involving Cytochrome P450 enzymes. *Toxicol Rev,* Vol. 23 pp. 239-249.

Delgoda, R.; Ellington, C. Barrett, S. Gordon, N. Clarke, N. Younger, N. (2004). The practice of polypharmacy involving herbal and prescription medicines in the treatment of diabetes mellitus, hypertension and gastrointestinal disorders in Jamaica. *West Indian Medical Journal,* Vol.53 pp. 400–404.

Delgoda, R.; Younger, N. Barrett, C. Braithwaite, J. and Davis, D. (2010). The prevalence of herbs use in conjunction with conventional medicines in Jamaica. Complementary Therapies in Medicine, Vol.18 pp. 13-20.

Forti, R.; and Wahlstrom, J. (2008). CYP2C19 inhibition: The impact of substrate probe selection on in vitro inhibition profiles. *Drug metabolism and disposition,* Vol.36 pp. 523-528.

Fuente, G.; Reina, M. and Timon, I. (1991). Chromene amides from *Amyris texana.* Phytochemistry, Vol. 30 pp. 2677-2684.

Guengerich, F.P; (1997) Role of cytochrome P450 enzymes in drug-drug interactions. *Adv Pharmacol,* Vol. 43 pp. 7–35.

Guengerich, F.P; (2003). Cytochromes P450, drugs, and diseases. *Mol Interv,* Vol. 3(4) pp. 194-204.

Guengerich, F.; and Shimada, T. (1991). Oxidation of toxic and carcinogenic chemicals by human cytochrome P-450 enzymes. Chem. Res. Toxicol, *Vol.* 4 pp. 391-407.

Gurwitz, J.; Field, T. Avorn, J. McCormick, D. Jain, S. Eckler, M. Benser, M. Edmondson, A. and Bates, D. (2000). Incidence and preventability of adverse drug events in nursing homes. *Am J Med,* Vol. 109 pp. 87-94.

Hartwell, J.; (1968). Plants used against cancer: a survey. *Loydia,* Vol.31 pp.171-179.

Ioannides, C.; (2002). Pharmacokinetic interactions between herbal remedies and medicinal drugs. *Xenobiotica,* Vol. 32 pp. 451-478. Read More: http://informahealthcare.com/doi/abs/10.1080/00498250210124147

King, H.W.S.; Osbourne, M.R. Beland, F.A. Harvey, R.G. and Brookes, P. (1976). (+)-7α,8β-Dihydroxy-9β,10β-epoxy-7,8,9,10-tetrahydrobenzo[a]-pyrene is an intermediate in the metabolism and binding to DNA of benzo[a]pyrene. *Proc. Nati. Acad. Sci.,* Vol. 73 pp. 2679-2681.

Klingenberg, M,; (1958). Pigments of rat liver microsome. Archives of biochemistry and biophysics, Vol.75 pp. 376-386.

Landi, M.; Sinha, R. Lang, N. and Kadlubar, F. (1999). Human cytochrome P4501A2. *IARC Sci Publ,* Vol.148 pp.173-195.

Leung, H.Y.; Wang, Y., Chan, H.Y., Leung, L.K. (2007) Developing a high throughput system for the screening of cytochrome P450 1A1- Inhibitory polyphenols. *Toxicol in Vitro,* Vol.21 pp. 996 –1002

Martinez, V.; O' Connor, R. Liang, Y. and Clynes, M. (2008). CYP1B1 expression is induced by docetaxel: effect on cell vialbility and drug resistance. *British journal of cancer,* Vol.98 pp. 564-570.

Meyer, B.; Pray-Grant, Vanden Heuvel, J. and Perdew, G (1998). Hepatitis B virus X-associated protein 2 is a subunit of the unliganded aryl hydrocarbon receptor core complex and exhibits transcriptional enhancer activity. *Mol Cell Biol,* Vol.18 pp. 978-988.

Mitchell, S.; and Ahmad, M. (2006). A Review of Medicinal Plant Research at the University of the West Indies, Jamaica, 1948-2001. *West Indies Medical Journal*, Vol. 55 pp. 243-269.

Moltke, L.L.; Weemhoff, J.L. Bedir, E. Khan, I.A. Harmatz, J.S. Goldman, P. Greenblatt, D.J. (2004). Inhibition of human cytochromes P450 by components of Ginkgo biloba. *Journal of Pharm. And Pharmacol*, Vol.56 pp. 1039-1044.

Mota, J.d.S.; Leite, A.C. Junior, M.B. Lopez, S.N. Ambrosio, D.L. Passerini, G.D. Kato, M.J. Bolzani, V.d.D. Cicarelli, B.R.M.. Furla, M. (2009). In vitro *Trypanocidal* Activity of Phenolic Derivatives from *Peperomia obtusifolia*. *Planta Med*, Vol. 75 pp. 620-623

Nelson, D.; (2009). "Cytochrome P450." Available at <http://drnelson.utmem.edu/CytochromeP450.html>.

Nelson, D.R.; Koymans, L. Kamataki, T. Stegeman, J.J. Feyereisen, R. Waxman, D.J. Waterman, M.R. Gotoh, O. Coon, M.J. Estabrook, R,W. Gunsalus, I.C. Nebert, D.W. (1996). P450 superfamily: update on new sequences, gene mapping, accession numbers and nomenclature. *Pharmacogenetics and genenomics*, Vol.6 pp. 1-42.

Nelson, D.; and Nebert , D.(2011). Cytochrome P450 gene (CYP) superfamily. Encyclopedia of Life Sciences (ELS) Chichester, John Wiley & Sons, Ltd: pp. 1-13.

Nebert, D.; Adesnik, M. Coon, M. Estabrook, R. Gonzalez, F. Guengerich, F. Gunsalus, I. Johnson, E. Kemper, B. Levin W. and et al. (1987). The P450 gene superfamily: recommended nomenclature. *DNA*, Vol. 6 pp. 1-11.

Osborne, M.R.; Brookes, P. Baland, F.A.,Harvey, R.G . (1976). The reaction of (±)-7α, 8β-dihydroxy-9β, 10β-epoxy-7,8,9,10-tetrahydrobenzo(a)pyrene with dna. *International journal of cancer*, Vol.18 pp. 362-368.

Picking, D.; Younger, N. Mitchell, S. Delgoda, R. (2011). The prevalence of herbal medicine home use and concomitant use with pharmaceutical medicines in Jamaica. *Journal of Ethnopharmacology*, Vol.137 pp. 305-311.

Shields, M..; Niazi, U. Badal, S. Yee, T. Sutcliffe, M. and Delgoda, R. (2009). Inhibition of CYP1A1 by Quassinoids found in *Picrasma excelsa*. *Planta Medica*, Vol.75 pp. 137-141.

Shimada, T.; Yun, C. Yamazaki, H. Gautier, J. Beaune, P. Guengerich, F. (1992). Characterization of human lung microsomal cytochrome P-450 1A1 and its role in the oxidation of chemical carcinogens. *Mol. Pharmacol.* Vol.41 pp. 586-864.

Shimada, T.; Yamazaki, H. Mimura, M. Inui, Y. and Guengerich, F. (1994b). Interindividual variations in human liver cytochrome P450 enzymes involved in the oxidation of drugs, carcinogens, and toxic chemicals: studies with liver microsomes of 30 Japanese and 30 Caucasians. *Pharmacol. Exp. Ther* , Vol.270 pp. 414-423.

Shukla, R.; Liu, T. Geacintov, N. Loechler, E. (1997). The Major, N2-dG Adduct of (+)-anti-B[a]PDE Shows a Dramatically Different Mutagenic Specificity (Predominantly, G → A) in a 5′-CGT-3′ Sequence Context. *Biochemistry*, Vol. 36 pp. 10256–10261.

Stoner, G.; Morse, M. and Kelloff, G. (1997). Perspectives in cancer chemoprevention." *Environ Health Perspect*, Vol.105 pp. 945-954.

Treasure, J.; (2000). Herbal Pharmacokinetics: A practitioner update with reference to St. Johns Wort (Hypericum perforatum) Herb-Drug Interactions. MNIMH Vol.1.4 pp. 6-7.

Zou, L.; Hrakey, M.R. Henderson, G.L. (2002). Effects of herbal components on cDNA-expressed cytochrome P450 enzyme catalytic activity. Life Sciences, Vol.71 pp. 1579-1589.

Yarbro, JW.; (1992). The scientific basis of cancer chemotherapy, In: Perry MG (Ed.): Chemotherapy source book (Ed3). Baltimore. Williams and Wilkins 2001. pp.3-18.

3

Pharmacomodulation of Broad Spectrum Matrix Metalloproteinase Inhibitors Towards Regulation of Gelatinases

Erika Bourguet[1], William Hornebeck[2], Janos Sapi[1], Alain Jean-Paul Alix[3] and Gautier Moroy[4]
[1]*CNRS UMR 6229, Institut de Chimie Moléculaire de Reims, IFR 53 Biomolécules, UFR de Pharmacie, Université de Reims-Champagne-Ardenne*
[2]*CNRS UMR 6237, Laboratoire de Biochimie Médicale, MéDyc, IFR 53 Biomolécules, UFR de Médecine, Université de Reims-Champagne-Ardenne*
[3]*Laboratoire de Spectroscopies et Structures Biomoléculaires (EA4303), IFR 53 Biomolécules, UFR Sciences, Université de Reims-Champagne-Ardenne*
[4]*INSERM UMR 973, Molécules thérapeutiques in silico (MTi), Université Paris Diderot France*

1. Introduction

Matrix metalloproteinases (MMP) constitute a family of 23 zinc- and calcium-dependent endopeptidases that play pivotal functions in several physiological processes such as embryogenesis, wound healing, vasculogenesis or stem cell mobilization (Nagase et al., 2006). These enzymes were originally defined as matrix-degrading proteases, but a myriad of other substrates have been discovered including cytokines, chemokines, growth factors and their receptors, cell adhesion molecules and angiogenic factors. MMP were first described to exert their degradative function extracellularly against matrix macromolecules or at the pericellular microenvironment. Recently, MMP proved to cleave intracellular substrates belonging to any subcellular compartments (Cauwe & Opdenakker, 2010). Among them were notably apoptotic regulators, signal transducers, molecular chaperones or transcriptional and translational regulators. Therefore, MMP can be considered as proteases mainly controlling signaling events through processing cytokines, chemokines and degrading matrix, liberating matrikines in the extracellular space, or in turn cleaving enzymes involved in signal transduction inside the cells. MMP are regulated at distinct levels including gene expression, compartmentalization, proenzyme activation, enzyme inhibition, endocytosis, and finally substrate availability and affinity. MMP up-regulation participates in tumor progression and metastasis, inflammatory disorders, cardiovascular and autoimmune diseases (Hu et al., 2007; López-Otín & Matrisian, 2007; Mandal et al., 2003; Murphy & Nagase, 2008).

All MMP are produced as proenzymes *i.e.* zymogen; enzyme latency is due to the formation of a coordinated bond between the zinc atom in the active site and an amino acid residue cysteine present in a consensus PRCGXPD sequence in MMP prodomain. Proteolysis of the prodomain, action of reactive oxygen species (O_2^-, NO) on the amino acid residue cysteine and allosteric perturbation (Sela-Passwell et al., 2010) of the prodomain can disrupt this Cys-Zn bond, a process named "cysteine switch" (Van Wart & Birkedal-Hansen, 1990). In the active enzyme, the zinc atom is linked to three histidine residues and a water molecule. A conserved glutamic acid residue (Glu) in the catalytic domain HEBXHXBGBXHS polarizes the water molecule (Gomis-Rüth, 2009; Lovejoy et al., 1994). This ligated water molecule attacks the carbonyl carbon of the scissile bond and transfers a proton to Glu and then to the scissile nitrogen atom. Then Glu releases the second proton from the water molecule to the scissile nitrogen atom and the peptide bond is cleaved (Figure 1).

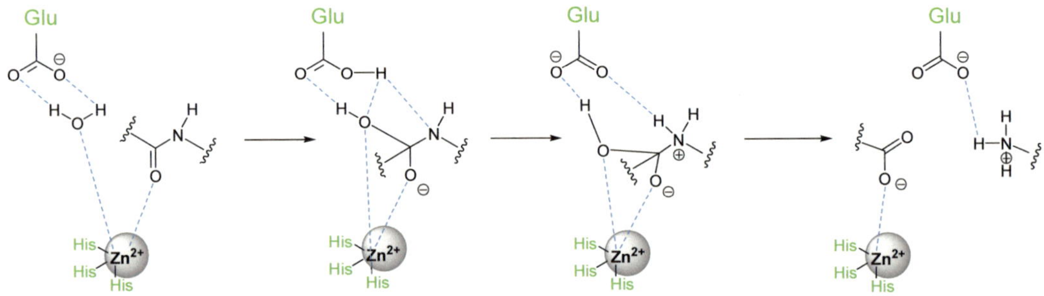

Fig. 1. Mechanism of action of MMP (adapted from Lovejoy et al., 1994).

Historically, MMP were named according to their preferential action on matrix components: collagenases (MMP-1, MMP-8, MMP-13), gelatinases (MMP-2, MMP-9), proteoglycanases or stromelysins (MMP-3, MMP-10), macrophage elastase (MMP-12).

To date, a classification based on their domain organization is favoured: five of them are secreted and the others are transmembrane proteins (MT-MMP) based on their structure similarities (Table 1) (Egeblad & Werb, 2002).

MMP family is constituted by: a pre-domain involved in enzyme secretion, a pro-domain including a cysteine residue interacting with the zinc atom in the catalytic domain that maintains the inactive enzyme form. The catalytic domain is responsible of the MMP activity. All MMP, except MMP-7, MMP-26 (28 kDa) and MMP-23 (56 kDa) possess a hemopexin-like domain involved in the substrate interactions. The gelatinases (MMP-2 (72 kDa) and MMP-9 (92 kDa)) contain a gelatin-binding type II domain with three fibronectin (Fn(II))-like repeats. MMP-11 (51 kDa) and MMP-28 (59 kDa) contain a furin motif for recognition by furin-like serine proteinases. This motif is also present in MMP containing a vitronectin-like domain (MMP-21 (70 kDa)) and membrane-type MMP (MT-MMP). In addition, MT-MMP have a transmembrane domain and a short cytoplasmic domain or a glycosylphosphatidylinositol anchored (MMP-17 (57 kDa) and MMP-25 (63 kDa)). Finally, MMP-23 is a type II transmembrane MMP with a cysteine array and immunoglobulin-like domain.

Designation	Main structure	Name
MMP-7 MMP-26	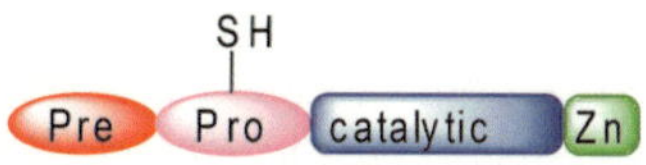 Minimal-domain	Matrilysin Matrilysin-2
MMP-1 MMP-3 MMP-8 MMP-10 MMP-12 MMP-13 MMP-18 MMP-19 MMP-20	 Simple hemopexin-domain	Collagenase-1 Stromelysin-1 Collagenase-2 Stromelysin-2 Metalloelastase Collagenase-3 Collagenase-4 Enamelysin
MMP-2 MMP-9	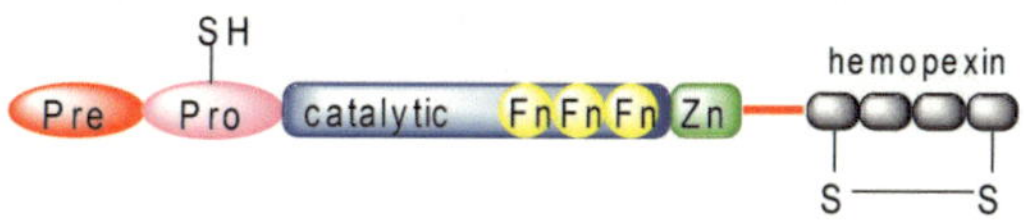 Gelatin binding	Gelatinase A Gelatinase B
MMP-11 MMP-28	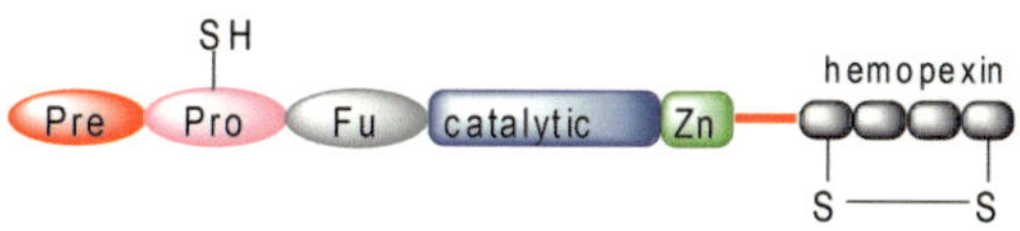 Furin activated	Stromelysin-3 Epilysin
MMP-14 MMP-15 MMP-16 MMP-24	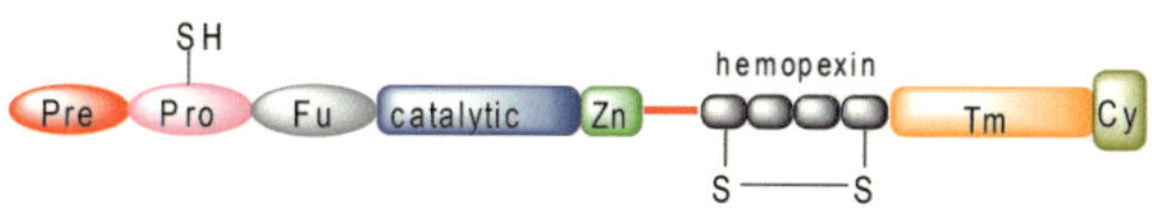 Transmembrane	MT1-MMP MT2-MMP MT3-MMP MT5-MMP
MMP-17 MMP-25	 GPI-anchored	MT4-MMP MT6-MMP
MMP-21	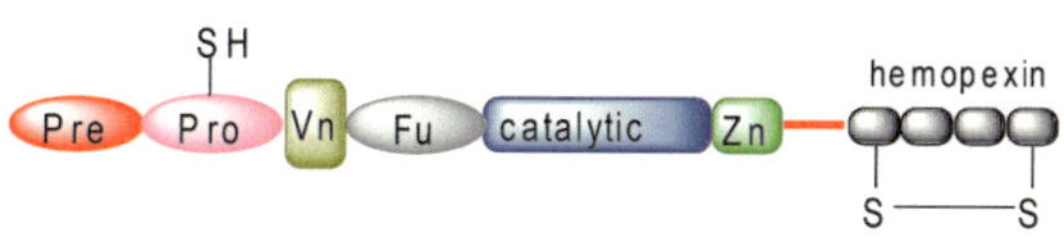 Vitronectin-like	XMMP (*Xenopus*)
MMP-23	 Type II transmembrane	Femalysin

Table 1. MMP family. Pre: signal peptide, Pro: propeptide, Fn: fibronectin type II domain, Fu: furin recognition site, Vn: vitronectin-like domain, TM: transmembrane domain, Cy: cytoplasmic domain, GPI: glycosylphosphatidylinositol, CA: cysteine array, Ig-like: immunoglobulin-like domain.

2. Structures and properties of gelatinases

2.1 Structure of gelatinases active sites

Gelatinases A (MMP-2) and B (MMP-9), as classified as both collagenases and elastases, are involved to a great extent in pathologies affecting major elastic tissues (lung, arteries). Among the MMP family members, gelatinases subclan, MMP-2 and MMP-9, do exhibit several originalities that could be taken into account for the design of inhibitors.

MMP family proved to have a great homology of sequence and the zinc-containing catalytic site is surrounded by subsite pockets named S1, S2, S3 for non-primed and S'1, S'2, S'3 for the primed side (Terp et al., 2002).

The conserved amino acid residues in gelatinases active-site region (Cuniasse et al., 2005; Kontogiorgis et al., 2005; Nicolotti et al., 2007; Rao, 2005) are given in Table 2.

The structural amino acid sequence of MMP is mainly similar except for the loop region (S'1 pocket), which displays different length and is composed of distinct amino acid composition. The similarities are ordered as S'1 > S2 > S'3 > S1, S3 > S'2.

Selective and/or combined occupancy of these pockets were believed to direct selectivity of inhibitor. More generally, it has been determined that such subsites display distinct potency in driving selectivity in order S'1 > S2, S'3, S3 > S1 > S'2.

S'1 pocket located immediately to "the right" of the catalytic site differs notably in size and shape among MMP and has been named specific pocket.

The S'1 pocket is deep, presenting an elongated and hydrophobic shape with an amino acid residue Leu at position 197 for all MMP except MMP-1 and MMP-7. The variation of amino acid residues among MMP, within this pocket, might be important. It adopts an extended shape in both gelatinases, but S'1 pocket in MMP-2 forms a large channel nearly bottomless, while it is slightly flexible in MMP-9 presenting a real pocket-like subsite.

The S'2 pocket is shallow, partly solvent-exposed and delimited on the top face by the amino acid residue 158 and on the bottom face by the amino acid residue 218. Its size is affected by the amino acid residues 162 (Asn), which is a Leu for both gelatinases, 163 (Val) which is an Ala for both gelatinases and 164 (Leu).

The S'3 pocket is neutral and partly solvent-exposed and delimited by the amino acid residues 222 (Leu) and 223 (Tyr). The size of this pocket is dependent on the amino acid residue 193, which is a Tyr for both gelatinases (Table 2).

As a rule, the substrates bind weakly with the unprimed subsites; however, some differences could be assigned between gelatinases.

The S1 pocket is shallow and hydrophobic. The same triad is pinpointed for MMP-2 and MMP-9 (His166-Phe168-Tyr155 and His183-Phe185-Tyr172, respectively). The amino acid residue 163 and to a lower extent the amino acid residue 155 influence the S1 subsite interactions with an inhibitor. The amino acid residue 163 is a Leu for both gelatinases.

The S2 pocket is solvent-exposed and the amino acid residues 86, 169 and 210 are poorly conserved in MMP family and affect the shape and the properties of this pocket. Its shape is

N°	S3		S2				S'2				S1							S'1 bottom			S'3
			Catalytic domain (S2 + S'2)																		
	155	168	86	165	169	210	158	162	163	218	155	163	166	168	172	183	185	197	220	222	193
MMP-2	Y	F	F	E	A	A	G	L	A	I	Y	L	H	F				L	T	R	Y
	FAPGTVGGDS		HEFGHAMGLEH								AHA							PIY-TYT-KNFRLSQ			DGLL
	Small		Small, + charge /H bond acceptor				Shallow Solvent-exposed				Hydrophobic							Channel with no bottom			Neutral Partly solvent-exposed
MMP-9	Y	F	-	D	P	P	G	L	A	M	-	L	H	Y	Y	H	F	L	T	T	Y
	FPPGPGIGGDA		HEFGHALGLAH								AHA							PMY-RFT-EGFPLHK			DGLL
	Small		Big, hydrophobic				Shallow Solvent-exposed				Hydrophobic							Pocket-like subsite			Neutral Partly solvent-exposed

Table 2. Overview of favorable ligand properties and conserved domains of gelatinases (adapted from Cuniasse et al., 2005; Nicolotti et al., 2007; Terp et al., 2002).

dependent on the amino acid residue Pro at the position 87, then the MMP-2 has its Phe87 leading to a small and hydrophobic pocket and MMP-2 interacts with positive charge probes. When the Pro87 is lacking, another conformation was observed. The amino acid residue at the position 169 is a Pro for MMP-9 and defines a large hydrophobic pocket.

Finally, the amino acid residue 210, Asn in MMP-9 and Glu in MMP-2, leads to a less exposed pocket and notably plays a crucial role in enzyme selectivity. The S2 pocket is important and differentiates both gelatinases.

The S3 pocket is composed of a hydrophobic cleft delimited by the amino acid residues 155 and 168. The pocket shape and size are influenced by the amino acid residue 155 which is a Tyr and 168 which is a Phe for both gelatinases.

Although these subsites might direct enzyme specificities, interaction of gelatinases with macromolecular substrates also relies on the presence of remote binding sites named exosites that also notably act in driving enzyme action (Figure 2).

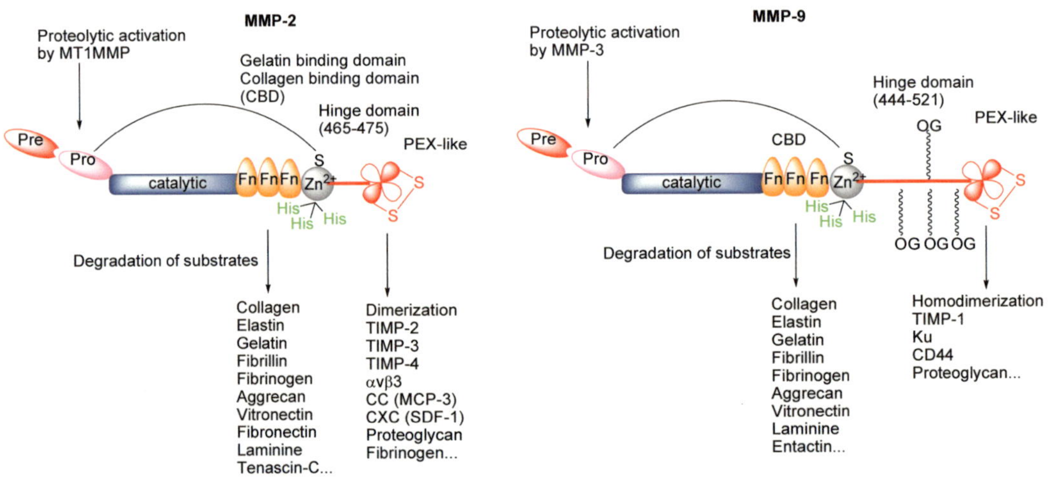

Fig. 2. Functions of gelatinases domains.

2.2 Biological properties of gelatinases

Gelatinases are most often associated to the cell plasma membrane of normal or transformed cells, thus targeting the proteolytic activity of invasive cells. Enzyme tethering to cell periphery requires the carboxy terminal hemopexin-like domain, designated as PEX of 200 amino acid residues on average forming a four bladed β-propeller structure.

In pro-MMP-2, the PEX domain interacts with the C-terminal domain of TIMP-2 (Tissue Inhibitors of Matrix MetalloProteinase-2 (Brew & Nagase, 2010)) which allowed the complex to tether to plasma membrane through interaction of the N-terminal part of the inhibitor to a MT1-MMP homodimer: one molecule of MT1-MMP acting as a docking molecule, the other catalyzing pro-MMP-2 activation (Itoh et al., 2006, 2011; Sato & Takino, 2010).

Of note, this PEX domain also reacts with TIMP-3 and TIMP-4 but no MMP-2 activation was noted in such case.

It needs to be emphasized that interaction of PEX domain in MMP-2 with MT3-MMP also leads to enzyme activation that can be enhanced by the prior binding of pro-MMP-2 to chondroitin-4 sulfate chains-containing proteoglycan. PEX domain of MMP-2 was also reported to bind to $\alpha_v\beta_3$, CC chemokine as monocyte chemoattractant protein-3 (MCP-3), CXC chemokine as stromal cell derived factor-1 (SDF-1) and fibrinogen.

Such PEX domain is also important in driving the pro-MMP-9 activation; in human neutrophils formation of pro-MMP-9-lipocalin complex favours enzyme activation by kallikrein. It is also important for localizing the enzyme at the cell periphery through interaction with low density lipoprotein-receptor related protein (LRP), CD-91 or different isomer forms of CD-44, protein Ku, and is involved in the formation of covalent complexes with proteoglycans (Malla et al., 2008; Monferran et al., 2004).

An important property of this carboxy terminal domain, in association with an unstructured, hydrophilic and flexible long-O-glycosylated domain (OG) in pro-MMP-9 which contains 11 repeats of the sequence T/SXXP (Figure 2), relies on its ability to catalyze the intracellular formation of enzyme dimers (Van den Steen et al., 2001). Importantly, the dimer form of pro-MMP-9 is more resistant to MMP-3 activation.

Gelatinases also appear unique in MMP family in exhibiting fibronectin type II domains which are also designated as collagen binding domains (CBD) (Shipley et al., 1996). Indeed, deletion of these domains in both enzymes led to protease devoided of collagen(s)-gelatin(s)- or elastin-degrading capacity (Allan et al., 1995). Recent data also indicated that OG could also mediate MMP-9 gelatin interaction by allowing the independent movement of enzyme terminal domain (Vandooren et al., 2011).

3. Design of MMP inhibitors (MMPI)

Up to now, a myriad of MMPI has already been synthesized (Table 3).

The most important studies focus on the combinations of diverse structural modifications; three classes of compounds have been developed: combined inhibitors, right hand side and left hand side inhibitors based on the scissile bond in the catalytic site (Skiles et al., 2004; Whittaker et al., 1999). Some of them have been used as potential therapeutic agents to limit tumor progression. Instead of using MMP inhibitors as therapeutic treatment, they might be also useful as preventive drugs or as biomarkers in early stage of cancer. Up to now, most of the clinical trials in cancer were rather disappointing (Abbenante & Fairlie, 2005; Dormán et al., 2010; Fingleton, 2007; Gialeli et al. 2011).

The first generation of MMPI was based on peptidomimetic skeleton containing a succinic acid motif and a hydroxamic acid as zinc binding group (ZBG) (batimastat®, marimastat®, solimastat®, galardin®, trocade®). Hydroxamic acid is a bidentate chelator, but it has also a good binding affinity to other ions (Cu^{2+}, Fe^{2+} and Ni^{2+}). They are broad spectrum inhibitors and led to musculoskeletal syndrome side effect mainly due to the presence of hydroxamic acid.

Prinomastat® contains a ring embedded sulfo-succinic acid motif, which increases oral availability and water solubility. Nevertheless, clinical trials have been discontinued after phase III for musculoskeletal toxicity and poor survival rate.

Tanomastat® has a thioether function increasing the oral activity and selectivity for MMP-2, MMP-3 and MMP-9. Unfortunately, clinical trials are discontinued for haematological toxicity and poor survival rate.

MMP Inhibitors	Indication	Clinical trials development	Structure
Batimastat® (BB-94) British Biotech	Cancer Broad spectrum (MMP-1, -2, -3, -7, -9)	Phase I (Discontinued)	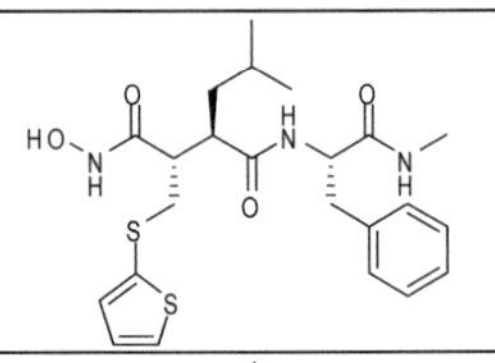
Marimastat® (BB-2516) British Biotech	Cancer Broad spectrum (MMP-1, -2, -7, -9)	Phase III (Discontinued)	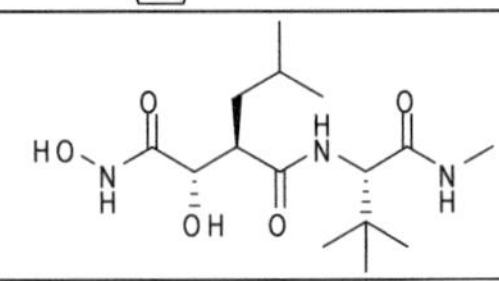
Solimastat® (BB-3644) British Biotech	Cancer Broad spectrum (MMP-1, -7)	Phase I (Discontinued)	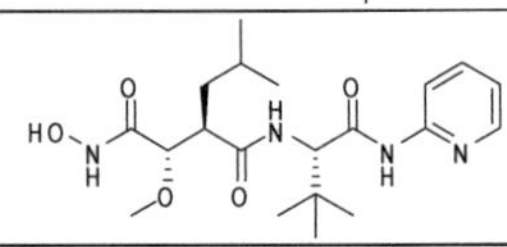
Galardin® (GM-6001) Glycomed	Eye disease, COPD Broad spectrum (MMP-1, -2, -9, -12)	Phase I (Discontinued)	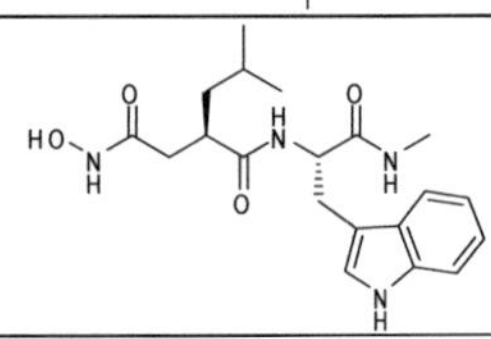
Prinomastat® (AG-3340) Agouron	Cancer Macular degeneration Broad spectrum (MMP-2, -3, -7, -9, -13)	Phase III (Discontinued)	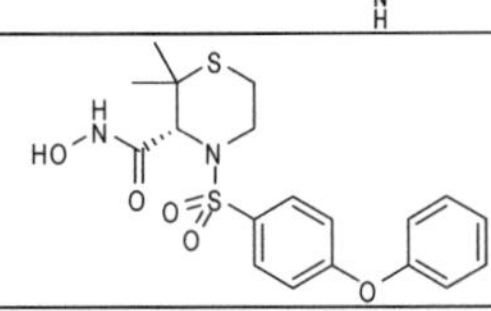
Trocade® (Ro32-3555) Roche	Rheumatoid arthritis (MMP-1, -8, -13)	Phase III (Discontinued)	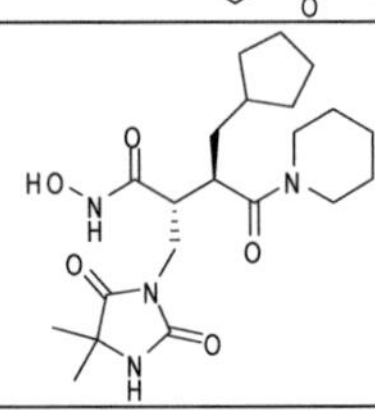
Tanomastat® (BAY 12-9566) Bayer	Cancer Arthritis (MMP-2, -3, -8, -9)	Phase II Phase III (Discontinued)	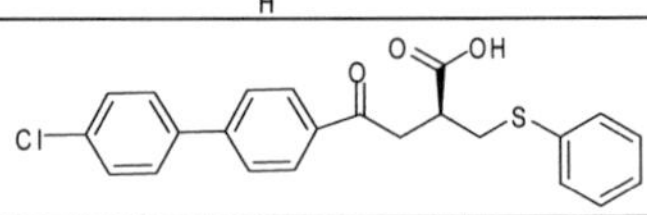
Rebimastat® (BMS-275291) Bristol-Myers Squibb	Cancer Broad spectrum (MMP-1, -2, -7, -9, -14)	Phase III (Discontinued)	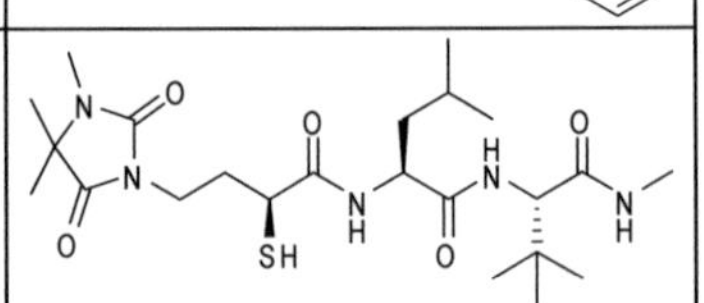
Metastat® (CMT-3) Collagenex	Cancer (MMP-2, -9)	Phase II	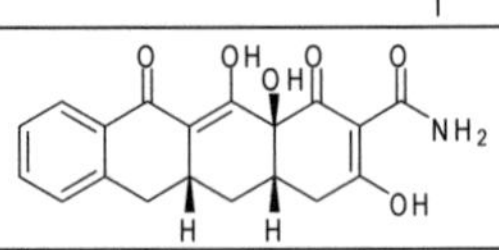

Table 3. Main MMPI in clinical development.

The thiol zinc binding group of rebimastat® is a weak monodentate ligand and the musculoskeletal toxicity and its poor response led clinicians to stop treatment.

Finally, metastat® is a second generation of tetracyclines still in phase II clinical trials (Acne, AIDS-related Kaposi's sarcoma and mainly used in cancer). The only detected side effect is its photosensibility. This inhibitor is selective of MMP-2 and MMP-9 and crosses the blood-brain barrier. Actually, it is the most promising MMPI.

These failures are mainly due to their broad spectrum MMP inhibitory activity, the similarity of their active sites with those of other metalloproteinases (ADAM, ADAMT...), the poor selectivity of the chelating group, the administration of MMPI in late disease stage, their poor pharmacokinetics, unavoidable side effects (musculoskeletal pain), toxicity and limited oral bioavailability.

However, these efforts led to pinpoint the importance of MMPI selectivity and allowed the identification of MMP as target and anti-target in various diseases progression (Overall & Kleifeld, 2006). Certain MMP are both targets and anti-targets depending on the stage of the disease.

4. Galardin® pharmacomodulation as a tool for designing specific gelatinases inhibitors

For a decade, we have been involved in the pharmacomodulation of galardin®, a powerful broad spectrum MMPI (Figure 3).

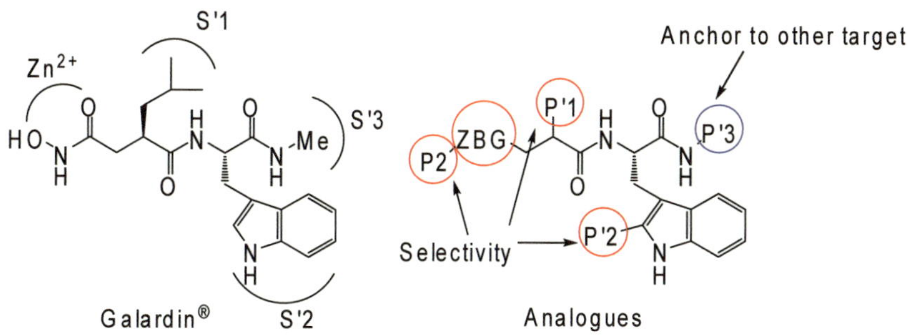

Fig. 3. Galardin® and analogues.

By the beginning of 1994, galardin® was in phase I clinical trials for age related macular degeneration (ARMD) and as chronic obstructive pulmonary disease (COPD) by Glycomed.

In order to increase selectivity, the synthesis of analogues of galardin®, has been achieved. The modifications have been focused on the P'1, P'2, P'3 groups and ZBG.

4.1 Influence of the S'1 subsite: Modulation of galardin® in gelatinases inhibition

Our first experiments started with the insertion of one unsaturation in P'1 position to increase the hydrophobicity of the new compounds and to study the effect of substitution on S'1 pocket specificity (Figure 4). For this purpose, replacement of isobutyl group by an isobutylidene group of *E* geometry enhanced by 100-fold MMP-2 selectivity *versus* MMP-3

(IC_{50} = 1.3 and 179 nM, respectively) (Marcq et al., 2003). The double bond geometry was found important for potency and selectivity as shown with the equimolar *E/Z* mixture which displayed lower activity.

Pursuing these pharmacomodulations aiming at better MMP-2 selectivities, we planned to increase hydrophobicity and rigidity with the dehydro and didehydro analogues which were synthesized (analogue **2a-d** and **3a-h**).

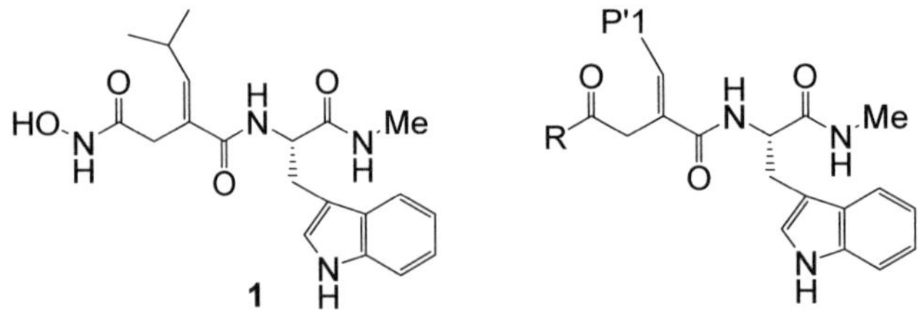

2a-d R = -NH-OH
P'1 = $-(CH_2)_5-CH_3$, $-(CH_2)_2-Ph$, -CH=CH-Me, -CH=CH-Ph

3a-h R = -OH
P'1 = $-(CH_2)n-CH_3$

Fig. 4. Pharmacomodulation of the P'1 group.

Introduction of either one or two unsaturations decreased their potent MMP inhibitory activity as compared to parent molecule over all MMP (Moroy et al., 2007). However, the presence of a phenyl group at the end of alkyl chain (**2b** and **2d**) led to inhibitors with a good activity and selectivity for MMP-9 (IC_{50} = 38 and 45 nM, respectively).

In parallel, C7 long alkyl chain containing galardin® **2a** displayed a MMP-2 selectivity comparing to MMP-9 (IC_{50} = 123 nM) (Table 4).

Compounds	ZBG	P'1	MMP-1	MMP-2	MMP-9	MMP-14
Galardin®	-NH-OH		1.5	1.1	0.5	13.4
1	-NH-OH	-*i*Pr	18	9.2	17	10
2a	-NH-OH	$-(CH_2)_5$-Me	32000	123	> 10^4	2660
2b	-NH-OH	$-(CH_2)_2$-Ph	9130	280	45	53100
2c	-NH-OH	-CH=CH-Me	984	78500	974	913
2d	-NH-OH	-CH=CH-Ph	1240	120	38	2490
3a	-OH	$-(CH_2)_6$-Me	> 10^5	7570	838	> 10^5
3b	-OH	$-(CH_2)_7$-Me	> 10^5	458	241	> 10^5
3c	-OH	$-(CH_2)_8$-Me	> 10^5	247	173	> 10^5
3d	-OH	$-(CH_2)_9$-Me	> 10^5	249	450	> 10^5
3e	-OH	$-(CH_2)_{10}$-Me	> 10^5	351	211	> 10^5
3f	-OH	$-(CH_2)_{11}$-Me	> 10^5	655	673	> 10^5
3g	-OH	$-(CH_2)_{14}$-Me	> 10^5	762	582	> 10^5
3h	-OH	$-(CH_2)_{18}$-Me	> 10^5	> 10^5	> 10^5	> 10^5

Table 4. Influence of the P'1 chain length on the selectivity and potency of MMP inhibitors. IC_{50} values are expressed in nM.

AutoDock 4.0 program (Huey et al., 2007; Morris et al., 1998) was used to perform the computational molecular docking. AutoDockTools package was employed to prepare the input files necessary to the docking procedures and to analyze the docking results. The figures have been done using Pymol program (DeLano, W. L. 2002. PyMol Molecular Graphics System, Palo Alto, CA. http://www.pymol.org).

Molecular modelling experiments confirmed the importance of the insertion of the alkyl chains in the S'1 pocket, supporting the observed biological data for **2a** (Figure 5).

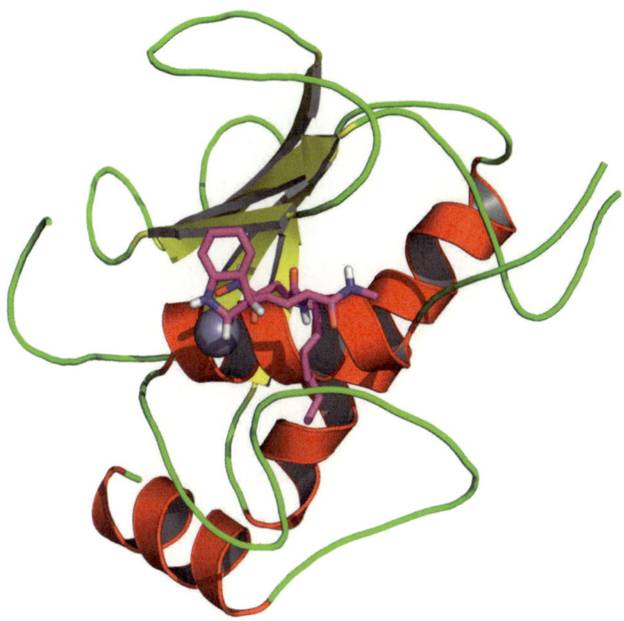

Fig. 5. Complex between MMP-2 and analogue **2a**. The MMP-2 secondary structure was represented in cartoon. The analogue **2a** is shown with sticks in which the C atoms are colored in magenta. Zn atom is displayed as a grey sphere.

The S'1 pocket of MMP-2 is sufficiently deep to accommodate the long alkyl chain. On the contrary, the S'1 pocket of MMP-9 at the end of the tunnel is restrained, like a funnel by the amino acid residues Glu233, Arg241, Thr246 and Pro247. The S'1 pocket of MMP-9 is large enough to accept a phenyl group at the entrance of the pocket such as compounds **2b** and **2d**.

In order to increase hydrophobicity, aiming at a better MMP-2 selectivity, the alkyl chain was elongated with n = 8 to 20 carbon atoms as described in the literature (Levy et al., 1998; Miller et al., 1997; Whittaker et al., 1999). Using galardin® as template, the activity and selectivity were not really increased with the length of the linear chain. Batimastat® inhibitors increase activity for MMP-2 with C8 long alkyl chain (IC_{50} = 0.6 nM) and a C12 analogue displays a MMP-2 selectivity comparing to MMP-1 (IC_{50} = 1 and 50000 nM, respectively). Finally, a good activity for MMP-2 is obtained for C9 long chain marimastat® analogues (IC_{50} < 0.15 nM) and a maximal selectivity occurs with a C16 for MMP-2 *versus* MMP-1 (IC_{50} = 0.6 and 5000 nM, respectively).

Nevertheless, in our case no better IC_{50} were found for MMP-2 with increasing chain length when an unsaturation was incorporated. However, a good selectivity for MMP-2 *versus* MMP-1 and MMP-14 is observed. Also, carboxylic derivatives **3c** and **3e** displayed a good selectivity for MMP-9 *versus* MMP-1 and MMP-14.

As stated previously, S'1 pocket is generally quite large in all MMP. However, the amino acid residue Arg214 redefines the bottom of the pocket in MMP-1 leading to a small and restricted S'1 pocket. The amino acid residue Arg214 can be flexible but in most cases, a large P'1 group inhibitor is expected to bind weakly to MMP-1. Thus, it is not surprising, to find that most of the MMP-1 inhibitors have relatively small P'1 groups.

MMP-14 appears as one key-proteolytic enzyme to promote cancer invasion and metastasis (Hernandez-Barrantes et al., 2002). Up to now, only one pentacyclic sterol sulphate MMP-14 inhibitor was described to display MMP-14 selectivity while it exhibits only low potency (Fujita et al., 2001).

S'1 pocket of MMP-14 was found to be two amino acid residues longer than those of gelatinases. The amino acid residue Met237 allows favourable interactions with the hydrophobic substituents at the bottom of the pocket, but unfavourable interaction with the positively charged substituents.

Therefore, lack of inhibition of this enzyme by long alkyl chain (Table 4) is rather surprising. The unsaturation might disturb the entrance of the S'1 pocket, but that is purely speculative. Finally, it seems to be more difficult to fit MMP-14 pockets, perhaps in keeping with its transmembrane localization and domain structure.

4.2 Influence of the S'2 subsite on gelatinases inhibition

Introduction of an alkyl chain in the P'2 position of the indole ring leads to lower gelatinase inhibitory activity. Nevertheless, the selectivity for MMP-2 was pinpointed (Marcq et al., 2003).

On the contrary, the introduction of a phenyl group at the P'2 position (analogue **4**) enhanced the selectivity towards MMP-2 maintaining a high potency (IC_{50} = 0.092 nM). To the best of our knowledge, the large and solvent-exposed S'2 pocket could accommodate large and hydrophobic groups (LeDour et al., 2008). A good activity was found for MMP-1 and MMP-14 (IC_{50} = 0.244 and 0.601 nM, respectively).

4.3 Influence of the unprimed subsites on gelatinases inhibition

It is well documented that the hydroxamic acid is one of the most powerful ZBG, but its toxicity and low bioavailability triggered tremendous efforts to design other ZBG (Jacobsen et al., 2007, 2010). Of note, the zing binding group affinity is as follows: hydroxamate > retrohydroxamate > sulfhydryl > phosphinate > carboxylate > heterocyclic core. Nevertheless, a zinc binding group with lower affinity may be advantageous.

In this line, we have proposed various hydrazide and sulfonylhydrazide-type functions as potential ZBG. The sulfonylhydrazide derivative is responsible for the increased acidity of the NH close to SO_2 function allowing the H-bond to be formed with the catalytic glutamate residue (Augé et al., 2003, 2004).

The hydroxamate acts as a bidentate ligand with the zinc ion to form the distorded trigonal-bipyramidal coordination geometry. With respect to the design of new ZBG, a DFT (Density Functional Theory) study revealed different modes of chelation of the sulfonylhydrazide group (Rouffet et al., 2009).

The zinc ion was found to be ligated to three 4-Me-imidazoles used as mimetics of histidine imidazole moieties located in the MMP catalytic site in physiological conditions. The sulfonylhydrazide group could chelate the zinc ion in two different manners either in a bi- or tri-dentate mode.

In all cases, interaction between

i. the Zn^{2+} ion and the sulfonamide nitrogen was observed as well as,
ii. the Zn^{2+} ion with the oxygen atom of the sulfonylhydrazide carbonyl.

In the case of the tridentate mode, the third interaction involves one of the sulfonyl oxygen atom and the Zn^{2+} ion. Consequently, the bidentate conformation was more favourable (4 to 5 kcal/mol) and the sulfonylhydrazide function seemed to possess ideal zinc binding properties and also biodisponibility and stability.

Following these investigations, the sulfonylhydrazide group was incorporated into the galardin® backbone as zinc binding group. Among the synthesized subtituents, the *p*-bromobiphenyl group displayed a good potency for MMP-2 (LeDour et al., 2008). Then, based on our preliminary results further modifications of the P'1 (long alkyl chain) and the P'2 (phenyl group) substituents were introduced to increase the selectivity for MMP-2 (Figure 6).

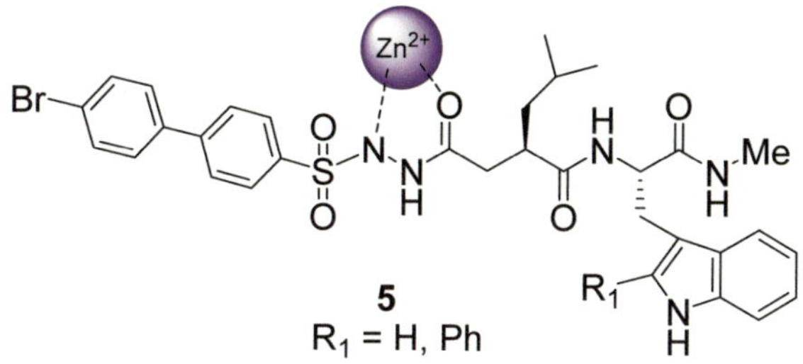

5
R_1 = H, Ph

Fig. 6. Modifications of the P'2 group and sulfonylhydrazide function as ZBG.

Finally, a high potency for MMP-9 was obtained with the compound **5a** (or **5b**) with a small group such as an isobutyl (Table 5). Introduction of a phenyl group at the position 2 of the indole ring did not modify the activity. This could easily be explained taking into account the S'2 solvent-exposed pocket.

Compounds	R_1	MMP-1	MMP-2	MMP-3	MMP-9	MMP-14
5a	H	30	98	5800	3	20000
5b	Ph	350	247	53	18	1237

Table 5. Influence of the P'2 group and ZBG on the selectivity and potency of MMP inhibitors. IC_{50} values are expressed in nM.

Docking studies of **5a** confirmed the occupancy of the MMP-9 S'1 pocket by the isobutyl group, the S2 subsite by the *p*-bromobiphenyl group and the chelation of the sulfonylhydrazide to the catalytic site (Figure 7). It is known that the amino acid residues Glu412 and Asp410, located in the S2 subsite control the selectivity of MMP-2 and MMP-9, respectively.

In MMP-2, the amino acid residue Glu412 is able to form an H-bond with the substrate which could not be formed in MMP-9 presenting the amino acid residue Asp410 (Chen et al., 2003).

In our case, no H-bond can be formed with the hydrophobic *p*-bromobiphenyl group and the compound **5a** displayed selectivity for MMP-9.

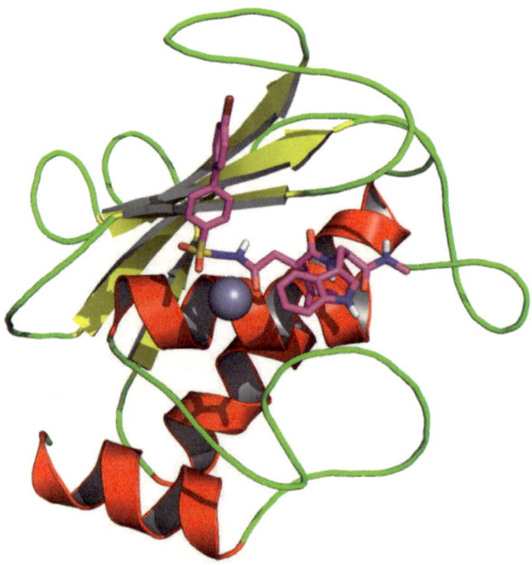

Fig. 7. Complex between MMP-9 and analogue **5a.** The MMP-9 secondary structure was represented in cartoon. The analogue **5a** is shown with sticks in which the C atoms are colored in magenta. Zn atom is displayed as a grey sphere.

Unfortunately, none of the P'1 (with an unsaturation and long alkyl chain or bulky substituent) and P'2 (with a phenyl group) modified galardin® derivatives exhibited increased inhibitory capacity and selectivity. Our docking data indicated that these compounds adopted a conformation in which sp2-hybridized carbon atom of the alkylidene side-chain led to steric hindrance impeding the entrance in the S'1 subsite. Consequently, when the S2 pocket is occupied, the primed subsites could not tolerate any large substituent and no synergistic effect could be obtained.

5. Control of elastolytic cascade by oleoyl-galardin®

Elastin degradation is at the genesis of cardiovascular disease as athero-arteriosclerosis and aneurysm formation, and pulmonary diseases as chronic obstructive pulmonary disease or lung cancer (Moroy et al., 2012; Muroski et al., 2008; Thompson & Parks, 1996).

Elastolysis requires the participation of serine- and metallo-elastases (Figure 8) which act through proteolytic cascades.

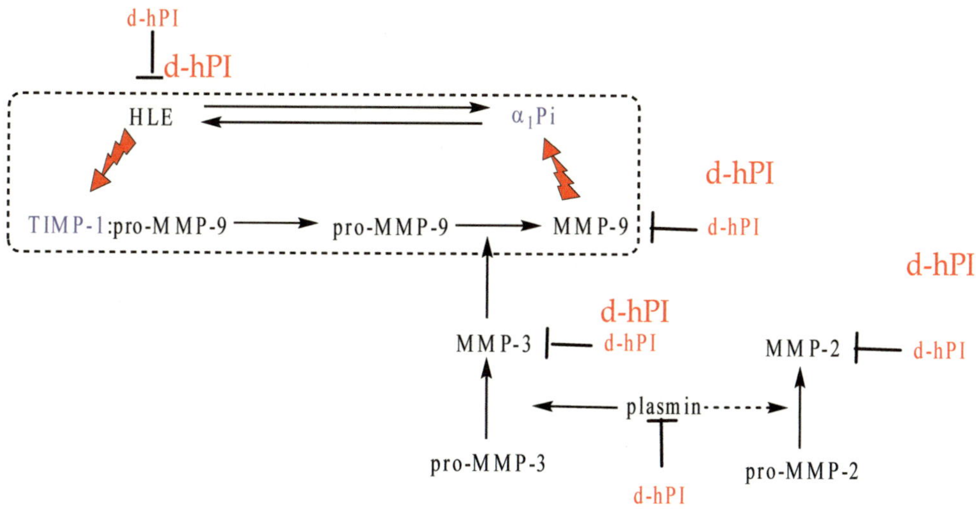

Fig. 8. Control of the serine (HLE)- and metallo-elastases (MMP-2, MMP-9) crosstalk in elastolysis by double-headed protease-MMP inhibitor (d-hPI).

Besides, a serine elastase as human leucocyte elastase (HLE) can degrade TIMP, and reversely MMP can hydrolyse serine protease inhibitor as α1 proteinase inhibitor (Nunes et al., 2011).

To control elastolysis, we thus attempted to design substances that could interfere with all actors of the depicted cascade. For that purpose, long chain-unsaturated fatty acids, as oleic acid, have been described to inhibit HLE (Hornebeck et al., 1985; Shock et al., 1990; Tyagi & Simon, 1990) and to impede plasmin-mediated prostromelysin-1 activation (Huet et al., 2004) as well as gelatinases activities (Berton et al., 2001). To that respect, we envisaged the synthesis of a double-headed protease-MMP inhibitor (d-hPI) able to block elastase and MMP activities. To that end, an oleoyl group was incorporated to galardin® at the P'3 position (Figure 9) (Moroy et al., 2011).

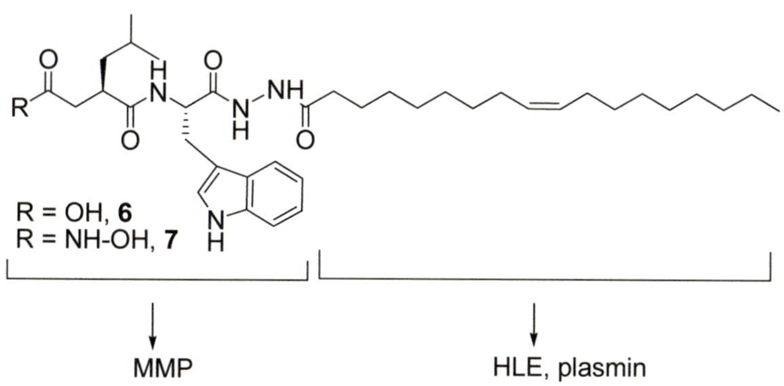

Fig. 9. Double-headed protease-MMP inhibitor (d-hPI).

Oleoyl analogues (carboxylic **6** or hydroxamic **7** acids) are more potent than oleic acid to inhibit MMP (Table 6). The hydroxamic acid **7** was found to improve the inhibitory capacity toward MMP-2 comparing to oleic acid.

Compounds	MMP-2	MMP-3	MMP-7	MMP-9	HLE	Plasmin
Oleic acid	4.3	-	6.5	6.4	3.0	3.5
6	0.5	0.5	0.07	0.7	0.6	0.7
7	0.1	0.05	0.6	0.04	8.7	8.3

Table 6. Inhibition of MMP and serine elastases by oleic acid and oleoyl-galardin® derivatives. MMP (*Kis* values are expressed in μM), HLE and plasmin (IC_{50} values are expressed in μM).

The molecular docking computations indicated that compound **7** is able to bind MMP-2 active site and unable to chelate Zn^{2+} ion in the active site (Figure 10).

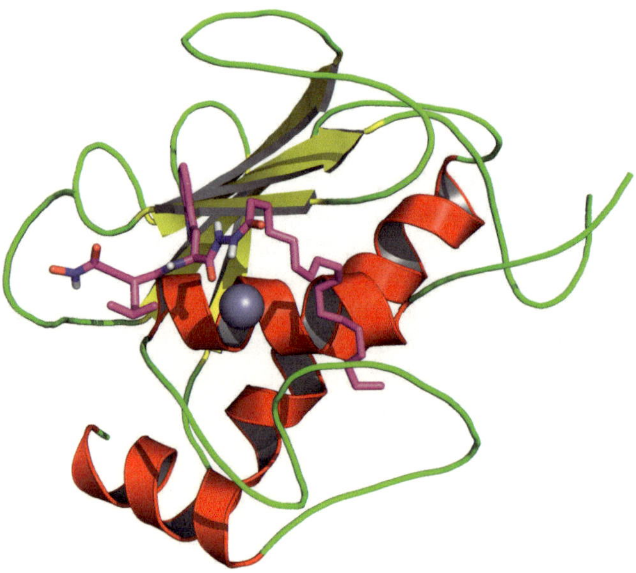

Fig. 10. Complex between MMP-2 and analogue **7**. The MMP-2 secondary structure was represented in cartoon. The analogue **7** is shown with sticks in which the C atoms are colored in magenta. Zn atom is displayed as a grey sphere.

Instead, the hydroxamic acid function forms a salt bridge with the N-terminal end of the amino acid residue Tyr110, while the long alkyl chain was inserted into the S'1 pocket as was already demonstrated. The heterocycle is inserted into the S1 subsite where it interacts *via* an H-bond with the carbonyl group of the amino acid residue Ala194 peptide bond.

We analyzed the inhibitory capacity of oleic acid, analogues **6** and **7** towards HLE and plasmin activities. The compound **6** displayed high potency (IC_{50} = 0.6 μM) against HLE, but lower inhibition was observed with oleic acid and the analogue **7** (IC_{50} = 3.0 and 8.7 μM, respectively). Molecular docking computations indicated that the carboxylic function of compound **6** and oleic acid can form a salt bridge with the amino acid residue Arg217, but not compound **7** (not shown).

Almost the same values are found in the same order for the plasmin-mediated pro-MMP-3 activation. The lowest energy model of oleic acid with the kringle 5 domain is characterized by the presence of a salt bridge between the carboxylic function and the amino acid residue Arg512 of plasmin (not shown).

6. Control of gelatinases through impeding enzyme-substrate interaction

Another approach to control the activity of those enzymes consists in the regulation of exosite protein-ligand interaction. To that respect both Fn(II) [or CBD] and PEX domains are involved and, in keeping with data presented on Figure 2, blocking either Fn(II) or PEX function will prevent the catalytic function of gelatinases on protein substrates selectively. For instance, the proteolysis of collagen and elastin might be inhibited while maintaining intact the potential of gelatinases to cleave proteoglycans or several growth factors.

6.1 Fn(II) domains

Using recombinant Fn(II) domain as bait, a one bead one-peptide combinatorial peptide library was screened (Xu et al., 2007). A peptide displaying high sequence identity with the segment 715-721 in human α1(I) collagen chain was identified and proved to inhibit by > 90% gelatinolysis catalyzed by MMP-2 (Xu et al., 2009).

The unsaturated fatty acid such as oleic acid inhibited MMP-2 with K_i = 4.3 μM. Molecular modelling studies focus on the interactions localized at two sites on MMP-2: the fatty chain filled the S'1 pocket while the carboxylic acid group was exposed to the solvent. This result is in agreement with our previous works showing that the S'1 pocket could accommodate long alkyl side chains (LeDour et al., 2008; Moroy et al., 2007). The molecular docking computations identified the second site of the oleic acid interactions as the 3rd Fn(II) domain. The carboxylic acid function interacts *via* an H-bond with the phenolic group from the amino acid residue Tyr381, *via* a salt bridge with the guanidinium group of the amino acid residue Arg385 and *via* van der Waals interactions with the amino acid residue Leu356 while the unsaturated bond forms van der Waals interaction with the amino acid residues Phe355, Trp374, Tyr381 and Trp387 (Figure 11).

Another approach could be the use of a more specific inhibitor directed against the 3rd Fn(II) domain.

For this latter, an inhibitor with two carboxyl groups at each end of the alkyl chain should be efficient for targeting respectively the amino acid residue Arg385, as it was observed for oleic acid, and the amino acid residue Arg368 that is also present on the rim of the hydrophobic pocket.

Interestingly, in the full MMP-2, the amino acid residue Arg368 forms a salt bridge with the amino acid residue Asp40 belonging to the propeptide domain. According to the binding mode of oleic acid, the size of the alkyl chain should be composed by 15 or 16 atoms of carbon such as (7Z)-hexadec-7-enedioic acid and (6Z)-pentadec-6-enedioic acid.

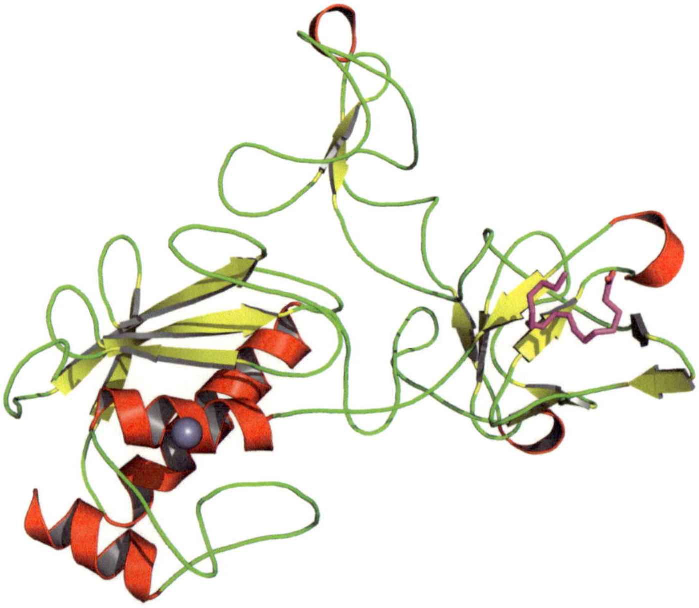

Fig. 11. Complex between MMP-2 and oleic acid. Oleic acid is anchored in the binding pocket at the surface of the 3rd Fn(II) domain.

6.2 Dual occupancy of the enzyme active site and Fn(II) domains

Another more complex strategy relies on the design of a dual occupancy of the enzyme active site and the exosite as Fn(II). That has been originally attempted with a coupled hydroxamate-based inhibitor to gelatin-like structures (Jani et al., 2005). No increase in selectivity or potency of those compounds towards gelatinases could be attained; it was attributed to the possibility that the Fn(II) and catalytic domains of enzyme are tumbling independently.

Thus, we have built a hypothetical inhibitor from the inhibitors previously studied. We have added alkyl groups until the hydrophobic pocket of the 3rd Fn(II) was reached: 19 are needed to interact with its rims, *i.e.* the amino acid residues Arg368 and Trp387. If we want to reproduce the binding mode of the oleic acid, 32 alkyl groups should be added (Figure 12). However, the size and the high flexibility of this kind of inhibitor could be problematic.

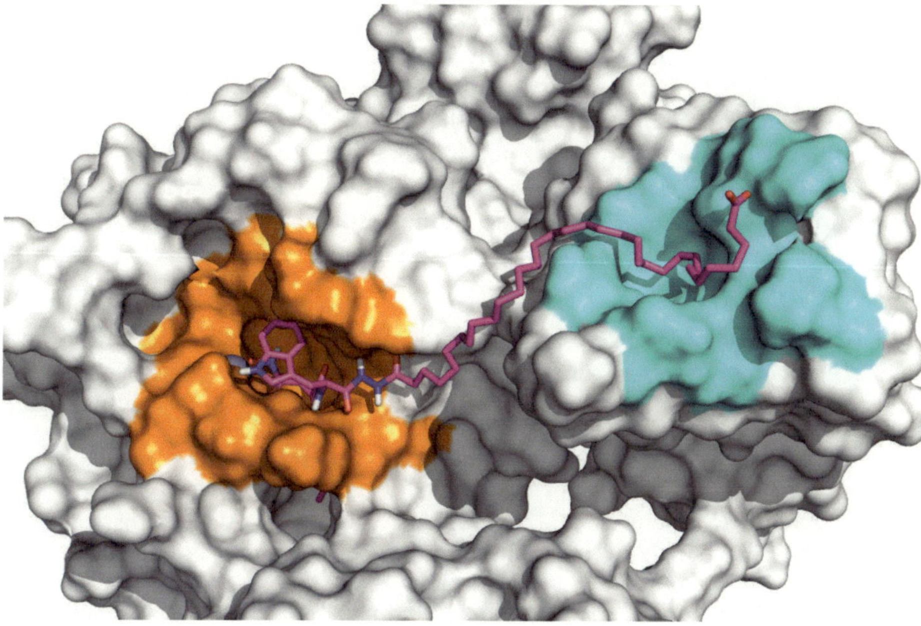

Fig. 12. Hypothetical model of an inhibitor able to bind both the catalytic site and the Fn(II) domain of the human MMP-2. The MMP-2 is displayed with its accessible surface area. The active site and the binding pocket on the 3rd Fn(II) domain are coloured in orange and in cyan, respectively.

6.3 PEX domains

Since gelatinases are critically involved in directing cellular invasion, interfering with PEX-integrins (receptors) interaction might be a nice alternative. As an example, the use of phage display identified a peptide that inhibits the association of MMP-9 PEX domain with the $\alpha_v\beta_5$ integrin, preventing proenzyme activation and cell migration (Björklund et al., 2004). More recently, a 20 mers peptide encompassing the PEX-binding tail region of C-TIMP-2 was found to inhibit the membrane-mediated activation in HT-1080 cells (Xu et al., 2011).

PEX domains play a crucial function in the non proteolytic function of gelatinases. As example, the PEX domain of MMP-9 is directly involved in the modulation of epithelial cell migration in a transwell chamber assay (Dufour et al., 2008). This domain also promotes B cell survival by interacting with $\alpha_4\beta_1$ and CD-44 receptors (Redondo-Muñoz et al., 2010).

7. Conclusion

General considerations need to be pinpointed at aims to give a novel expansion to the design of substances able to regulate MMP activity.

First, MMP as gelatinases are produced by nearly all cell types, but their cellular source may intervene in their function and activity. Proteolytic activity liberated by activated neutrophils is one pivotal element in the genesis and progression of aneurysms or chronic obstructive pulmonary diseases (Muroski et al., 2008; Thompson & Parks, 1996).

It has been demonstrated that pro-MMP-9 is produced by neutrophils as a free form *i.e.* not associated with TIMP-1 molecule, more readily activatable by enzyme as stromelysin-1 (Ardi et al., 2007). In addition, following activation, those cells release extracellular traps *i.e.* neutrophil extracellular traps (NET) formed by the association of chromatin and granule proteins; NET are enriched in neutral endopeptidases as neutrophil elastase and MMP-9 (Brinkmann et al., 2004). To that respect, the use of double-headed (HLE-MMP-9) inhibitors as oleoyl-galardin® might be of therapeutic value. Advantageously, as we recently documented, oleoyl moiety might be replaced by β-lactam (Moroy et al., 2012), a more potent and selective HLE inhibitor.

Although inflammation can orchestrate cancer (Kessenbrock et al., 2010), MMP-2 and MMP-9 intervene in several other stages of cancer progression. Both enzymes have been involved in promoting cell growth; MMP-2 is more linked to cancer cell invasiveness while MMP-9 may contribute to cell survival.

Up to now, one selective MMP-9 inhibitor is a monoclonal antibody binding to N-terminal part of catalytic domain (Martens et al., 2007). Thus, intuitively, in keeping with those distinct functions, the concept of selective inhibitor among gelatinases is emerging.

Their role in establishing a "metastatic niche" has also been delineated and their contribution in angiogenesis has been widely underlined.

However, paradoxically, MMP-9 can generate either pro- or anti-angiogenic signals (Figure 13).

On one side, it can

i. regulate VEGF bioavailability for VEGF-R2 receptor (Bergers et al., 2000),
ii. activate the basic fibroblast growth factor-2 (FGF-2) pathway (Ardi et al., 2009),
iii. generate elastin fragments *i.e.* elastokines with potent angiogenic activity (Robinet et al., 2005).

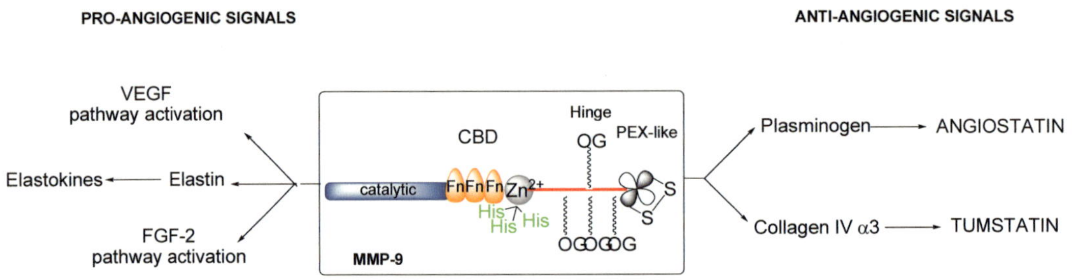

Fig. 13. The paradoxical function of MMP-9 in angiogenesis.

At the opposite, proteolysis of plasminogen and α_3 chain of collagen IV leads to the formation of angiostatin and tumstatin (Cornelius et al., 1998; Kessenbrock et al., 2010). Importantly, mice deficient in MMP-9 evidenced an increased-tumor growth which was attributed to lack of tumstatin formation (Hamano et al., 2003).

As mentioned, both gelatinases exerted their action at the pericellular environment, following binding of their PEX domain to receptors as $\alpha_v\beta_3$ for MMP-2 or CD-44 for MMP-9.

Peptide or chemical libraries can be developed at aims to impede MMP-2(PEX)-$\alpha_v\beta_3$ or MMP-9(PEX)-CD-44 interactions.

As one example, a bivalent derivatized dilysine tetraamide was isolated which proved to interfere with MMP-2/$\alpha_v\beta_3$ interaction and inhibit angiogenesis (Silletti et al., 2001). Possibly, this compound could be chemically modified to confer it additionally MMP-2 inhibitory activity (Bourguet et al., 2009).

One main problem related with the control of MMP in general and gelatinases in particular relies on the kinetics of production of those enzymes during the cancer course from initiation to metastasis formation. In other words, at what stage for one particular type of tumor do we need to incorporate MMPI in cancer treatment?

The development of imaging MMP activity using derivatized selective inhibitor will probably answer to this question. Several techniques have been already developed using Positron Emission Tomography (PET) with ^{18}F-labelled MMP-2 inhibitor (Furumoto et al., 2003), Single Photon Emission Computed Tomography (SPECT) with a ^{123}I gelatinases inhibitor (Schaffers et al., 2004), or the use of fluorogenic substrates bearing self quenched and near infrared FRET pairs (Scherer et al., 2008).

In our Federal Research Institute (IFR), we aim to develop hybrid nanoprobes build from MMPI and fluorescent nanocrystal quantum dots (QDs). Design and chemical synthesis of derivatives of galardin®, selective inhibitors of MMP-2, will be followed by their tagging with QDs. Photo- and chemical stability of QDs will enable long-term spatiotemporal tracking of the process of inhibition of MMP-2 with developed nanoprobe thus permitting understanding of physiological process of invasion of melanoma for example.

8. Acknowledgment

Authors thank *Université de Reims Champagne-Ardenne, IFR53 "Biomolécules", Région Champagne-Ardenne, EU* (Fonds Feder) and *CNRS* for financial support. Technical assistance of Mrs. M. Decarme is greatfully acknowledged.

9. References

Abbenante, G. & Fairlie, D.P. (2005). Protease inhibitors in the clinic. *Medicinal Chemistry*, Vol.1, No.1, (January 2005), pp. 71-104, ISSN 1573-4064

Allan, J.A.; Docherty, A.J.; Barker, P.J.; Huskisson, N.S.; Reynolds, J.J. & Murphy, G. (1995). Binding of gelatinases A and B to type-I collagen and other matrix components. *The Biochemical Journal*, Vol.309, Pt.1, (July 1995), pp. 299-306, ISSN 0264-6021

Ardi, V.C.; Kupriyanova, T.A.; Deryugina, E.I. & Quigley, J.P. (2007). Human neutrophils uniquely release TIMP-free MMP-9 to provide a potent catalytic stimulator of angiogenesis. *Proceedings of the National Academy of Sciences of the United States of America*, Vol.104, No.51, (December 2007), pp. 20262-20267, ISSN 0027-8424

Ardi, V.C.; Van den Steen, P.E.; Opdenakker, G.; Schweighofer, B.; Deryugina, E.I. & Quigley, J.P. (2009). Neutrophil MMP-9 proenzyme, unencumbered by TIMP-1, undergoes efficient activation in vivo and catalytically induces angiogenesis via a basic fibroblast growth factor (FGF-2)/FGFR-2 pathway. *The Journal of Biological Chemistry*, Vol.284, No.38, (September 2009), pp. 25854-25866, ISSN 0021-9258

Augé, F.; Hornebeck, W.; Decarme, M. & Laronze, J.-Y. (2003). Improved gelatinase A selectivity by novel zinc binding groups containing galardin derivatives. *Bioorganic & Medicinal Chemistry Letters*, Vol.13, No.10, (May 2003), pp. 1783-1786, ISSN 0960-894X

Augé, F.; Hornebeck, W. & Laronze, J.-Y. (2004). A novel strategy for designing specific gelatinase A inhibitors: potential use to control tumor progression. *Critical reviews in Oncology/Hematology*, Vol.49, No.10, (March 2004), pp. 277-282, ISSN 1040-8428

Bergers, G.; Brekken, R.; McMahon, G.; Vu, T.H.; Itoh, T.; Tamaki, K.; Tanzawa, K.; Thorpe, P.; Itohara, S.; Werb, Z. & Hanahan, D. (2000). Matrix metalloproteinase-9 triggers the angiogenic switch during carcinogenesis. *Nature Cell Biology*, Vol.2, No.10, (October 2000), pp. 737-744, ISSN 1465-7392

Berton, A.; Rigot, V.; Huet, E.; Decarme, M.; Eeckhout, Y.; Patthy, L.; Godeau, G.; Hornebeck, W.; Bellon, G. & Emonard, H. (2001). Involvement of fibronectin type II repeats in the efficient inhibition of gelatinases A and B by long-chain unsaturated fatty acids. *The Journal of Biological Chemistry*, Vol.276, No.23, (June 2001), pp. 20458-20465, ISSN 0021-9258

Björklund, M.; Heikkilä, P. & Koivunen, E. (2004). Peptide inhibition of catalytic and noncatalytic activities of matrix metalloproteinase-9 blocks tumor cell migration and invasion. *The Journal of Biological Chemistry*, Vol.279, No.28, (July 2004), pp. 29589-29597, ISSN 0021-9258

Bourguet, E.; Sapi, J.; Emmonard, H. & Hornebeck, W. (2009). Control of melanoma invasiveness by anticollagenolytic agents: a reappraisal of an old concept. *Anti-cancer Agents Current Medicinal Chemistry*, Vol.9, No.5, (June 2009), pp. 576-597, ISSN 1871-5206

Brew, K. & Nagase, H. (2010). The tissue inhibitors of metalloproteinases (TIMPs): an ancient family with structural and functional diversity. *Biochimica et Biophysica Acta*, Vol.1803. No.1, (January 2010), pp. 55-71, ISSN 0006-3002

Brinkmann, V.; Reichard, U.; Goosmann, C.; Fauler, B.; Uhlemann, Y.; Weiss, D.S.; Weinrauch, Y. & Zychlinsky, A. (2004). Neutrophil extracellular traps kill bacteria. *Science*, Vol.303, No.5663, (March 2004), pp. 1532-1535, ISSN 0036-8075

Cauwe, B. & Opdenakker, G. (2010). Intracellular substrate cleavage: a novel dimension in the biochemistry, biology and pathology of matrix metalloproteinases. *Critical Reviews in Biochemistry and Molecular Biology*, Vol.45, No.5, (October 2010), pp. 351-423, ISSN 1040-9238

Chen, E.I.; Li, W.; Godzik, A.; Howard, E.W. & Smith, J.W. (2003). A residue in the S2 subsite controls substrate selectivity of matrix metalloproteinase-2 and matrix metalloproteinase-9. *The Journal of Biological Chemistry*, Vol.278, No.19, (May 2003), pp. 17158-17163, ISSN 0021-9258

Cornelius, L.A.; Nehring, L.C.; Harding, E.; Bolanowski, M.; Welgus, H.G.; Kobayashi, D.K.; Pierce, R.A. & Shapiro, S.D. (1998). Matrix metalloproteinases generate angiostatin: effects on neovascularization. *Journal of Immunology*, Vol.161, No.12, (December 1998), pp. 6845-6852, ISSN 0022-1767

Cuniasse, P.; Devel, L.; Makaritis, A.; Beau, F.; Georgiadis, D.; Matziari, M.; Yiotakis, A. & Dive, V. (2005). Future challenges facing the development of specific active-site-directed synthetic inhibitors of MMPs. *Biochimie*, Vol.87, No.3-4, (March-April 2005), pp. 393-402, ISSN 0300-9084

Dormán, G.; Cseh, S.; Hajdú, I.; Barna, L.; Kónya, D.; Kupai, K.; Kovács, L. & Ferdinandy, P. (2010). Matrix metalloproteinase inhibitors: a critical appraisal of design principles and proposed therapeutic utility. *Drugs*, Vol.70, No.8, (May 2010), pp. 949-964, ISSN 0012-6667

Dufour, A.; Sampson, N.S.; Zucker, S. & Cao, J. (2008). Role of the hemopexin domain of matrix metalloproteinases in cell migration. *Journal of Cellular Physiology*. Vol.271, No.3, (December 2008), pp. 643-651, ISSN 0021-9541

Egeblad, M. & Werb, Z. (2002). New functions for the matrix metalloproteinases in cancer progression. *Nature Reviews*, Vol.2, No.3, (March 2002), pp. 161-174, ISSN 1474-1768

Fingleton, B. (2007). Matrix Metalloproteinases as Valid Clinical Targets. *Current Pharmaceutical Design*, Vol.13, No.3, pp. 333-346, ISSN 1381-6128

Fujita, M.; Nakao, Y.; Matsunaga, S.; Seiki, M.; Itoh, Y.; van Soest, R.W.M.; Heubes, M.; Faulkner, D.J. & Fusetani, N. (2001). Isolation and structure elucidation of two phosphorylated sterol sulfates, MT1-MMP inhibitors from a marine sponge *Cribrochalina* sp.: revision of the structures of haplosamates A and B. *Tetrahedron*, Vol.57, No.18, (April 2001), pp. 3885-3890, ISSN 0040-4020

Furumoto, S.; Takashima, K.; Kubota, K.; Ido, T.; Iwata, R. & Fukuda, H. (2003). Tumor detection using 18F-labeled matrix metalloproteinase-2 inhibitor. *Nuclear Medicine and Biology*, Vol.30, No.2, (February 2003), pp.119-125, ISSN 0969-8051

Gialeli, C.; Theocharis, A.D. & Karamanos N.K. (2011). Roles of matrix metalloproteinases in cancer progression and their pharmacological targeting. *The FEBS Journal*, Vol.278, No.1, (January 2011), pp. 16-27, ISSN 1742-4658

Gomis-Rüth, F.X. (2009). Catalytic domain architecture of metzincin metalloproteases. *The Journal of Biological Chemistry*, Vol.284, No.23, (June 2009), pp. 15353-15357, ISSN 0021-9258

Hamano, Y.; Zeisberg, M.; Sugimoto, H.; Lively, J.C.; Maeshima, Y.; Yang, C.; Hynes, R.O.; Werb, Z.; Sudhakar, A. & Kalluri, R. (2003). Physiological levels of tumstatin, a fragment of collagen IV alpha3 chain, are generated by MMP-9 proteolysis and suppress angiogenesis via alphaV beta3 integrin. *Cancer Cell*, Vol.3, No.6, (June 2003), pp. 589-601, ISSN 1535-6108

Hernandez-Barrantes, S.; Bernardo, M.; Toth, M. & Fridman, R. (2002). Regulation of membrane type-matrix metalloproteinases. *Seminars in Cancer Biology*, Vol.12, No.2, (April 2002), pp. 131-138, ISSN 1044-579X

Hornebeck, W.; Moczar, E.; Szecsi, J. & Robert, L. (1985). Fatty acid peptide derivatives as model compounds to protect elastin against degradation by elastases. *Biochemical Pharmacology*, Vol.34, No.18, (September 1985), pp. 3315-3321, ISSN 0006-2952

Hu, J.; Van den Steen, P.E.; Sang, Q.X. & Opdenakker, G. (2007). Matrix metalloproteinase inhibitors as therapy for inflammatory and vascular diseases. *Nature Reviews Drug Discovery*, Vol.6, No.6, (June 2007), pp. 480-498, ISSN 1474-1776

Huet, E.; Cauchard, J.H.; Berton, A.; Robinet, A.; Decarme, M.; Hornebeck, W. & Bellon, G. (2004). Inhibition of plasmin-mediated prostromelysin-1 activation by interaction of long chain unsaturated fatty acids with kringle 5. *Biochemical Pharmacology*, Vol.67, No.4, (March 2004), pp. 643–654, ISSN 0006-2952 (Erratum in: ibid 2004, Vol.67, No.5, p 1011)

Huey, R.; Morris, G.M.; Olson, A.J. & Goodsell, D.S. (2007). A semiempirical free energy force field with charge-based desolvation. *Journal of Computational Chemistry*, Vol.28, No.6, (April 2007), pp. 1145-1152, ISSN 0192-8651

Itoh, Y.; Ito, N.; Nagase, H.; Evans, R.D.; Bird, S.A. & Seiki, M. (2006). Cell surface collagenolysis requires homodimerization of the membrane-bound collagenase MT1-MMP. *Molecular Biology of the Cell*, Vol.17, No.12, (December 2006), pp. 5390-5399, ISSN 1059-1524

Itoh, Y.; Palmisano, R.; Anilkumar, N.; Nagase, H.; Miyawaki, A. & Seiki, M. (2011). Dimerization of MT1-MMP during cellular invasion detected by flourescence resonance energy transfer. *The Biochemical Journal*, Vol.440, Part.3, (August 2011), pp. 319-326, ISSN 0264-6021

Jacobsen, F.E.; Lewis, J.A. & Cohen, S.M. (2007). The design of inhibitors for medicinally relevant metalloproteins. *Chem Med Chem*, Vol.2, No.2, (December 2006), pp. 152-171, ISSN 1860-7187

Jacobsen, J.A.; Jourden, J.L.M.; Miller, M.T. & Cohen, S.M. (2010). To bind zinc or not to bind zinc: An examination of innovative approaches to improved metalloproteinase inhibition. *Biochimica et Biophysica Acta (BBA) - Molecular Cell Research*, Vol.1803, No.1, (January 2010), pp. 72-94, ISSN 0167-4889

Jani, M.; Tordai, H.; Trexler, M.; Bányai, L. & Patthy, L. (2005). Hydroxamate-based peptide inhibitors of matrix metalloprotease 2. *Biochimie*, Vol.87, No.3-4, (March-April 2005), pp. 385-392. ISSN 0300-9084

Kessenbrock, K.; Plaks, V. & Werb, Z. (2010). Matrix metalloproteinases: regulators of the tumor microenvironment. *Cell*, Vol.141, No.1, (April 2010), pp. 52-67, ISSN 0092-8674

Kontogiorgis, C.A.; Papaioannou, P. & Hadjipavlou-Litina, D.J. (2005). Matrix metalloproteinase inhibitors: a review on pharmacophore mapping and (Q)SARs results. *Current Medicinal Chemistry*, Vol.12, No.3, pp. 339-355, ISSN 0929-8673

LeDour, G.; Moroy, G.; Rouffet, M.; Bourguet, E.; Guillaume, D.; Decarme, M.; ElMourabit, H.; Augé, F.; Alix, A.J.P.; Laronze, J.Y.; Bellon, G.; Hornebeck, W. & Sapi, J. (2008). Introduction of the 4-(4-bromophenyl)benzenesulfonyl group to hydrazide analogs of Ilomastat leads to potent gelatinase-B (MMP-9) inhibitors with improved selectivity. *Bioorganic & Medicinal Chemistry*, Vol.16, No.18, (September 2008), pp. 8745-8759, ISSN 0968-0896

Levy, D.E.; Lapierre, F.; Liang, W.; Ye, W.; Lange, C.W.; Li, X.; Grobelny, D.; Casabonne, M.; Tyrrell, D.; Holme, K.; Nadzan, A. & Galardy, R.E. (1998). Matrix Metalloproteinase Inhibitors: A Structure−Activity Study. *Journal of Medicinal Chemistry*, Vol.41, No.2 , (January 1998), pp. 199-223, ISSN 0022-2623

López-Otín, C. & Matrisian, L.M. (2007). Emerging roles of proteases in tumour suppression. *Nature Reviews Cancer*, Vol.7, No.10, (October 2007), pp. 800-808, ISSN 1474-175X

Lovejoy, B.; Hassell, A.M.; Luther, M.A.; Weigl, D. & Jordan, S.R. (1994). Crystal structures of recombinant 19-kDa human fibroblast collagenase complexed to itself. *Biochemistry*, Vol.33, No.27, (July 1994), pp. 8207-8217, ISSN 0006-2960

Malla, N.; Sjøli, S.; Winberg, J.O.; Hadler-Olsen, E. & Uhlin-Hansen, L. (2008). Biological and pathobiological functions of gelatinase dimers and complexes. *Connective Tissue Research*, Vol.49, No.3, pp. 180-184, ISSN 0300-8207

Mandal, M.; Mandal, A.; Das, S.; Chakraborti, T. & Chakraborti, S. (2003). Clinical implications of matrix metalloproteinases. *Molecular and Cellular Biochemistry*, Vol.252, No.1-2, (October 2003), pp. 305-329, ISSN 0300-8177

Marcq, V.; Mirand, C.; Decarme, M.; Emonard, H. & Hornebeck, W. (2003). MMPs inhibitors: New succinylhydroxamates with selective inhibition of MMP-2 over MMP-3. *Bioorganic & Medicinal Chemistry Letters*, Vol.13, No.17, (September 2003), pp. 2843-2846, ISSN 0960-894X

Martens, E.; Leyssen, A.; Van Aelst, I.; Fiten, P.; Piccard, H.; Hu, J.; Descamps, F.J.; Van den Steen, P.E.; Proost, P.; Van Damme, J.; Liuzzi, G.M.; Riccio, P.; Polverini E. & Opdenakker G. (2007). A monoclonal antibody inhibits gelatinase B/MMP-9 by selective binding to part of the catalytic domain and not to the fibronectin or zinc binding domains. *Biochimica et Biophysica Acta*, Vol.1770, No.2, (February 2007), pp. 178-186, ISSN 0006-3002

Miller, A.; Askew, M.; Beckett, R.P.; Bellamy, C.L.; Bone, E.A.; Coates, R.E.; Davidson, A.H.; Drummond, A.H.; Huxley, P.; Martin, F.M.; Saroglou, L.; Thompson, A.J.; van Dijk, S.E. & Whittaker, M. (1997). Inhibition of Matrix Metalloproteinases: An examination of the S1' pocket. *Bioorganic & Medicinal Chemistry Letters*, Vol.7, No.2, (January 1997), pp. 193-198, ISSN 0960-894X

Monferran, S.; Paupert, J.; Dauvillier, S.; Salles, B. & Muller C. (2004). The membrane form of the DNA repair protein Ku interacts at the cell surface with metalloproteinase 9. *The EMBO Journal*, Vol.23, No.19, (October 2004), pp. 3758-3768, ISSN 0261-4189

Moroy, G.; El Mourabit, H.; Toribio, A.; Denhez, C.; Dassonville, A.; Decarme, M.; Renault, J.H.; Mirand, C.; Bellon, G.; Sapi, J.; Alix, A.J.P.; Hornebeck, W. & Bourguet, E. (2007). Simultaneous presence of unsaturation and long alkyl chain at P'1 of Ilomastat confers selectivity for gelatinase A (MMP-2) over gelatinase B (MMP-9) inhibition as shown by molecular modelling studies. *Bioorganic & Medicinal Chemistry*, Vol.15, No.14, (July 2007), pp. 4753-4766, ISSN 0968-0896

Moroy, G.; Bourguet, E.; Decarme, D.; Sapi, J.; Alix, A.J.P.; Hornebeck, W. & Lorimier, S. (2011). Inhibition of Human Leukocyte Elastase, Plasmin and Matrix Metalloproteinases by Oleic acid and Oleoyl-Galardin derivative(s). *Biochemical Pharmacology*, Vol.81, No.5, (March 2011), pp. 626–635, ISSN 0006-2952

Moroy, G.; Alix, A.J.P.; Sapi, J.; Hornebeck, W. & Bourguet, E. (2012). Neutrophil Elastase as a Target in Lung Cancer. *Anti-cancer Agents Current Medicinal Chemistry*, In press, ISSN 1871-5206

Morris, G.M.; Goodsell, D.S.; Halliday, R.S.; Huey, R.; Hart, W.E.; Belew, R.K. & Olson, A.J. (1998). Automated Docking Using a Lamarckian Genetic Algorithm and an Empirical Binding Free Energy Function. *Journal of Computational Chemistry*, Vol.19, No.14, (November 1998), pp. 1639-1662, ISSN 0192-8651

Muroski, M.E.; Roycik, M.D.; Newcomer, R.G.; Van den Steen, P.E.; Opdenakker, G.; Monroe, H.R.; Sahab, Z.J. & Sang, Q.X. (2008). Matrix metalloproteinase-9/gelatinase B is a putative therapeutic target of chronic obstructive pulmonary disease and multiple sclerosis. *Current Pharmaceutical Biotechnology*, Vol.9, No.1, (February 2008), pp. 34-46, ISSN 1389-2010

Murphy, G. & Nagase, H. (2008). Progress in matrix metalloproteinase research. *Molecular Aspects of Medicine*, Vol.25, No.5, (October 2008), pp. 290-308, ISSN 0098-2997

Nagase, H.; Visse, R. & Murphy, G. (2006). Structure and function of matrix metalloproteinases and TIMPs. *Cardiovascular Research*, Vol.69, No.3, (February 2006), pp. 562-573, ISSN 0008-6363

Nicolotti, O.; Miscioscia, T.F.; Leonetti, F.; Muncipinto, G. & Carotti, A. (2007). Screening of matrix metalloproteinases available from the protein data bank: insights into biological functions, domain organization, and zinc binding groups. *Journal of Chemical Information and Modeling*, Vol.47, No.6, (November 2007), pp. 2439-2448, ISSN 1549-9596

Nunes, G.L.; Simões, A.; Dyszy, F.H.; Shida, C.S.; Juliano, M.A.; Juliano, L.; Gesteira, T.F.; Nader, H.B.; Murphy, G.; Chaffotte, A.F.; Goldberg, M.E.; Tersariol, I.L. & Almeida, P.C. (2011). Mechanism of heparin acceleration of tissue inhibitor of metalloproteases-1 (TIMP-1) degradation by the human neutrophil elastase. *PLoS One*, Vol.6, No.6, (June 2011), pp. e21525, ISSN 1932-6203

Overall, C.M. & Kleifeld, O. (2006). Tumour microenvironment - opinion: validating matrix metalloproteinases as drug targets and anti-targets for cancer therapy. *Nature Reviews Cancer*, Vol.6, No.3, (March 2006), pp. 227-239, ISSN 1474-1768

Overall, C.M. & Kleifeld, O. (2006). Towards third generation matrix metalloproteinase inhibitors for cancer therapy. *British Journal of Cancer*, Vol.94, No.7, (April 2006), pp. 941-946, ISSN 0007-0920

Rao, B.G. (2005). Recent Developments in the Design of Specific Matrix Metalloproteinase Inhibitors aided by Structural and Computational Studies. *Current Pharmaceutical Design*, Vol.11, No.3, pp. 295-322, ISSN 1381-6128

Redondo-Muñoz, J.; Ugarte-Berzal, E.; Terol, M.J.; Van den Steen, P.E.; Hernández del Cerro, M.; Roderfeld, M.; Roeb, E.; Opdenakker, G.; García-Marco, J.A. & García-Pardo, A. (2010). Matrix metalloproteinase-9 promotes chronic lymphocytic leukemia b cell survival through its hemopexin domain. *Cancer Cell*, Vol.17, No.2, (February 2010), pp. 160-172, ISSN 1535-6108

Robinet, A.; Fahem, A.; Cauchard, J.H.; Huet, E.; Vincent, L.; Lorimier, S.; Antonicelli, F.; Soria, C.; Crepin, M.; Hornebeck, W. & Bellon, G. (2005). Elastin-derived peptides enhance angiogenesis by promoting endothelial cell migration and tubulogenesis through upregulation of MT1-MMP. *Journal of Cell Science*, Vol.118, Pt.2, (January 2005), pp. 343-356, ISSN 0021-9533

Rouffet, M.; Denhez, C.; Bourguet, E.; Bohr, F. & Guillaume, D. (2009). In silico study of new inhibitors of MMPs. *Organic & Biomolecular Chemistry*, Vol.7, No.18, (September 2009), pp. 3817-3825, ISSN 1477-0520

Sato, H. & Takino, T. (2010). Coordinate action of membrane-type matrix metalloproteinase-1 (MT1-MMP) and MMP-2 enhances pericellular proteolysis and invasion. *Cancer Science*, Vol.101, No.4, (April 2010), pp. 843-847, ISSN 1347-9032

Schäfers, M.; Riemann, B.; Kopka, K.; Breyholz, H.J.; Wagner, S.; Schäfers, K.P.; Law, M.P.; Schober, O. & Levkau, B. (2004). Scintigraphic imaging of matrix metalloproteinase activity in the arterial wall in vivo. *Circulation*, Vol.109, No.21, (June 2004), pp. 2554-2559, ISSN 0009-7322

Scherer, R.L.; McIntyre, J.O. & Matrisian, L.M. (2008). Imaging matrix metalloproteinases in cancer. *Cancer Metastasis Reviews*, Vol.27, No.4, (December 2008), pp. 679-990, ISSN 0167-7659

Sela-Passwell, N.; Rosenblum, G.; Shoham, T. & Sagi, I. (2010). Structural and functional bases for allosteric control of MMP activities: can it pave the path for selective inhibition? *Biochimica et Biophysica Acta*, Vol.1803, No.1, (January 2010), pp. 29-38, ISSN 0006-3002

Shipley, J.M.; Doyle, G.A.; Fliszar, C.J.; Ye, Q.Z.; Johnson, L.L.; Shapiro, S.D.; Welgus, H.G. & Senior, R.M. (1996). The structural basis for the elastolytic activity of the 92-kDa and 72-kDa gelatinases. Role of the fibronectin type II-like repeats. *The Journal of Biological Chemistry*, Vol.271, No.8, (February 1996), pp. 4335-4341, ISSN 0021-9258

Shock, A.; Baum, H.; Kapasi, M.F.; Bull, F.M. & Quinn, P.J. (1990). The susceptibility of elastin fatty acid complexes to elastolytic enzymes. *Matrix*, Vol.10, No.3, (July 1990), pp. 179-185, ISSN 0934-8832

Silletti, S.; Kessler, T.; Goldberg, J.; Boger, D.L. & Cheresh, D.A. (2001). Disruption of matrix metalloproteinase 2 binding to integrin alpha v beta 3 by an organic molecule inhibits angiogenesis and tumor growth in vivo. *Proceedings of the National Academy of Sciences of the United States of America*, Vol.98, No.1, (January 2001), pp. 119-124, ISSN 0027-8424

Skiles, J.W.; Gonnella, N.C. & Jeng, A.Y. (2004). The design, structure, and clinical update of small molecular weight matrix metalloproteinase inhibitors. *Current Medicinal Chemistry*, Vol.11, No.22, (November 2004), pp. 2911-2977, ISSN 0929-8673

Terp, G.E.; Cruciani, G.C.; Christensen, I.T. & Jørgensen, F.S. (2002). Structural differences of matrix metalloproteinases with potential implications for inhibitor selectivity examined by the GRID/CPCA approach. *Journal of Medicinal Chemistry*, Vol.45, No.13, (June 2002), pp. 2675-2684, ISSN 0022-2623

Thompson, R.W. & Parks, W.C. (1996). Role of matrix metalloproteinases in abdominal aortic aneurysms. *Annals of the New York Academy of Sciences*, Vol.800, (November 1996), pp. 157-174, ISSN 0077-8923

Tyagi, S.C. & Simon, S.R. (1990). Inhibitors directed to binding domains in neutrophil elastase. *Biochemistry*, Vol.29, No.42, (October 1990), pp. 9970-9977, ISSN 0006-2960

Van den Steen, P.E.; Opdenakker, G.; Wormald, M.R.; Dwek, R.A. & Rudd, P.M. (2001). Matrix remodelling enzymes, the protease cascade and glycosylation. *Biochimica et Biophysica Acta*, Vol.1528, No.2-3, (October 2001), pp. 61-73, ISSN 0006-3002

Vandooren, J.; Geurts, N.; Martens, E.; Van den Steen, P.E.; Jonghe, S.D.; Herdewijn, P. & Opdenakker, G. (2011). Gelatin degradation assay reveals MMP-9 inhibitors and function of O-glycosylated domain. *World Journal of Biological Chemistry*, Vol.2, No.1, (January 2011), pp. 14-24, ISSN 1949-8454 (Electronic)

Van Wart, H.E. & Birkedal-Hansen, H. (1990). The cysteine switch: a principle of regulation of metalloproteinase activity with potential applicability to the entire matrix metalloproteinase gene family. *Proceedings of the National Academy of Sciences of the United States of America*, Vol.87, No.14, (July 1990), pp. 5578-5582, ISSN 0027-8424

Whittaker, M.; Floyd, C.D.; Brown, P. & Gearing, A.J.H. (1999). Design and therapeutic application of matrix metalloproteinase inhibitors. *Chemical Reviews*, Vol.99, No.9, (September 1999), pp. 2735-2776, ISSN 0009-2665

Xu, X.; Chen, Z.; Wang, Y.; Bonewald, L. & Steffensen, B. (2007). Inhibition of MMP-2 gelatinolysis by targeting exodomain-substrate interactions. *Biochemical Journal*, Vol.406, No.1, (August 2007), pp. 147-155, ISSN 0264-6021

Xu, X.; Mikhailova, M.; Ilangovan, U.; Chen, Z.; Yu, A.; Pal, S.; Hinck, A.P. & Steffensen, B. (2009). Nuclear magnetic resonance mapping and functional confirmation of the collagen binding sites of matrix metalloproteinase-2. *Biochemistry*, Vol.48, No.25, (June 2009), pp. 5822-5831, ISSN 0006-2960

Xu, X.; Mikhailova, M.; Chen, Z.; Pal, S.; Robichaud, T.K.; Lafer, E.M.; Baber, S. & Steffensen, B. (2011). Peptide from the C-terminal domain of tissue inhibitor of matrix metalloproteinases-2 (TIMP-2) inhibits membrane activation of matrix metalloproteinase-2 (MMP-2). *Matrix biology: journal of the International Society for Matrix Biology, Biology*, Vol.30, No.7-8, (September 2011), pp. 404-412, ISSN 0945-053X

4

Non-Enzymatic Glycation of Aminotransferases and the Possibilities of Its Modulation

Iva Boušová, Lenka Srbová and Jaroslav Dršata
Department of Biochemical Sciences, Charles University in Prague,
Faculty of Pharmacy in Hradec Králové
Czech Republic

1. Introduction

Enzymes are catalytic protein molecules performing specific functions in their native form. Native structure is one of basic conditions for normal function of proteins including enzymes. Catalytic activity of enzyme may be decreased by both non-covalent modulation by true inhibitors and covalent modifications by metabolites in the body or natural products. One example of such modulation is non-enzymatic glycation of proteins by sugars.

Enzymes are very good models for studies of protein interactions with other molecules, including sugars (e.g., Arai et al., 1987; Dolhofer & Wieland, 1978; Okada et al., 1994; Okada et al., 1997; Sakurai et al., 1987). Advantage of these studies is the fact that enzyme interactions may be investigated not only by common methods of studies of protein properties like changes in spectral characteristics, molecular weight, charge, solubility etc. but also by measurement of their catalytic activity as the most characteristic property of enzyme molecules. For such studies, it is important to use the enzyme with known structure and reaction mechanism, available in sufficient purity, and measurable by simple assay method. Aminotransferases belong to such enzymes. That is the reason why research activities of our laboratory devote to inhibition of aminotransferases by low-molecular compounds for many years (e.g. Dršata & Veselá, 1984; Netopilová et al., 1991; Netopilová et al., 2001). Our studies of aminotransferase glycation belong to the efforts made also by several other research groups (see e.g. Okada & Ayabe, 1997; Fitzgerald et al., 2000; Seidler & Kowalewski 2003; Hobart et al., 2004).

1.1 Aminotransferases

Aspartate aminotransferase (AST, EC 2.6.1.1) and alanine aminotransferase (ALT, EC 2.6.1.2), enzymes frequently assessed in clinical laboratories, catalyse reversible transfer of α-amino group from amino acids aspartate and alanine to the acceptor 2-oxoglutarate, respectively. The resulting products are oxaloacetate, pyruvate and glutamate. In metabolism, AST activity mediates the connection between the metabolism of amino acids and saccharides and participates in the transportation of reduced equivalents across the membrane of mitochondria as a part of the malate–aspartate shuttle. The catalytic activities of AST in cells are implemented by two isoenzymes—the cytosolic and the mitochondrial one, which differ in primary structure and in some properties (Yagi et al., 1985; Metzler et al., 1979).

Porcine heart cytosolic AST, which was used in further described experiments of our research group, is a homodimer with molecular mass of about 93,150 Da. Both subunits are composed of 412 amino acid residues. This dimer contains 38 lysine and 52 arginine residues with six Lys–Arg and four Arg–Lys sequence pairs (Seidler & Kowalewski, 2003). Presence of these amino acid residues makes glycation of aminotransferases *in vitro* as well as *in vivo* possible. The Lys258 binding coenzyme PLP could be one of the possible targets of glycating agents as can be seen from the loss of enzymatic activity due to the effect of methylglyoxal as well as other glycating agents (e.g. Boušová et al., 2009; Seidler & Kowalewski, 2003).

Aminotransferases are characterised by the presence of coenzyme pyridoxal-5′-phosphate (PLP) and its direct participation in catalysis. The protein part of aminotransferase molecules consists of two identical, non-covalently bound subunits, which are composed of one small and one large domain. PLP coenzyme forms a Schiff base with ε-amino group of lysine residue (Lys313 in ALT and Lys258 in AST) located in an active site of the larger domain of each subunit (Kirsch et al., 1984). The PLP form of AST shows, depending on pH, a major absorption peak at 360 nm (an active, unprotonized form of the coenzyme, prevailing at lower pH values), and/or a peak at 430 nm (an inactive, protonized form, increasing at lower pH values) (Kirsch et al., 1984). After a reaction with L-aspartate during the first part of a ping-pong transaminating reaction, the pyridoxamine-5′-phosphate (PMP) form of the coenzyme appears and the original absorption maximum shifts to 325–330 nm (Fig. 1A). While free PLP or PMP are not optically active substances, the coenzyme bound in the active site of AST shows circular dichroism (CD) spectra in the range of 300–500 nm, which are similar to absorption spectra (Fig. 1C). The CD effect is caused by the change in the electronic configuration of the molecule (Kelly & Price, 2000). Circular dichroism clears away absorption characteristics of optically inactive components, which facilitates identification of the specific coenzyme signal and its changes (Dršata et al., 2005). This method is also a powerful tool for characterizing secondary (190-240 nm, chromophore is peptide bond) and tertiary (250-320 nm, chromophores are aromatic amino acids and disulfide bonds) structures of studied proteins as well as determining whether the protein is folded (Kelly & Price, 2000).

1.2 Non-enzymatic glycation of proteins

Non-enzymatic glycation, also called Maillard reaction, was first described in 1912 by Louis Camille Maillard (Monnier, 1989). This non-enzymatic browning process had been first extensively studied by food chemists and later has become the center of attention of geological or agricultural sciences. Much later it was recognized that the process is important for medical science (Monnier, 1989; Singh et al., 2001). A very common but serious human disease is diabetes mellitus, which in its untreated or unsuccessfully treated form is accompanied with development of chronic complications due to increased intra- and

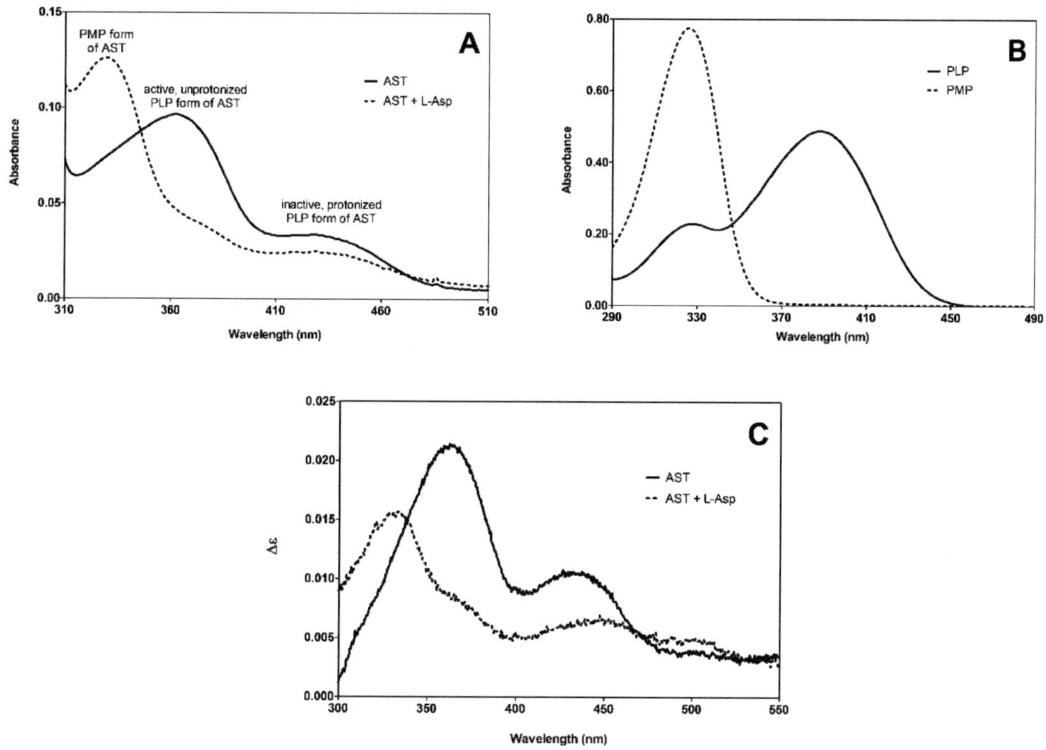

Fig. 1. Characteristic UV-VIS absorption and CD spectra of AST (panel A and C) and of the free coenzyme (panel B). (A) UV-VIS spectra of AST (1 mg/mL; 0.1 M sodium phosphate buffer, pH 7.4) alone or in the presence of L-Asp (1 mM), (B) UV-VIS spectra of free PLP (0.1 mM) and PMP (0.1 mM) in sodium phosphate buffer (0.1 M; pH 7.4) were recorded using a diode array spectrophotometer HP 8453 (Hewlett Packard, USA) in a 0.5 cm quartz cuvette against sodium phosphate buffer (0.1 M; pH 7.4). (C) CD spectra of AST (1 mg/mL; 0.1 M sodium phosphate buffer, pH 7.4) alone or in the presence of L-Asp (1 mM) were recorded using a dichrograph CD6 (Jobin Yvon, France) in a 0.5 cm quartz cuvette against sodium phosphate buffer (0.1 M; pH 7.4). For further details on experimental work see (Dršata et al., 2005)

extracellular concentration of glucose including a hallmark of the disease - hyperglycemia. There is increasing evidence that Maillard reaction plays an important role in the onset and progression of some other diseases, such as atherosclerosis and Alzheimer's disease (Nursten, 2005; Yegin et al., 1995). Plenty of studies have been devoted to investigation of protein structural and functional changes caused by glycation process (e.g. Jabeen & Saleemuddin, 2006; Seidler & Seibel, 2000; Yan & Harding 1997, 2006; Zeng et al., 2006; Zhao et al., 2000).

1.2.1 Formation of advanced glycation end-products

The early stage of the Maillard reaction is initiated by non-enzymatic condensation of a reducing sugar or a certain related compound (e.g. ascorbate) with a free amino group of a protein, a lipid or a nucleic acid. In the case of glucose, the reaction first leads to the formation of acid-labile Schiff base, which undergoes a rearrangement to a relatively stable Amadori product, e.g. fructosamine. Only a small portion of these Amadori-adducts experiences further rearrangements leading to an irreversible formation of advanced glycation end-products (AGEs) (Monnier, 1989). The reaction with fructose proceeds in a similar way, but it is called Heyns rearrangement and two separate Heyns products are generated (Suarez et al., 1989). The formation of Schiff base proceeds in the range of hours and it is fully reversible, while Amadori rearrangement takes days and is reversible only to a certain extent.

In the intermediate stage, Amadori product subsequently degrades and various reactive intermediates are formed. These products are known as α-dicarbonyls or α-oxoaldehydes and are represented by products like methylglyoxal (MGO), 3-deoxyglucosone (3-DG), and glyoxal (GO). Also a Schiff base is a potential source of reactive α-dicarbonyls, because it can be fragmented to MGO and GO. These dicarbonyls possess higher reactivity towards proteins than the parent monosaccharide. They are capable of forming various cross-links as well as chromo/fluorophoric adducts called AGEs, upon reaction with proteins (Schalkwijk et al., 2004; Wolff et al., 1991). Both MGO and 3-DG form adducts with proteins and nucleic acids up to 10,000 times more readily than glucose (Beisswenger et al., 2003). The accumulation of α-dicarbonyl compounds is termed carbonyl stress (Miyata et al., 1999). The other process proceeding during the intermediate stage of glycation is metal catalyzed autoxidation of glucose, in which the carbonyl compounds (arabinose and glyoxal), H_2O_2 and free radicals are formed (Hunt et al., 1988; Wolff & Dean, 1987). The generated free radicals initiate further oxidative steps. The glycation process accompanied by oxidation steps is called glycoxidation (Baynes, 1991). Various pathways incorporated in the formation of AGEs are shown in Fig. 2.

The advanced glycation end-products are formed during the late stage of glycation over a period of weeks, thereby affecting predominantly long-lived proteins, such as collagen and lens crystallins. They represent a heterogeneous group of compounds rising from different precursors. The chemical structures of AGEs have not been fully described yet. These compounds are formed either by oxidative pathway (pentosidine and CML) or by non-oxidative pathway (pyrraline, DOLD, GOLD, MOLD, and CEL) as can be seen in Fig. 2. Proteins modified by advanced glycation are characterized by a much higher molecular weight than the original protein, a yellow-brown pigmentation, a typical fluorescent spectra (λex/λem: 370/440 nm), an ability to form various cross-links, and by their biological half-life, which is comparable to the half-lives of parent proteins (Lapolla et al., 2005; Singh et al., 2001).

Glucose is the least reactive of the common sugars and that is probably the reason for its evolutionary selection as the principal sugar *in vivo* (Bunn et al., 1978). Because of its low reactivity towards proteins, AGEs have been thought to form only at long-lived extracellular proteins, such as collagen, crystallines, and myelin. Recently also rapid intracellular AGE formation by various intracellular sugars (e.g. fructose, ribose, glyceraldehyde, dihydroxyacetone phosphate, glyceraldehyde-3-phosphate, glyoxal, methylglyoxal, and 3-deoxyglucosone) *in vivo* has been described. The rate of glycation is directly proportional to the percentage of sugar in the open-chain form and the rate for fructose (0.7% open-chain) is 7.5-fold faster than that of glucose (0.002% open-chain). More strikingly, the glycolytic intermediate glyceraldehyde-3-phosphate (100% open-chain) forms 200-fold more glycated proteins than do equimolar amounts of glucose (Schalkwijk et al., 2004).

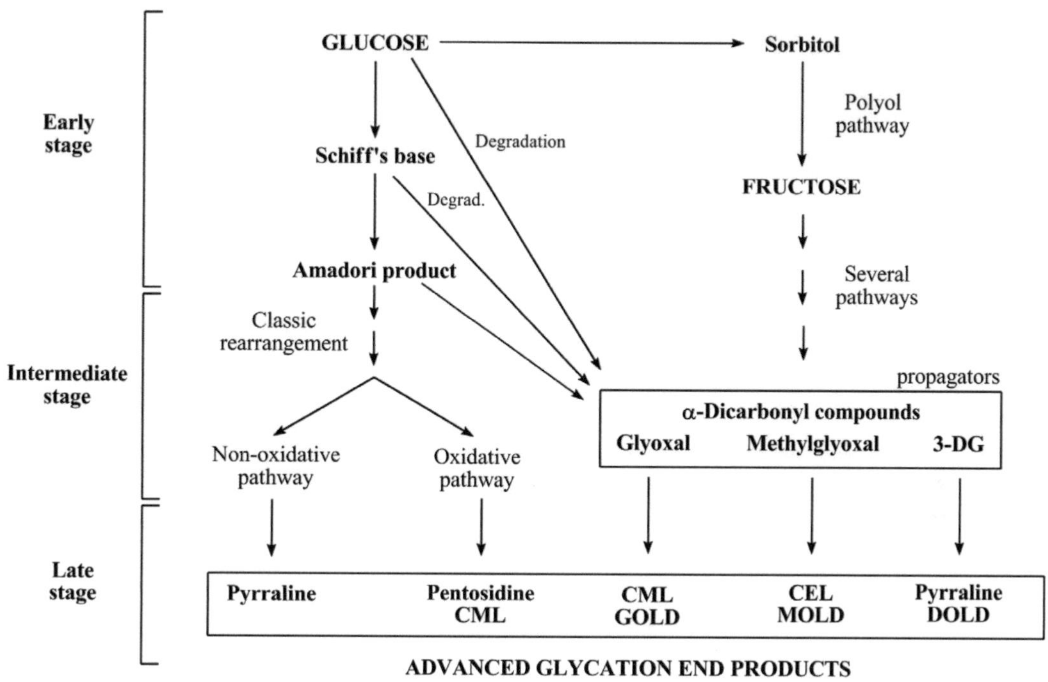

Fig. 2. Three stages of non-enzymatic glycation reaction. CML, *N*-ε-(carboxymethyl)lysine; GOLD, glyoxal-lysine dimer; CEL, *N*-ε-(carboxyethyl)lysine; MOLD, methylglyoxal-lysine dimer; 3-DG, 3-deoxyglucosone; DOLD, deoxyglucosone-lysine dimer

1.2.2 Structure of AGEs

The advanced glycation end-products found under physiological conditions can be classified according to their fluorescent properties and their ability to form cross-links. The first group is represented by fluorescent AGE cross-links, which are thought to be responsible for a major share of the deleterious effects of AGEs in diabetes and aging. Fluorescence is a good qualitative indicator used to estimate AGEs formation. Pentosidine, crossline, and various vesperlysines are members of this group. However, also non-fluorescent AGE cross-links are found *in vivo*. Their isolation and identification is more complicated than in the case of fluorescent AGE cross-links. It is thought that they account just for 1% of all cross-links rising under physiological conditions. Various imidazolium dilysine cross-links (GOLD, MOLD), arginine-lysine cross-links, and glucosepan belong to this group. Last but not least, a group of non-cross-linking protein bound AGE structures have been identified *in vivo*. These structures may exert deleterious effects as precursors of cross-links or as biological receptor ligands inducing a variety of adverse cellular and tissue changes. The well-known members of this group are pyrraline, carboxyalkyllysines (CML, CEL), imidazolones, and argpyrimidine (Ulrich & Cerami, 2001). Classification and examples of each above mentioned group are shown in Fig. 3.

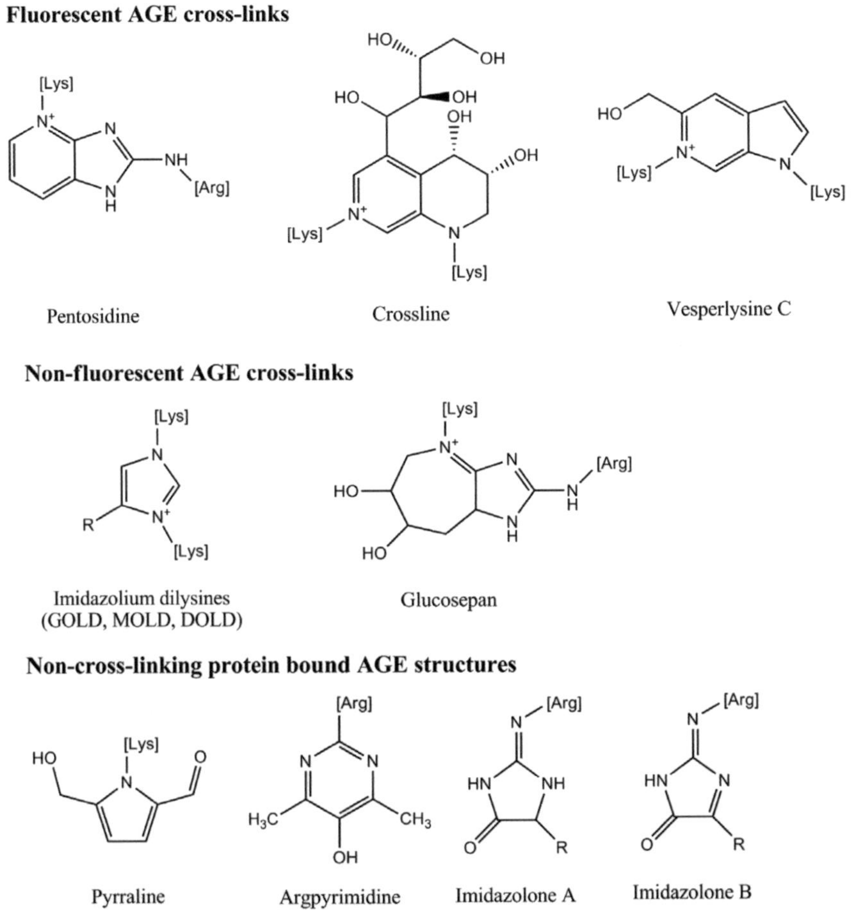

Fig. 3. Classification of AGEs formed under physiological conditions including several examples to each group. [Lys] represents a desamino-lysine residue; [Arg] stands for a desguanidino-arginine residue; R represents either hydrogen atom (GOLD), methyl group (MOLD), 1,2,3-trihydroxypropyl (DOLD) or 2,3,4-trihydroxybutyl group (imidazolone A and B).

1.2.3 Therapeutic strategies targeting the AGEs

The therapeutic intervention to the glycation process has followed three main approaches. A first approach follows inhibition of RAGE by neutralizing antibodies or suppression of post-receptor signaling using antioxidants. A second one is inhibition of AGE formation process by carbonyl-blocking agents (aminoguanidine) or by antioxidants. The last approach is reducing AGE deposition by using cross-link breakers or by enhancing cellular uptake and degradation.

Interactions of AGEs with the receptor for AGEs (RAGE) have been implicated in the development of diabetic vascular complications, which cause various disabilities and shortened life expectancy, and reduced quality of life in patients with diabetes. These undesirable effects can be suppressed by the use of specific antibodies to RAGE, soluble

RAGE or by suppression of post-receptor signaling using antioxidants (Hudson et al., 2003; Stuchbury & Münch, 2005). The secreted RAGE form, named soluble RAGE (sRAGE), acts as a decoy to trap ligands and prevent interaction with cell surface receptors (Bucciarelli et al., 2002). Soluble RAGE was shown to have important inhibitory effects in several cell culture and transgenic mouse models, in which it prevented or reversed full-length RAGE signaling. The administration of sRAGE has been shown to suppress accelerated diabetic atherosclerosis (Park et al., 1998).

Aminoguanidine, also known by its trade name Pimagedine (Alteon Inc.), is a prototype therapeutic agent for prevention of the AGEs formation. It is a low-molecular, highly nucleophilic hydrazine compound that rapidly reacts with α-dicarbonyl compounds such as MGO, GO, and 3-DG to prevent formation of AGE cross-links. The products of the scavenging reaction are substituted 3-amino-1,2,4-triazines. Aminoguanidine does not affect the formation of the Schiff base and Amadori products (Thornalley et al., 2000). Clinical trials of aminoguanidine in overt diabetic nephropathy (ACTION) were performed, but they were early terminated due to safety concerns. Reported side effects of aminoguanidine in clinical therapy were gastrointestinal disturbance, abnormalities in liver function tests, flu-like symptoms, and a rare vasculitis (Bolton et al., 2004; Thornalley, 2003). Other nucleophilic compounds, which are designed to trap reactive carbonyl intermediates in AGE formation, are for example OPB-9195, diaminophenazine, tenilsetam, and pyridoxamine (Baynes & Thorpe, 2000; De La Cruz et al., 2004). With regard to the presence of free radicals and oxidative steps in the course of glycoxidation, compounds with antioxidant effect such as α-lipoic acid, α-tocopherol, ascorbic acid, and ß-carotene were tested. Dipeptide carnosine, pyridoindole derivative stobadine, hypolipidemic drug probucol, and mucolytic remedy *N*-acetylcysteine are just a few more examples of the compounds with described antioxidant properties, which were tested in order to estimate their potential protective effect in the process of glycation. Also some antioxidant enzymes such as superoxide dismutase, catalase, and selenium-dependent glutathione peroxidase may protect proteins against impairment caused by non-enzymatic glycation (De La Cruz et al., 2004; Kyselova et al., 2004).

Aminoguanidine and other compounds mentioned before can inhibit the formation of new AGE cross-links, but they are not able to cleave those already formed. Vasan et al. (1996) reported the first of cross-link breakers, phenyl thiazolium bromide (PTB). This anti-AGE agent chemically breaks α-dicarbonyl compounds by cleaving the carbon-carbon bond between the carbonyls. Under physiological conditions, PTB is not stable and therefore its analogs were tested and alagebrium chloride (ALT-711), a highly potent cross-link breaker with higher stability, has been discovered. This compound successfully completed preclinical studies and Phase II clinical study on healthy volunteers (Yamagishi et al., 2008). Unfortunately, the specific types of AGEs affected by alagebrium are more important in rats than humans; hence the promising results in animals were never repeated in human studies. Randomized Phase II clinical trial (BENEFICIAL) in patients with chronic heart failure (Willemsen et al., 2010) has been terminated early due to financial constraints.

The objective of current study was to evaluate potential antiglycation activity of two mitochondrial antioxidants, α-phenyl *N-tert*-butyl nitrone (PBN) and *N-tert*-butyl hydroxylamine (NtBHA). PBN is a nitrone that traps free radicals and protects against damage in different models such as inflammation, ischemia reperfusion, and aging. Its decomposition product NtBHA mimics PBN and is much more effective in delaying senescence of human lung fibroblasts IMR90 (Atamna et al., 2000). NtBHA appears to act on mitochondria to delay alterations in function (Atamna et al., 2001). Supplementation with NtBHA improved the respiratory control ratio of mitochondria from liver of old rats (Atamna et al., 2001).

1.3 Summary of existing results

Our laboratory deals with research of the protein glycation *in vitro* for many years. Aminotransferases were chosen as suitable model proteins, because they are commercially available in a highly purified and stable form, their structures and mechanism of catalysis are well known, and at least two simple methods for their enzyme activity assessment are in use. This subchapter summarizes results, which have been obtained by our research group up to now.

In our laboratory, we found decrease in alanine aminotransferase (ALT, EC 2.6.1.2) activities in the presence of several reducing monosaccharides *in vitro*. The decrease in the catalytic activity of ALT from porcine heart after 20 days of incubation with D-glucose, D-fructose, D-ribose or D,L-glyceraldehyde varied and was dependent on the nature of glycating agent (the percentage of sugar present in open-chain form). The dependence of enzyme inactivation on the presence of these sugars and time of incubation is presented in Fig. 4 (Beránek et al., 2001). As was described earlier, the rate of glycation is directly proportional to the percentage of sugar in the open chain form and the rate for fructose (0.7% open-chain) is 7.5-fold faster than that of glucose (0.002% open-chain). More strikingly, the glycolytic intermediate glyceraldehyde 3-phosphate (100% open-chain) forms 200-fold more glycated proteins than do equimolar amounts of glucose (Schalkwijk et al., 2004).

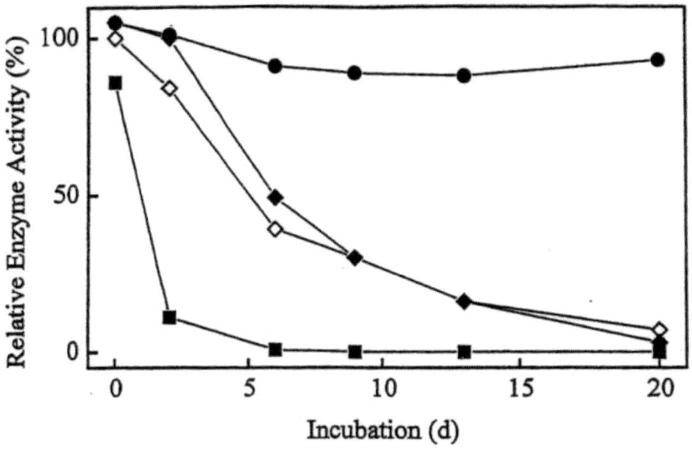

Fig. 4. An effect of glycation on ALT activity. ALT was incubated with 50 mM sugars in sodium phosphate buffer (0.1 M, pH 7.4) at 25 °C. Aliquots of samples were taken at days 0, 2, 6, 9, 13, 20 and remaining enzyme activities in samples were determined. Activity was expressed as a percentage of the activity of the control sample (without sugars). ● ALT with D-glucose; ◊ ALT with D-ribose; ♦ ALT with D-fructose; ■ ALT with D,L-glyceraldehyde

Fructose, ribose and glyceraldehyde have been found more potent glycating agents than glucose. The strongest glycation effect was exerted by D,L-glyceraldehyde. Complete enzyme inhibition was reached after 6 days but most enzymatic activity (about 75%) was reduced in the course of the first 2 days of incubation. Moreover, a decrease in ALT activity to 88% of the relevant control was apparent at the zero-time determination of the enzymatic activity in the glyceraldehyde sample. Similar results were obtained with AST (Beránek et al., 2002). These data are presented in Fig. 5.

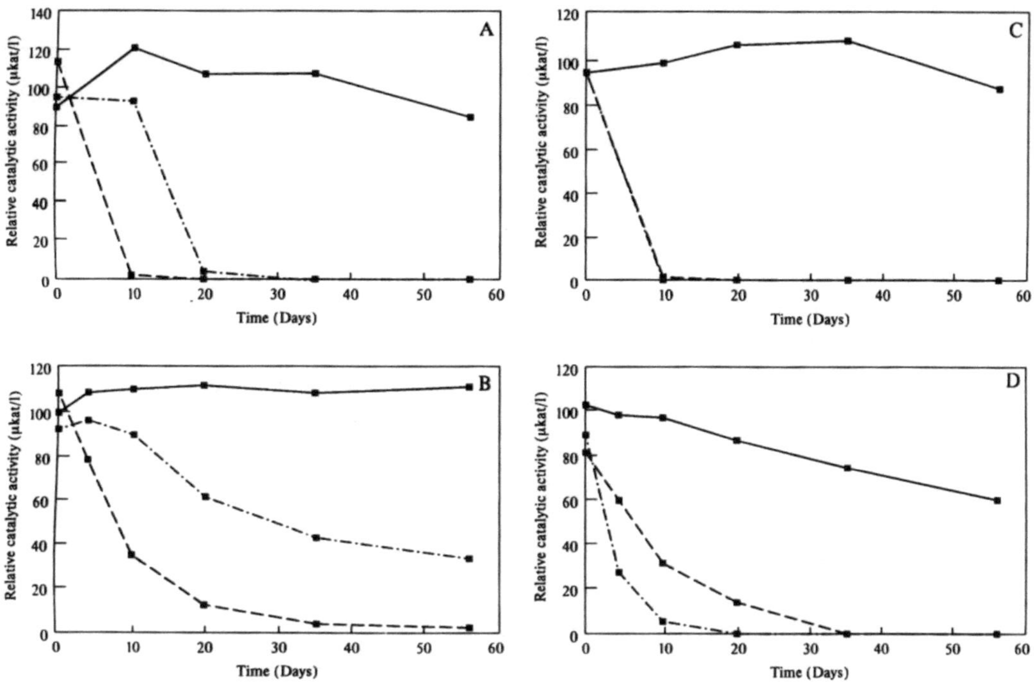

Fig. 5. Inhibitory effect of glycation on the catalytic activity of aminotransferases incubated in sodium phosphate buffer (50 mM, pH 7.4) at 4 °C (——), 25 °C (----) and 37 °C (-·-·-) for up to 56 days. The values show residual catalytic activities of the enzymes related to the appropriate control. (A) ALT incubated with 50 mM D-fructose; (B) AST with 50 mM D-fructose; (C) ALT with 500 mM D-fructose; (D) AST with 500 mM D-fructose.

In our further experiments, aspartate aminotransferase (AST, EC 2.6.1.1) has been chosen as a model protein due to its relative stability during *in vitro* incubation and availability of the enzyme preparations of high purity (for further experimental details see Dršata et al., 2005).

The purity of the enzyme preparation was crucial for experiments with prolonged incubation. While AST activity of the rat liver 20 000 x g supernatant was very unstable *in vitro* and declined rapidly independently of presence or absence of sugar during incubation even at 25 °C, the purified preparation enabled experiments using incubation for many days (Tupcová, 1996). Data are presented in Fig. 6.

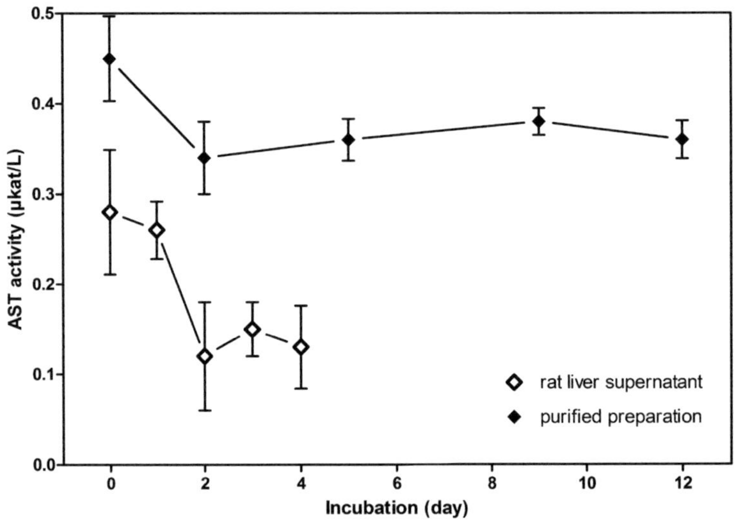

Fig. 6. Comparison of stability of AST in rat liver supernatant and in purified preparation from pig heart during *in vitro* incubation at 25 °C. The incubation mixture was diluted before the assay in order to obtain activities within the analytical range of the method. Values are mean ± S.D. of three independent samples. Each sample was measured three-times and the mean was used to calculate the value presented.

Moreover, glycation of AST was accompanied by a decrease in its catalytic activity in dependence on the concentration and activity of the glycating agent, while the concentration effect was not clearly demonstrated in case of ALT (Dršata et al., 2002). The effect of substrates (2-oxoglutarate and L-aspartate) on AST stability and on glycation process was assessed as well. There was no effect of 2-oxoglutarate on the control AST activity throughout the experiment, which demonstrated a high stability of the pyridoxal form both in the presence and absence of this substrate. On the other hand, the presence of 25 mM L-aspartate (inducing the pyridoxamine form of the enzyme) caused a rapid decrease in AST activity even in the control reaction (Dršata et al., 2002). The results of AST incubation with D-ribose and D-fructose in presence of 0.5 mM 2-oxoglutarate suggested that AST glycation could be partly prevented by this substrate. This finding supports the idea that Lys258 in the active center of AST is involved in glycation of the free enzyme (Fig. 7, taken from Dršata et al., 2002).

The *in vitro* model of protein glycation (AST) by D-fructose has been then established and experimentally used. (Boušová et al., 2005a). As mentioned above, attempts have been made by researchers to investigate various chemical compounds as potential antiglycating agents. With this model, influence of potential antiglycating compounds with antioxidant activities was investigated. In an attempt to reduce glycoxidation process and formation of AGEs, influence of endogenous antioxidant uric acid (0.2-1.2 mM) on glycoxidation process of AST by 50 mM and 500 mM D-fructose *in vitro* was studied. Uric acid at 1.2 mM concentration reduced AST activity decrease caused by incubation of the enzyme with 50 mM sugar up to 25 days at 37 °C (Fig. 8), as well as formation of total fluorescent AGE products (Fig. 9). The results obtained supported the hypothesis that uric acid has beneficial effects in controlling protein glycoxidation (Fig. 8 and 9).

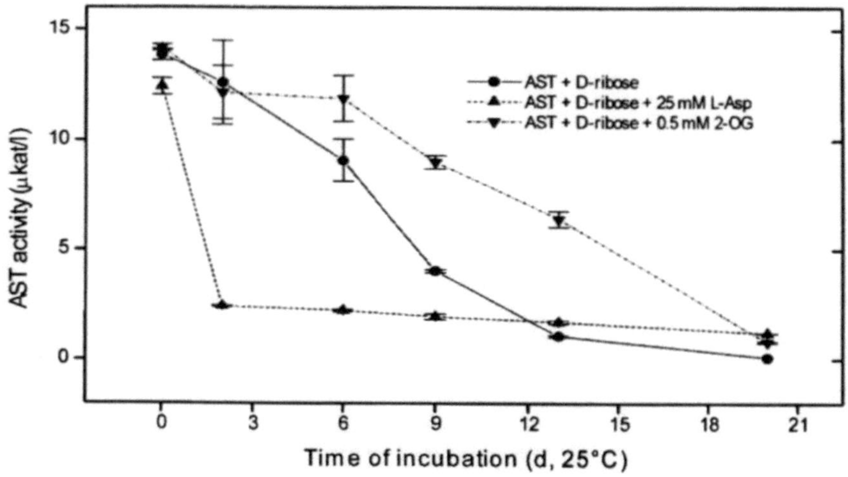

Fig. 7. Influence of substrates on AST activity during glycation by D-ribose *in vitro*. Concentration of the enzyme (purified Serva preparation) in the incubation mixture: 0.413 mg/ml, final catalytic concentration in the mixture 7.83 μkat/mg. Values are mean ± S.D. of 6 (control) or 3 (with D-ribose) independent samples. Each sample was measured three-times and the mean was used to calculate the value presented.
L-Asp = L-aspartate; 2-OG = 2-oxoglutarate

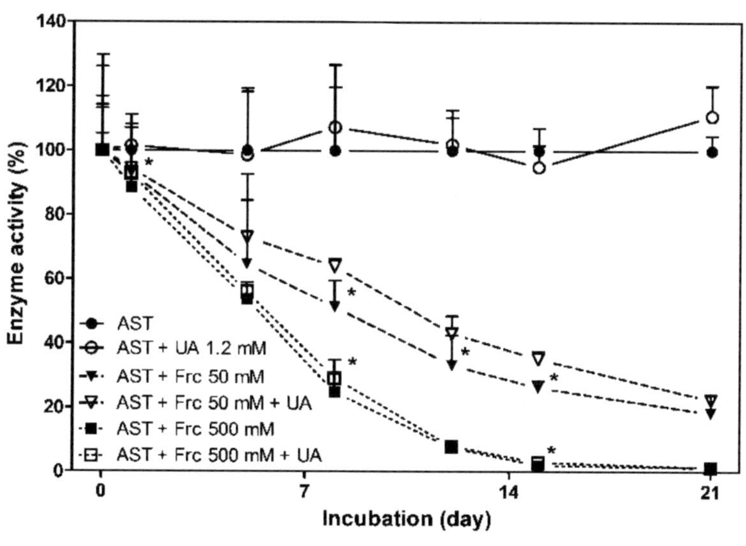

Fig. 8. Effect of glycation on AST activity and intervention of uric acid (UA) in the process. AST (1.33 mg/ml) was incubated with and without fructose (Frc) in 0.1 mM phosphate buffer, pH 7.4 at 37 °C in the presence or absence of 1.2 mM uric acid. AST activity was expressed as percentage of that of control sample (without sugar), which was considered as 100% ± S.D. (%) at every interval. Each point represents an average of three experiments in interval 0-12 days, and an average of two experiments on the 15th and 21st day. Each experiment was performed in triplicate (*data with $P<0.05$, Student's *t*-test).

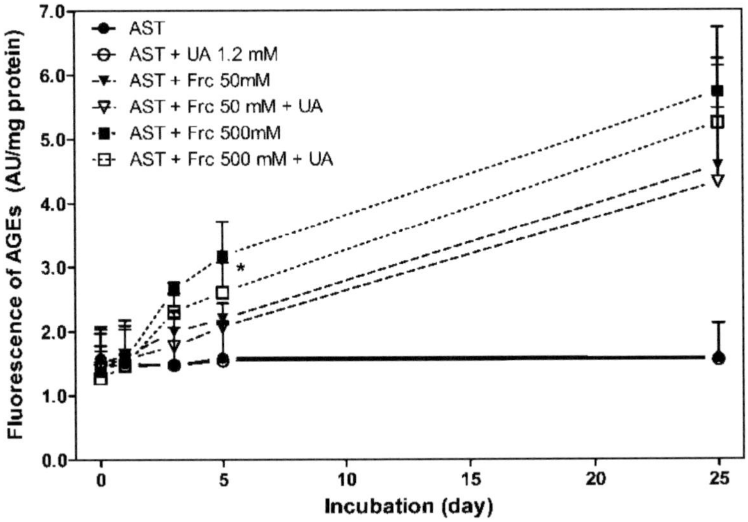

Fig. 9. Formation of fluorescent products of glycation under the conditions of long lasting incubation. AST (1.33 mg/ml) was incubated with or without D-fructose in sodium phosphate buffer (0.1 M, pH 7.4) at 37 °C in the presence or absence of 1.2 mM uric acid up to 25 days. Aliquots of samples were taken on days 0, 1, 3, 5, 25, and arising fluorescent AGE products in samples were determined at specific wavelengths of excitation and emission (λex/λem) corresponding to total AGEs (370/440 nm). Data of relative fluorescence were expressed in arbitrary units per mg of protein ± S.D., with 1 AU corresponding to the fluorescence of BSA 1.0 mg/ml. Every point in days 0 and 5 represents an average of four experiments (10 samples), in days 1 and 3 an average of three experiments for mixtures with 50 mM fructose (7 samples) and of two experiments for mixtures with 500 mM fructose (4 samples), and in day 25 an average of three experiments (7 samples), (*data with $P<0.05$, Student's *t*-test).

Also further studies of our research group have been pointed at possible protective activity of selected natural compounds (e.g. hydroxycinnamic acids, flavonoids, arbutin, hydroxycitric acid) on AST glycation by fructose. The results have shown that these compounds can exert also negative effects on enzyme activity but some of them have been able to slow down the course of AST modification by glycating agent. The compound with overall positive activity has been hydroxycitric acid (Fig. 10), which is major active component in the fruit rinds of certain species of the plant *Garcinia*. The effect of this compound was concentration-depended and its positive activity was most pronounced at 2.5 mM concentration. On the other hand, flavonoid baicalin (Fig. 11) and its aglycone baicalein rapidly decreased the *in vitro* activity of the enzyme in all concentrations used (0.5–3 mM), and no beneficial effects of these compounds on glycation of the enzyme by fructose were found (Boušová et al., 2005b).

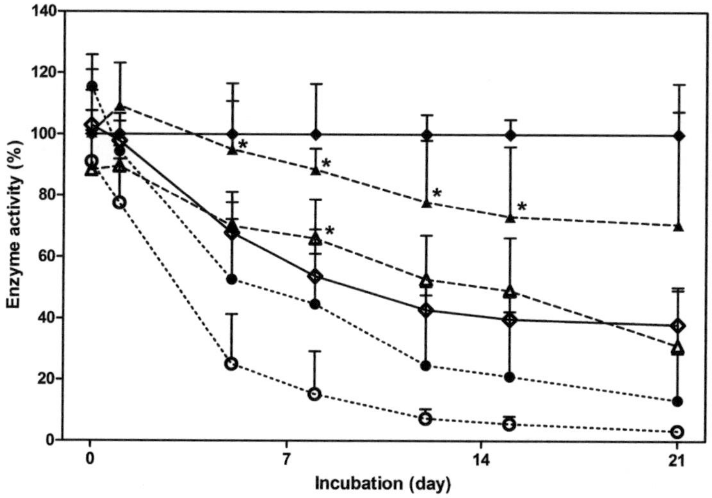

Fig. 10. The effect of hydroxycitric acid (HCA) on the glycation of AST by fructose *in vitro*. Concentration of AST in incubation mixtures was 1.33 mg/ml. AST was incubated with or without D-fructose 50 mM in sodium phosphate buffer (0.1 M, pH 7.4, 0.05% sodium azide) at 37 °C in the presence or absence of HCA up to 21 days. The samples were diluted before the assay to fit to the analytical range of the method. Values are expressed as a percentage of the activity of respective control (AST) ± S.D. of six independent samples (*data with $P<0.05$, Student's *t*-test). ♦ AST; ◊ AST + Frc 50 mM; ▲ AST + Frc + HCA 2.5 mM; Δ AST + Frc + HCA 5.0 mM; • AST + Frc + HCA 7.5 mM; o AST + Frc + HCA 10.0 mM.

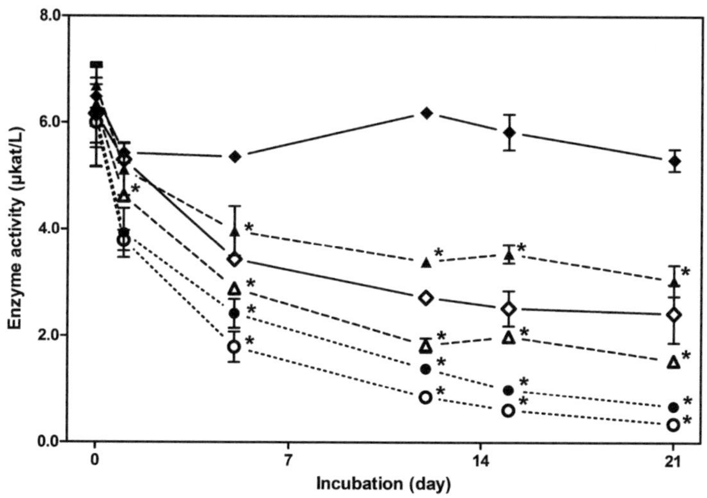

Fig. 11. Direct effect of baicalin on the AST activity *in vitro*. Concentration of AST in incubation mixtures was 1.33 mg/ml in 0.1M sodium phosphate buffer (pH 7.4; 0.05% sodium azide). Incubation at 37 °C. The samples were diluted before the assay to fit to the analytical range of the method. Values are expressed as mean ± S.D. of six (control and AST + Frc 50 mM) or three (with baicalin) independent samples (*data with $P<0.05$, Student's *t*-test). ♦ AST; ◊ AST + Frc 50 mM; ▲ AST + baicalin 0.5 mM; Δ AST + baicalin 1.0 mM; • AST + baicalin 1.5 mM; o AST + baicalin 3.0 mM.

Following experiments were conducted using methylglyoxal (MGO) as a glycating agent, because this compound has higher glycating potential than reducing monosaccharides and the incubation period has been shortened to one week only. Changes in the catalytic activity of AST caused by MGO were observable even after 120 min of incubation at 37 °C. Antiglycating activity of hydroxycitric (Fig. 12) and uric acid has been studied in this modified model (Boušová et al., 2009).

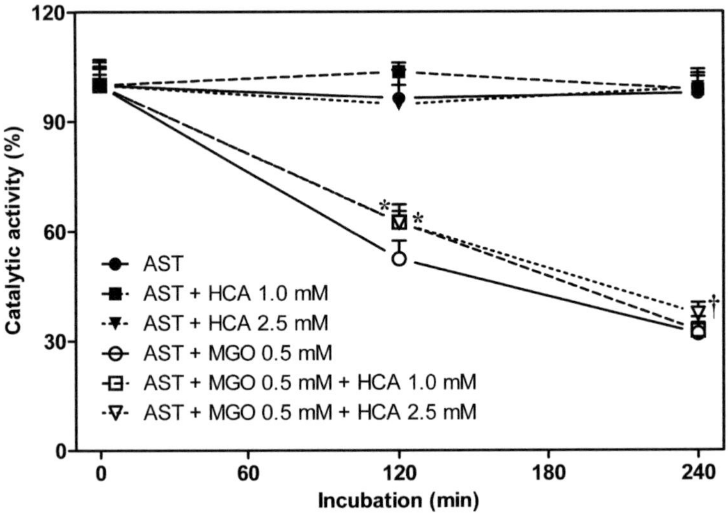

Fig. 12. Effect of glycation on AST activity and its intervention by hydroxycitric acid. AST (5 µg/ml) was incubated with or without methylglyoxal (0.5 mM) in 0.1 M sodium phosphate buffer (pH 7.4) at 37 °C in the presence or absence of hydroxycitric acid (1.0 and 2.5 mM). Catalytic activity of AST was expressed as percentage of each sample activity at the time 0, which was 100% ± S.D. (%). Every point represents an average of two independent experiments, in which assays were performed in triplicates († data with $P<0.05$ and *data with $P<0.01$, Student's *t*-test).

2. Methods

Following parameters have been assessed: enzyme activity, fluorescence (total AGEs and argpyrimidine), amount of primary amino groups, molecular charge of AST (using native polyacrylamide gel electrophoresis), and protein cross-linking and aggregation (using polyacrylamide gel electrophoresis under reducing conditions with subsequent western blotting). Structures of all tested compounds are shown in Fig. 13.

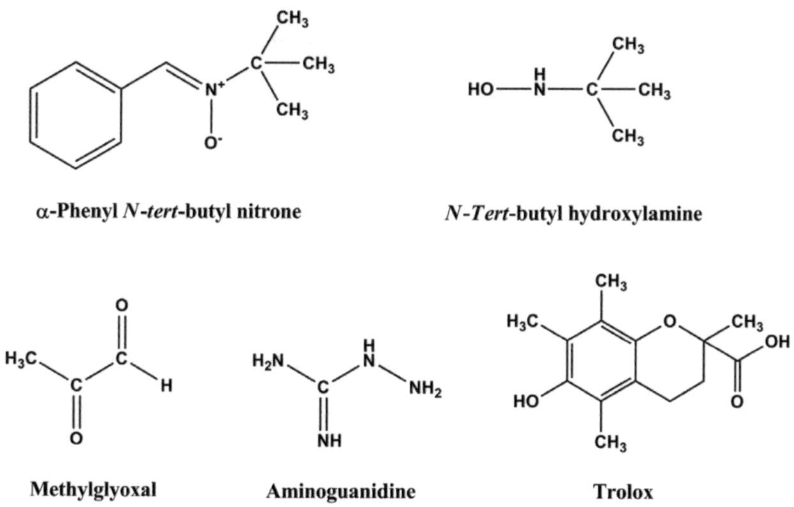

Fig. 13. Structures of tested compounds

2.1 Sample preparation and incubation

The enzyme suspension was centrifuged at 5000 rpm at 4 °C for 20 minutes, the supernatant was removed, and protein pellet was reconstituted in 0.1 M sodium phosphate buffer (pH 7.4, 0.05% sodium azide) and the stock solution of 1.0 mg/ml was prepared. This stock solution was used for the preparation of four different types of incubation mixtures: (a) control samples (with buffer only), (b) methylglyoxal-modified samples (with MGO in a final concentration of 0.5 mM), (c) direct protein-antioxidant interaction samples (with individual antioxidants in a final concentration 0.5-10 mM), (d) antiglycation samples (with individual antioxidants in a final concentration of 0.5-10 mM and MGO in a final concentration 0.5 mM). The inhibitory effect of α-phenyl *N-tert*-butyl nitrone and *N-tert*-butyl hydroxylamine on protein glycation was compared to the effect of aminoguanidine (AG) in a concentration of 1.0 mM and Trolox in a concentration of 2.5 mM. The final concentrations of the enzyme were 5 μg/ml for catalytic activity assessment and 0.5 mg/ml for electrophoresis, amine content and fluorescence measurements. All incubation mixtures were incubated in the dark at 37 °C for up to 14 days. The low-molecular compounds were removed using Amicon centrifugal filtration device with 0.1 M sodium phosphate buffer (pH 7.4), protein content was measured using Bradford assay, and adjusted to the concentration 0.5 mg/ml. Aliquots were stored frozen at -20 °C until analysis. All samples were assessed in triplicates and experiments were repeated twice if not stated otherwise.

2.2 Enzyme assay

Catalytic activity of AST was assessed spectrophotometrically using kinetic UV method with addition of PLP (Bergmeyer et al., 1986). Sample aliquots were diluted by 0.1 M sodium phosphate buffer (pH 7.4) to obtain enzyme activities within the analytical range of the method used. Sampling and measuring was carried out at 37 °C in the intervals 0, 120, and 240 minutes using Helios ß spectrophotometer. Absorbance changes at 340 nm were used to calculate enzyme activities. All results of enzyme assays were expressed in μkat/l and usually recalculated as activities relative to those of the value of individual sample at time 0.

2.3 Fluorescence measurements

Formation of fluorescent AGEs and argpyrimidine were measured using the method of Wu & Yen (2005) with some modifications. Briefly, samples were incubated for 7 days at 37 °C. The aliquots were taken away at time 0, 3 and 7 days and stored frozen at -20 °C. Aliquots of time 0 were used as unincubated blanks. Fluorescence of samples was measured at excitation and emission wavelengths of 330 nm/410 nm (fluorescent AGEs) and 320 nm/380 nm (argpyrimidine) against corresponding blanks in 96-well-plate by microplate reader (Tecan Infinity M200) using 0.1 mg of protein per well. The percentage inhibition of AGEs and argpyrimidine formation were calculated according to following formula: % inhibition = [1 - (fluorescence of test group/fluorescence of glycated control)] x 100%.

2.4 Determination of primary amino groups

Amine content, which is a measure of protein glycation, was estimated spectrophotometrically with trinitrobenzenesulfonic acid (Steinbrecher, 1987) using ß-alanine as the standard. Sample containing 50 μg of AST was incubated with 0.1% trinitrobenzenesulfonic acid in alkaline conditions for 2 h at 37 °C. The reaction was stopped by acidification (1 M HCl) and addition of 10% sodium dodecyl sulfate. The absorbance of trinitrophenyl-amino acid complex was measured at 340 nm. The standard curve was linear in the range 5–100 nmol of NH_2.

2.5 Effect of glycation on molecular charge of AST

Native polyacrylamide gel electrophoresis (PAGE) was used to investigate the changes in the molecular charge of AST due to glycation. Electrophoresis was performed in discontinuous system with 4% stacking gel and 7.5% separating non-denaturating gel (Ornstein, 1964; Davies, 1964). All lanes were loaded with 9 μg of protein. Electrophoresis was performed at 30 mA for 2 hours using Mini ProteanIII apparatus. The gel was then stained by colloidal Coomassie Blue G250, scanned, and relative migration distances were calculated from Rf using Quantity One software. Electrophoretic mobilities were expressed as a rise in percentage mobility compared to the native enzyme (control).

2.6 SDS-PAGE and western blotting

Protein cross-linking and aggregation were assessed using a sodium dodecyl sulfate polyacrylamide gel electrophoresis (SDS–PAGE) on Mini ProteanIII apparatus (BioRad). SDS-PAGE was performed using discontinuous system with 4% stacking gel and 10% separating gel (Laemmli, 1970). Lanes were loaded with 4 μg of protein. Proteins after electrophoretic separation were transferred to PVDF membrane (0.2 μm, Bio-Rad) at a constant voltage 100 V for 90 minutes (Mini Trans-Blot Electrophoretic Transfer Cell, Bio-Rad). After blotting, membranes were blocked with 8% non-fat dry milk in Tris buffered saline-Tween-20 buffer (TBST) overnight at 4°C, then washed in TBST and reacted with primary antibody (dilution 1:1000) for 45 minutes at room temperature. Subsequently, membranes were washed six times with TBST and incubated with secondary antibody for 45 minutes (dilution 1:1000). The blots were extensively washed in 0.1 M TRIS buffer containing 5 mM $MgCl_2.6H_2O$ (pH 9.5), covered with chemiluminescent substrate DuoLux (Vector Laboratories) and incubated for 5 minutes. The membranes were then exposed to X-ray film (CL-XPosure film, Thermo Fisher Scientific), developed by standard developing process, and images were recorded with a GelDoc XR system. The blots were densitometrically quantified using Quantity One software.

2.7 Statistical analysis

Values of catalytic activity are given as means ± S.D. and mostly expressed in % of the time 0 of individual samples ± relative S.D. Values of fluorescence (AU) are given as means ± S.D. Statistical significance was determined using Student's *t*-test and differences were regarded as significant when $P<0.05$ and $P<0.01$, respectively.

3. Results and discussion

The activity of tested compounds was compared to the effect of known carbonyl-blocking agent aminoguanidine and to the effect of Trolox (water-soluble derivative of vitamin E), which is often used in various methods for assessing antioxidant/antiradical properties of potential antioxidants as reference compound.

3.1 Enzyme assay

Activity of control sample (AST alone) was stable throughout the experiment. Some of tested compounds had a more or less pronounced negative direct effect on enzyme activity, which was probably due to a direct interaction of their molecules with the molecule of the enzyme. PBN itself had no harmful influence on stability and catalytic activity of AST in the concentrations used (Fig. 14B), while NtBHA caused concentration-dependent decrease in AST catalytic activity, which was statistically significant at concentrations of NtBHA 1 mM and higher (Fig. 14A). Aminoguanidine 1.0 mM caused significant decrease of enzyme activity by 21.7% after 240 min of incubation. This inhibitory effect of AG may be explained by its binding to PLP coenzyme forming a Schiff base, which disturbs tissue distribution of PLP *in vivo* and decreases its concentration in liver (Okada & Ayabe, 1995; Taguchi et al., 1998). Trolox 2.5 mM did not influence AST activity.

Following incubation of enzyme with MGO 0.5 mM, a rapid decline of AST activity was observed. The enzymatic activity decreased to 53.1 and 30.1% of control sample after 120 and 240 min, respectively. In addition, positive antiglycation effects were observed with some compounds. NtBHA exerted antiglycation influence only at 5 mM concentration after 120 and 240 min and at 1 mM concentration after 240 min of incubation (Fig. 15A). Negative direct effect of this compound observed at 10 mM concentration probably outweighed its positive antiglycation activity. The catalytic activity of AST was by 33.1% and 16.5% higher in the presence of PBN 10 mM and 1 mM after 120 min of incubation with MGO compared to the activity of sample containing AST + MGO only (Fig. 15B), respectively. PBN 1-10 mM protected AST against MGO-induced glycation also after 240 min of incubation, when all three concentrations increased activity of AST by 20%. In comparison, aminoguanidine 1.0

mM almost completely reversed negative effect of MGO and the AST activity of the sample containing AST + MGO + AG was 86.2 and 76.6% of control sample activity after 120 and 240 min, respectively. The effect of Trolox 2.5 mM was slightly higher than that of NtBHA 5 mM. The AST activity was by 15.1 and 13.7% higher in the presence of Trolox after 120 and 240 min, respectively.

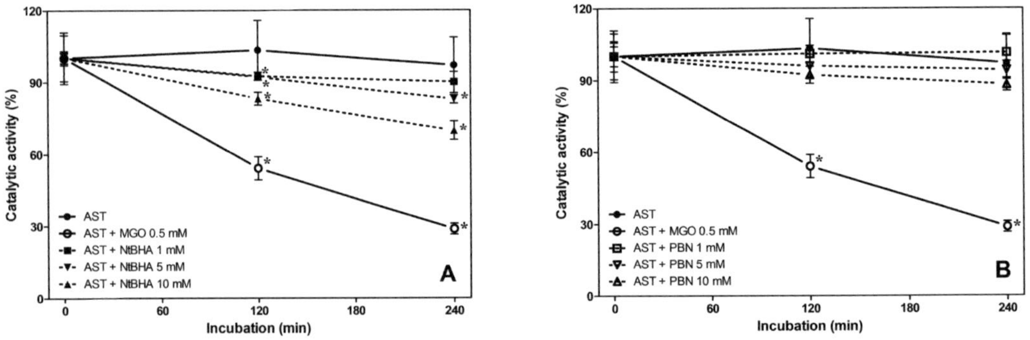

Fig. 14. Direct effect of NtBHA (panel A) and PBN (panel B) on AST catalytic activity. AST (0.5 µg/ml) was incubated with or without methylglyoxal (0.5 mM) in 0.1 M sodium phosphate buffer (pH 7.4) at 37 °C in the presence or absence of NtBHA (1-10 mM) or PBN (1-10 mM). Catalytic activity of AST was expressed as percentage of each sample activity at the time 0, which was 100% ± S.D. (%). Every point represents an average of eighteen (AST and AST + MGO) or six independent samples (*data with $P<0.01$, Student's *t*-test)

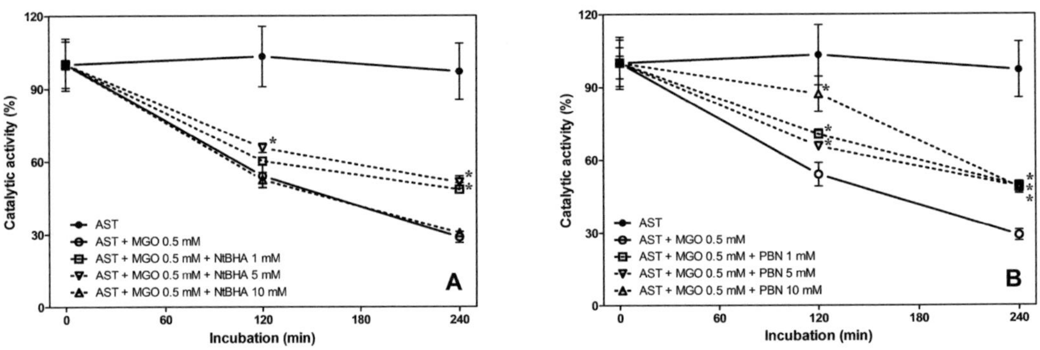

Fig. 15. Antiglycating activity of NtBHA (panel A) and PBN (panel B) towards methylglyoxal-induced deactivation of AST. AST (0.5 µg/ml) was incubated with or without methylglyoxal (0.5 mM) in 0.1 M sodium phosphate buffer (pH 7.4) at 37 °C in the presence or absence of NtBHA (1-10 mM) or PBN (1-10 mM). Catalytic activity of AST was expressed as percentage of each sample activity at the time 0, which was 100% ± S.D. (%). Every point represents an average of eighteen (AST and AST + MGO) or six independent samples (*data with $P<0.01$, Student's *t*-test)

3.2 Fluorescence measurements

The inhibition of MGO-mediated protein glycation by several antioxidants was determined by measuring of AGEs with fluorescent properties. Figure 16B shows formation of argpyrimidine and the effect of NtBHA on this process. Sample containing AST + MGO exerted 12.5 times higher fluorescence intensity than the control sample without MGO after 7 days of incubation. NtBHA caused statistically significant decrease in the formation of argpyrimidine during incubation (inhibition by 42.3–83.1%). The most remarkable decline in argpyrimidine formation was observed at 10 mM concentration of NtBHA, which exhibited inhibition by 83.1%. Effect of PBN on the argpyrimidine formation was less remarkable (inhibition by 35.2–55.8%) but still highly significant (Fig. 16A). The influence of AG 1.0 mM was well-pronounced (93.2%), while the effect of Trolox 2.5 mM was much weaker (53.2%) and comparable to the activity of NtBHA 1 mM and PBN 1-10 mM.

The effect of NtBHA on the formation of "non-specific" AGE products is presented in Fig. 16D. Methylglyoxal caused almost 13 fold increase in concentration of AGEs with fluorescent properties compared to the control sample (AST alone) after 7 days of incubation. Positive effect of NtBHA reached almost the same extent as in the case of argpyrimidine formation with statistically significant inhibition of glycation (34.3–76.6%). Little bit lower rate of inhibition (8.9–48.2%) was obtained also with PBN (Fig. 16C). Aminoguanidine and Trolox showed 88.7 and 44.4% suppressing effect on AGEs generation, respectively.

As for fluorescence measurement, control sample showed stable but not negligible fluorescence, since the start of the experiment. Most of this fluorescence is probably constituted by general fluorescence properties of this protein. Presence of pyridoxal-5'-phosphate coenzyme in the molecule of AST also contributes to basal fluorescence of the enzyme. Results of fluorescence measurements clearly show an inhibiting effect of NtBHA and PBN on the formation of AGE products. NtBHA was also quite effective in the inhibition of argpyrimidine generation. Apart from these findings, the use of fluorescence method for evaluation of protein glycation is limited by its imprecision. The measurement of some well-identified AGEs (e.g., pentosidine and carboxymethyllysine) by techniques as HPLC or ELISA could give more precise information on this matter (Boušová et al., 2009).

3.3 Determination of primary amino groups

Following incubation of AST with methylglyoxal, there was a decrease in amine content compared to control (Table 1). Unmodified AST exhibited 25.2 ± 0.4 nmol NH_2/mol AST versus 13.9 ± 1.1 nmol NH_2/mol AST for sample containing AST + MGO ($P<0.001$, using Student's *t*-test). This difference represents a 45% decrease in amine content due to chemical modification of the primary amines (α-amino group of *N*-terminal amino acids and ε-amino group of Lys residues). Native AST in the dimer form contains 40 primary amines (38 Lys residues and 2 *N*-terminal amino acids), suggesting that approximately 18 amines were modified by methylglyoxal. Modification of primary amino group of Lys258 in AST molecule by MGO may be also responsible for the loss of its catalytic activity, because this Lys residue binds coenzyme PLP in the active centre of AST and thus directly participates in the enzymatic catalysis.

PBN as well as NtBHA significantly protected AST against the loss of primary amino groups induced by MGO. Their effect was concentration-dependent and more pronounced in the case of NtBHA. Sample containing AST + MGO + NtBHA 10 mM exhibited 21.8 ± 0.7 nmol NH_2/mol AST ($P<0.001$, using Student's *t*-test) suggesting that about 5 amines were modified by MGO. In the sample containing AST + MGO + PBN 10 mM was found 19.8 ± 1.2 nmol NH_2/mol AST (P<0.001, using Student's t-test), which means that approximately 9 amines were lost. This result is comparable to the effect of NtBHA at 1 mM concentration. Moreover, AG 1 mM exerted the same effect as NtBHA 10 mM. In the sample containing AST + MGO + AG was detected 22.0 ± 0.4 nmol NH_2/mol AST suggesting that 5 amines was lost in its presence.

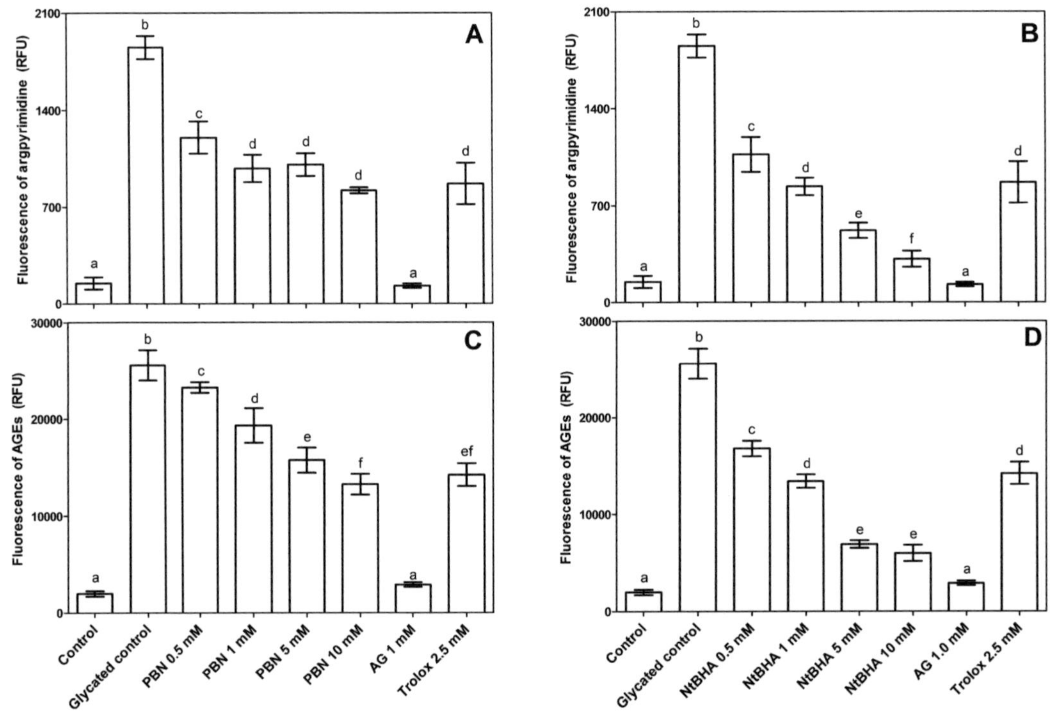

Fig. 16. Formation of argpyrimidine (panel A and B) and fluorescent AGEs (panel C and D). AST (0.5 mg/ml) was incubated with or without methylglyoxal (0.5 mM) in 0.1 M sodium phosphate buffer (pH 7.4) at 37 °C in the presence or absence of PBN (0.5-10 mM), NtBHA (0.5–10 mM), aminoguanidine (1 mM) or Trolox (2.5 mM) for 7 days. Aliquots of samples were taken on days 0 and 7 and stored frozen at -20 °C. Fluorescence of samples was measured at specific excitation and emission wavelengths (λex/λem) corresponding to argpyrimidine (335/385 nm) and AGEs (330/410 nm) versus the unincubated blanks. Data of fluorescence were expressed in relative fluorescence units ± S.D. Every point represents an average of two independent experiments (6 samples). Groups with different letters are significantly different (P<0.01, Student's t-test).

To determine whether the changes in fluorescence of argpyrimidine observed in the samples containing various concentrations of NtBHA (0.5–10 mM) and MGO 0.5 mM correlated with the loss of primary amino groups in these samples, the fluorescence intensity was plotted as a function of amine content (data not shown). The decrease in fluorescence emission (λex/λem = 320/380 nm) varied directly with the loss of primary amino groups (r = 0.963, P<0.019). Changes in fluorescence of argpyrimidine in these samples also correlated well with the formation of fluorescent AGEs (r = 0.991, $P<0.005$). In addition, the amine content measured in samples containing AST + MGO + NtBHA (0.5-10 mM) was plotted as a function of AGEs fluorescence, indicating that MGO-induced formation of AGEs was directly proportional to an irreversible loss of primary amino groups in AST molecule (r = 0.946, $P<0.027$). Similar results were obtained also in samples containing AST + MGO + PBN (0.5-10 mM).

Sample	Amine content	Number of primary NH_2	
	(nmol NH_2/mol AST)	remaining	modified
AST	25.2 ± 0.4^{a}	40	0
AST + MGO 0.5 mM	13.9 ± 1.1^{b}	22	18
AST + MGO + PBN 0.5 mM	17.9 ± 0.9^{c}	28	12
AST + MGO + PBN 1 mM	18.2 ± 1.1^{cd}	29	11
AST + MGO + PBN 5 mM	18.6 ± 1.7^{cd}	30	10
AST + MGO + PBN 10 mM	19.8 ± 1.2^{d}	31	9
AST + MGO + NtBHA 0.5 mM	17.3 ± 0.4^{c}	27	13
AST + MGO + NtBHA 1 mM	19.5 ± 0.4^{d}	31	9
AST + MGO + NtBHA 5 mM	20.8 ± 0.8^{de}	32	8
AST + MGO + NtBHA 10 mM	21.8 ± 0.7^{ef}	35	5
AST + MGO + AG 1 mM	22.0 ± 0.4^{f}	35	5

[a,b,c,d,e,f] Groups with different letters vary significantly ($P<0.05$, Student's *t*-test)

Table 1. Effect of PBN, NtBHA and AG on the changes in AST primary amine content induced by MGO 0.5 mM

3.4 Effect of glycation on molecular charge of AST

Native PAGE was run several times, and the representative native PAGE gel is presented in Fig. 17. Mobility of MGO-modified protein to the positive pole significantly increased (by 41%) after 7 days of incubation compared to the mobility of control sample (AST alone). This result indicates the progressive loss of the positive charge in the MGO-modified AST during the glycation reaction. NtBHA showed concentration-dependent protective effect against changes in AST molecular change induced by MGO, when the relative mobility of sample containing AST + MGO + NtBHA 10 mM was increased only by 9.2% compared to the mobility of control. The enzyme incubated in the presence of both MGO and PBN showed a smaller rise in mobility, up to 21.6% in the case of PBN 10 mM. The effect of this compound on the protein electrophoretic mobility ranged from 21.6 to 26.3%. Aminoguanidine 1 mM completely reversed effect of MGO and the relative mobility of sample containing AST + MGO + AG was only slightly increased (by 0.35%) against the mobility of control, whereas Trolox 2.5 mM showed similar effect on molecular charge of AST as PBN, i.e., the mobility was increased by 27% compared to control sample (data not shown). These data indicated that the molecule of enzyme became more anionic due to glycation and that PBN as well as NtBHA had significant inhibitory effect on the middle stage of glycation process.

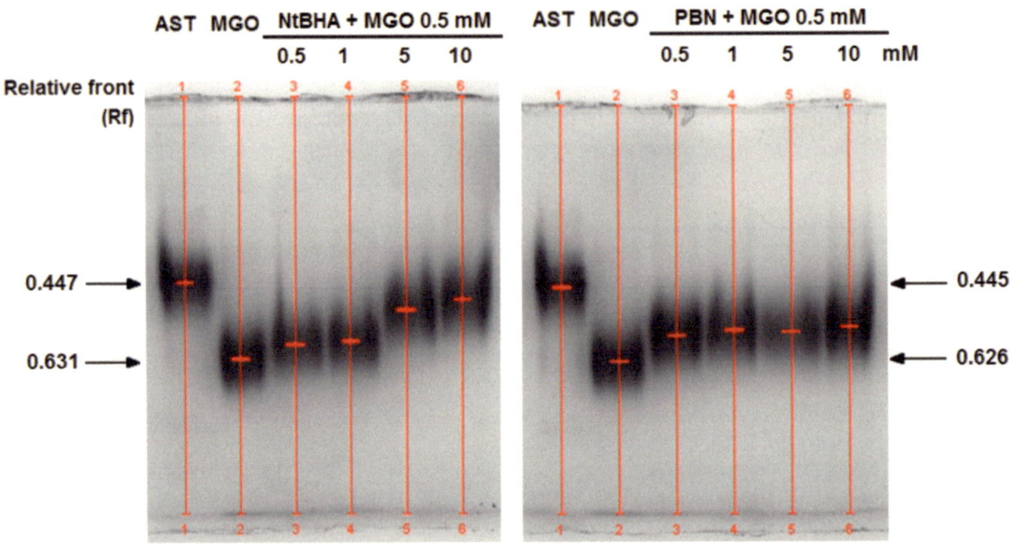

Fig. 17. Protective effect of NtBHA and PBN on changes in molecular change of AST caused by MGO-induced glycation. AST (0.5 mg/ml) was incubated with or without MGO (0.5 mM) and NtBHA (0.5-10 mM) or PBN (0.5-10 mM) in sodium phosphate buffer (0.1 M, pH 7.4) at 37 °C for 7 days and then subjected to native PAGE. Proteins were visualized by Coomassie Blue G250. Gels were scanned and Rf was obtained using Quantity One software

Methylglyoxal-induced chemical modifications led to a change in molecular charge of AST, which became more anionic as revealed by native PAGE. These results indicate the progressive loss of the positive charge in the glycation-modified AST molecule, which is caused by the irreversible modification of Arg and Lys residues (Kang, 2006; Nagai et al., 2000) as was confirmed by determination of amine content. Both PBN and NtBHA partially protected native AST against glycation by MGO. The antiglycation activity was more pronounced in the case of NtBHA mainly at higher concentrations tested. The antiglycation activity of NtBHA 10 mM was a little bit lower than that of AG 1 mM.

3.5 SDS-PAGE and western blotting

The ability of aggregation and cross-link formation of tested antioxidants was determined by SDS-PAGE under denaturing conditions (Fig. 18). MGO readily reacts with lysine and arginine residues to produce high molecular weight protein products. Incubation of AST with MGO 0.5 mM at 37 °C for 7 days resulted in the formation of protein aggregates with molecular weight about 85, 107, and 145 kDa corresponding to protein dimer, trimer, and tetramer, respectively. No presence of protein dimer and tetramer, and lower concentration of protein trimer were observed in samples containing AST alone (lane 2), AST + NtBHA 10 mM (lane 7), and AST + AG 1 mM (data not shown). Also lower concentrations of NtBHA were able to partially protect formation of protein tetramer, although they had no effect on formation of protein dimer and trimer. On the other hand, PBN as well as Trolox were not able to prevent formation of protein cross-links and high molecular weight aggregates. Additional bands with molecular weight 20–35, 57 and 63 kDa were constituted of several contaminating proteins present in commercial preparation (Fig. 18).

Western blotting with specific antibody against advanced glycation end products derived from MGO (anti-MGO [3C]) was used to confirm formation of protein aggregates as a result of MGO activity. The presence of high molecular weight protein cross-links in samples containing AST + MGO, AST + MGO + NtBHA, and AST + MGO + PBN was observed (data not shown). These protein aggregates had molecular weight about 85, 107, and 145 kDa corresponding to AST dimer, trimer, and tetramer, respectively. Quantitative differences between bands of samples with and without PBN or Trolox were not observed. These compounds are not able to prevent formation of protein cross-links. On the other hand, some reduction in the amount of AST tetramer was observed in samples containing NtBHA. These results suggest that NtBHA possesses, at least in part, antiglycation properties. Nevertheless, aminoguanidine 1 mM completely inhibited formation of protein aggregates, since no bands of AST dimer, trimer or tetramer were present.

The electrophoretic techniques confirmed the results obtained by other methods; i.e., changes in protein molecule caused by the presence of methylglyoxal and positive antiglycating effect of NtBHA. Methylglyoxal-induced chemical modifications led to a change in molecular charge of AST, which became more anionic as revealed by native PAGE. The SDS-PAGE and subsequent western blotting clearly showed formation of protein cross-links with higher molecular weight than native enzyme. NtBHA partially protected native AST from glycation by MGO and also exhibited mild anti-cross-linking activity.

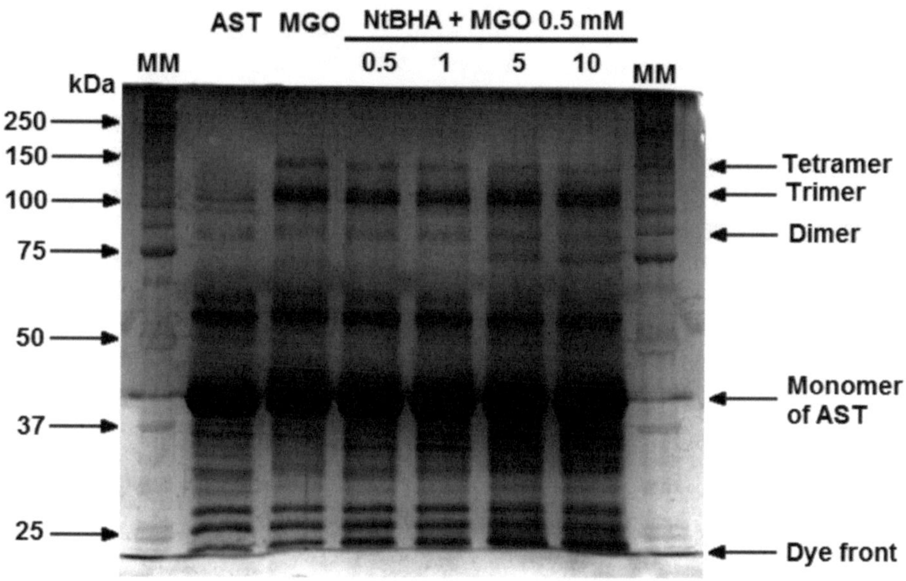

Fig. 18. Formation of protein cross-links on reaction of AST with methylglyoxal. AST (0.5 mg/ml) was incubated with or without methylglyoxal (0.5 mM) in sodium phosphate buffer (0.1 M, pH 7.4) at 37 °C in the presence or absence of *N-tert*-butyl hydroxylamine (0.5–10 mM) for 7 days and then subjected to SDS-PAGE. Electrophoretic separation was performed on 4% stacking and 10% resolving polyacrylamide gels under reducing conditions. Bands were visualized with silver staining. Each lane was loaded with 4 μg of protein. MM = Mw marker; AST = aspartate aminotransferase; MGO = methylglyoxal; NtBHA = *N-tert*-butyl hydroxylamine

Modification of proteins caused by methylglyoxal can be accompanied by formation of free radicals. Lee et al. (1998) identified three types of free radical species in samples containing methylglyoxal and bovine serum albumin by electron spin resonance spectroscopy. These radicals (methylglyoxal dialkylimine radical cation, methylglyoxal radical anion, and superoxide anion radical) were formed by direct 1-electron transfer process. Scavenging ability of NtBHA and PBN were already described (Lee et al., 2004; Atamna et al., 2001). It can be assumed that the positive antiglycation activity of these compounds may be at least partly attributed to their scavenging ability.

4. Conclusion

- Catalytic activity, which is biologically the most important property of all enzymes, is fully dependent on the native structure of the enzyme. Changes in the structure of enzymes usually lead to the progressive loss of their catalytic activities. These changes may be caused by reversible binding of various low-molecular inhibitors or by irreversible modification. Such example of the irreversible change is non-enzymatic glycation of proteins by various reducing monosaccharides or reactive α-dicarbonyl compounds (e.g., MGO).
- In our studies, aspartate aminotransferase (AST) was used in glycation studies as model protein, which possesses catalytic properties.
- Catalytic activity of aminotransferases *in vitro* has been found to be impaired by glycating agents. The extent of this effect depends on the activity and concentration of the agent, susceptibility of given enzyme to such modification as well as on the duration of action (see Fig. 4, Fig. 5 and Fig. 8).
- Among several glycating agents, effect of fructose is strong enough for investigation of changing AST properties during long time of incubation. Nevertheless, methylglyoxal, an intermediate of glycation process, is more reactive and permits to investigate the process of AST glycation *in vitro* in shorter course of time.
- Catalytic activity of AST as the protein function may serve as the most important criterion of glycation effect. Beside this, molecular properties of purified AST and character of glycoxidation permit using other methods of investigation of molecular changes during the process, like fluorescence of advanced glycation end-products, decrease in primary amino groups in the protein molecule, and protein cross-linking and aggregation.
- Several approaches of therapeutic intervention to the glycation process have been used (e.g., reduction in deposition of already formed AGEs, inhibition of new AGEs formation, and inhibition of the receptor for AGE).
- In our own experiments, compounds with described antioxidant and potential antiglycating activities have been studied. Among the compounds studied, hydroxycitric acid, uric acid, and two mitochondrial antioxidants α-phenyl *N-tert*-butyl nitrone and *N-tert*-butyl hydroxylamine had pronounced antiglycating activity against protein glycation by methylglyoxal (hydroxycitric acid, PBN and NtBHA) and by fructose (hydroxycitric and uric acids). On the other hand, flavonoids baicalin and baicalein exerted overall negative influence on the catalytic activity of AST alone and in the combination with fructose. Other studied compounds (i.e., ferulic, isoferulic, *o*-coumaric, and *p*-coumaric acids, arbutin and methylarbutin) showed no positive antiglycation activity.

- The goal of our research group is to participate in the search for compounds with potential antiglycating activity with a perspective of their use as remedies against diabetic complications.

5. Acknowledgment

This study was supported by the Charles University in Prague (Project SVV 263 004).

6. References

Arai, K.; Iizuka, S.; Tada, Y.; Oikawa, K. & Taniguchi, N. (1987). Increase in the glucosylated form of erythrocyte Cu-Zn-superoxide dismutase in diabetes and close association of the nonenzymatic glucosylation with the enzyme activity. *Biochimica et Biophysica Acta - General Subjects*, Vol.924, No.2, pp. 292-296, ISSN 0304-4165

Atamna, H.; Paler-Martínez, A. & Ames, B.N. (2000). N-t-butyl hydroxylamine, a hydrolysis product of alpha-phenyl-N-t-butyl nitrone, is more potent in delaying senescence in human lung fibroblasts. *Journal of Biological Chemistry*, Vol.275, No.9, pp. 6741-6748, ISSN 0021-9258

Atamna, H.; Robinson, C.; Ingersoll, R.; Elliott, H. & Ames, B.N. (2001). N-t-butyl hydroxylamine is an antioxidant that reverses age-related changes in mitochondria in vivo and in vitro. *The FASEB Journal*, Vol.15, No.12, pp. 2196-2204, ISSN 0892-6638

Baynes, J.W. (1991). Role of oxidative stress in development of complications in diabetes. *Diabetes*, Vol.40, No.4, pp. 405-412, ISSN 0012-1797

Baynes, J.W. & Thorpe, S.R. (2000). Glycoxidation and lipoxidation in atherogenesis. *Free Radical Biology & Medicine*, Vol.28, No.12, pp. 1708-1716, ISSN 0891-5849

Beisswenger, P.J.; Howell, S.K.; Nelson, R.G.; Mauer, M. & Szwergold, B.S. (2003). Alpha-oxoaldehyde metabolism and diabetic complications. *Biochemical Society Transactions*, Vol.31, No.Pt 6, pp. 1358-1363, ISSN 0300-5127

Beránek, M.; Dršata, J. & Palička, V. (2001). Inhibitory effect of glycation on catalytic activity of alanine aminotransferase. *Molecular and Cellular Biochemistry*, Vol.218, No.1-2, pp. 35-39, ISSN 0300-8177

Beránek, M.; Dršata, J. & Palička, V. (2002). In vitro glycation of aminotransferases: A process closely depending on the employed experimental conditions. *Acta Medica*, Vol.45, No.3, pp. 89-92, ISSN 1211-4286

Bergmeyer, H.U.; Horder, M. & Rej, R. (1986). International Federation of Clinical Chemistry (IFCC) Scientific Committee, Analytical Section: approved recommendation (1985) on IFCC methods for the measurement of catalytic concentration of enzymes. Part 2. IFCC method for aspartate aminotransferase (L-aspartate: 2-oxoglutarate aminotransferase, EC 2.6.1.1). *Journal of Clinical Chemistry and Clinical Biochemistry*, Vol.24, No.7, pp 497–510, ISSN 0340-076X

Bolton, W.K.; Cattran, D.C.; Williams, M.E.; Adler, S.G.; Appel, G.B.; Cartwright, K.; Foiles, P.G.; Freedman, B.I.; Raskin, P.; Ratner, R.E.; Spinowitz, B.S.; Whittier, F.C.; Wuerth, J.P. & ACTION I Investigator Group (2004). Randomized trial of an inhibitor of formation of advanced glycation end products in diabetic nephropathy. *American Journal of Nephrology*, Vol.24, No.1, pp. 32-40, ISSN 0250-8095

Boušová, I.; Vukasović, D.; Juretić, D.; Palička, V. & Dršata, J. (2005a). Enzyme activity and AGE formation in a model of glycoxidation of AST by D fructose in vitro. *Acta Pharmaceutica*, Vol.55, No.1, pp. 107-114, ISSN 1330-0075

Boušová, I.; Martin, J.; Jahodář, L.; Dušek, J.; Palička, V. & Dršata, J. (2005b). Evaluation of in vitro effects of natural substances of plant origin using a model of protein glycoxidation. *Journal of Pharmaceutical and Biomedical Analysis*, Vol.37, No.5., pp. 957-962, ISSN 0731-7085

Boušová, I.; Bakala, H.; Chudáček, R.; Palička, V. & Dršata, J. (2005c). Glycation-induced inactivation of aspartate aminotransferase, effect of uric acid. *Molecular and Cellular Biochemistry*, Vol.278, No.1-2, pp. 85-92, ISSN 0300-8177

Boušová, I.; Bacílková, E.; Dobrijević, S. & Dršata J. (2009). Glycation of aspartate aminotransferase by methylglyoxal, effect of hydroxycitric and uric acid. *Molecular and Cellular Biochemistry*, Vol.331, No.1-2, pp. 215-223, ISSN 0300-8177

Bucciarelli, L.G.; Wendt, T.; Rong, L.; Lalla, E.; Hofmann, M.A.; Goova, M.T.; Taguchi, A.; Yan, S.F.; Yan, S.D.; Stern, D.M. & Schmidt, A.M. (2002). RAGE is a multiligand receptor of the immunoglobulin superfamily: implications for homeostasis and chronic disease. *Cellular and Molecular Life Sciences*, Vol.59, No.7, pp. 1117-1128, ISSN 1420-682X

Bunn, H.F.; Gabbay, K.H. & Gallop, P.M. (1978). The glycosylation of hemoglobin: relevance to diabetes mellitus. *Science*, Vol.200, No.4337, pp. 21-27, ISSN 0036-8075

Davies, B.J. (1964). Disc electrophoresis. II. Method and application to human serum proteins. *Annals of the New York Academy of Sciences*, Vol.121, pp. 404–427, ISSN 0077-8923

De La Cruz, J.; González-Correa, J.; Guerrero, A. & De la Cuesta, F. (2004). Pharmacological approach to diabetic retinopathy. *Diabetes/Metabolism Research and Reviews*, Vol.20, No.2, pp. 91-113, ISSN 1520-7552

Dolhofer, R. & Wieland, O.H. (1978). In vitro glycosylation of hemoglobins by different sugars and sugar phosphates. *FEBS Letters*, Vol.85, No.1, pp. 86-90, ISSN 0014-5793

Dršata, J.; Beránek, M. & Palička, V. (2002). Inhibition of aspartate aminotransferase by glycation in vitro under various conditions. *Journal of Enzyme Inhibition and Medicinal Chemistry*, Vol.17, No.1, pp. 31-36, ISSN 1475-6366

Dršata, J.; Boušová, I. & Maloň, P. (2005). Determination of quality of pyridoxal-5´-phosphate enzyme preparations by spectroscopic methods. *Journal of Pharmaceutical and Biomedical Analysis*, Vol.37, No.5., pp. 1173-1177, ISSN 0731-7085

Dršata, J. & Veselá, J. (1984). Inhibition of liver aminotransferases with some potential cytostatic agents. *Cesko-Slovenska Farmacie*, Vol.33, No.9, pp. 372-375, ISSN 0009-0530

Fitzgerald, C.; Swearengin, T.A.; Yeargans, G.; McWhorter, D.; Cucchetti, B. & Seidler, N.W. (2000). Non-enzymatic glycosylation (or glycation) and inhibition of the pig heart cytosolic aspartate aminotransferase by glyceraldehyde 3-phosphate. *Journal of Enzyme Inhibition*, Vol.15, No.1, pp. 79-89, ISSN 8755-5093

Hudson, B.I.; Bucciarelli, L.G.; Wendt, T.; Sakaguchi, T.; Lalla, E.; Qu, W.; Lu, Y.; Lee, L.; Stern, D.M. & Naka, Y. (2003). Blockade of receptor for advanced glycation endproducts: a new target for therapeutic intervention in diabetic complications and inflammatory disorders. *Archives of Biochemistry and Biophysics*, Vol.419, No.1, pp. 80-88, ISSN 0003-9861

Hunt, J.V.; Dean, R.T. & Wolff, SP. (1988). Hydroxyl radical production and autoxidative glycosylation. Glucose autoxidation as the cause of protein damage in the experimental glycation model of diabetes mellitus and ageing. *The Biochemical Journal*, Vol.256, No.1, pp. 205-212, ISSN 0264-6021

Jabeen, R. & Saleemuddin, M. (2006). Polyclonal antibodies inhibit the glycation-induced inactivation of bovine Cu,Zn-superoxide dismutase. *Biotechnology and Applied Biochemistry*, Vol.43, No. Pt 1, pp. 49-53, ISSN 0885-4513

Kang, J.H. (2006). Oxidative modification of human ceruloplasmin by methylglyoxal: an in vitro study. *Journal of Biochemistry and Molecular Biology*, Vol.39, No.3, pp. 335–338, ISSN 1225-8687

Kelly, S.M. & Price, N.C. (2000). The Use of Circular Dichroism in the Investigation of Protein Structure and Function. *Current Protein and Peptide Science*, Vol.1, No.4, pp. 349-384, ISSN 1389-2037

Kirsch, J.F.; Eichele, G.; Ford, G.C.; Vincent, M.G.; Jansonius, J.N.; Gehring, H. & Christen, P. (1984). Mechanism of action of aspartate aminotransferase proposed on the basis of its spatial structure. *Journal of Molecular Biology*, Vol.174, No.3, pp. 497-525, ISSN 0022-2836

Kyselova, Z.; Stefek, M. & Bauer, V. (2004). Pharmacological prevention of diabetic cataract. Journal of Diabetes and its Complications, Vol.18, No.2, pp. 129-140, ISSN 1056-8727

Lapolla, A.; Traldi, P. & Fedele, D. (2005). Importance of measuring products of non-enzymatic glycation of proteins. *Clinical Biochemistry*, Vol.38, No.2, pp. 103-115, ISSN 0009-9120

Lee, J.H.; Kim, I.S. & Park, J.W. (2004). The use of N-t-butyl hydroxylamine for radioprotection in cultured cells and mice. *Carcinogenesis*, Vol.25, No.8, pp. 1435-1442, ISSN 0143-3334

Lee, C.; Yim, M.B.; Chock, P.B.; Yim, H.S. & Kang, S.O. (1998). Oxidation-reduction properties of methylglyoxal-modified protein in relation to free radical generation. *Journal of Biological Chemistry*, Vol.273, No.39, pp. 25272–25278, ISSN 0021-9258

Metzler, C.M.; Rogers, P.H.; Arnone, A.; Martin, D.S. & Metzler, D.E. (1979). Investigation of crystalline enzyme-substrate complexes of pyridoxal phosphate-dependent enzymes. *Methods in Enzymology*, Vol.62, pp. 551-558, ISSN 0076-6879

Monnier, V.M. (1989). Toward a Maillard reaction theory of aging. *Progress in Clinical and Biological Research*, Vol.304, pp. 1-22, ISSN 0361-7742

Nagai, R.; Matsumoto, K.; Ling, X.; Suzuki, H.; Araki, T. & Horiuchi, S. (2000). Glycolaldehyde, a reactive intermediate for advanced glycation end products, plays an important role in the generation of an active ligand for the macrophage scavenger receptor. *Diabetes*, Vol.49, No.10, pp. 1714–1723, ISSN 0012-1797

Netopilová, M.; Haugvicová, R.; Kubová, H.; Dršata, J. & Mareš, P. (2001). Influence of convulsants on rat brain activities of alanine aminotransferase and aspartate aminotransferase. *Neurochemical Research*, Vol.26, No.12, pp. 1285-1291, ISSN 0364-3190

Netopilová, M.; Veselá, J. & Dršata, J. (1991). Influence of 5-[2-N,N-dimethylamino)ethoxy]-7-oxo-7H-benzo(c)fluorene hydrochloride (benflurone) on the activity of rat liver aspartate and alanine aminotransferases. *Drug Metabolism and Drug Interactions*, Vol.9, No.3-4, pp. 301-309, ISSN 0792-5077

Nursten, H. (2005). *The Maillard Reaction : Chemistry, Biochemistry and Implications* (1st edition), The Royal Society of Chemistry, ISBN 0-85404-964-9, Cambridge

Okada, M.; Sogo, A. & Ohnishi, N. (1994). Glycation reaction of aspartate aminotransferase by various carbohydrates in an in vitro system. *Journal of Nutritional Biochemistry*, Vol.5, No.10, pp. 485-489, ISSN 0955-2863

Okada, M. & Ayabe, Y. (1995). Effects of aminoguanidine and pyridoxal phosphate on glycation reaction of aspartate aminotransferase and serum albumin. *Journal of Nutritional Science and Vitaminology*, Vol.41, No.4, pp. 43-50, ISSN 0301-4800

Okada, M.; Murakami, Y. & Miyamoto, E. (1997). Glycation and inactivation of aspartate aminotransferase in diabetic rat tissues. *Journal of Nutritional Science and Vitaminology*, Vol.43, No.4, pp. 463-469, ISSN 0301-4800

Ornstein, L. (1964). Disc electrophoresis. I. Background and theory. *Annals of the New York Academy of Sciences*, Vol.121, pp. 321–349, ISSN 0077-8923

Park, L.; Raman, K.G.; Lee, K.J.; Lu, Y.; Ferran, L.J.; Chow, W.S.; Stern, D. & Schmidt, A.M. (1998). Suppression of accelerated diabetic atherosclerosis by the soluble receptor for advanced glycation endproducts. *Nature Medicine*, Vol.4, No.9, pp. 1025-1031, ISSN 1078-8956

Sakurai, T.; Matsuyama, M. & Tsuchiya, S. (1987). Glycation of erythrocyte superoxide dismutase reduces its activity. *Chemical and Pharmaceutical Bulletin*, Vol.35, No.1, pp. 302-307, ISSN 0009-2363

Schalkwijk, C.G.; Stehouwer, C.D. & van Hinsbergh, V.W. (2004). Fructose-mediated non-enzymatic glycation: sweet coupling or bad modification. *Diabetes/Metabolism Research and Reviews*, Vol.20, No.5, pp. 369-82, ISSN 1520-7552

Seidler, N.W. & Kowalewski, C. (2003). Methylglyoxal-induced glycation affects protein topography. *Archives of Biochemistry and Biophysics*, Vol.410, No.1, pp. 149–154, ISSN 0003-9861

Seidler, N.W. & Seibel, I. (2000). Glycation of Aspartate Aminotransferase and Conformational Flexibility. *Biochemical and Biophysical Research Communications*, Vol.277, No.1, pp. 47-50, ISSN 1090-2104

Singh, R.; Barden, A.; Mori, T. & Beilin, L. (2001). Advanced glycation end-products: a review. *Diabetologia*, Vol.44, No.2, pp. 29-146, ISSN 1432-0428

Steinbrecher, U.P. (1987). Oxidation of human low density lipoprotein results in derivatization of lysine residues of apolipoprotein B by lipid peroxide decomposition products. *Journal of Biological Chemistry*, Vol.262, No.8, pp. 3603–3608, ISSN 0021-9258

Stuchbury, G. & Münch, G. (2005). Alzheimer's associated inflammation, potential drug targets and future therapies. *Journal of Neural Transmission*, Vol.112, No.3, pp. 429-453, ISSN 0300-9564

Suarez, G.; Rajaram, R.; Oronsky, A.L. & Gawinowicz, M.A. (1989). Nonenzymatic glycation of bovine serum albumin by fructose (fructation). Comparison with the Maillard reaction initiated by glucose. *Journal of Biological Chemistry*, Vol.264, No.7, pp. 3674-3679, ISSN 0021-9258

Taguchi, T.; Sugiura, M.; Hamada, Y. & Miwa, I. (1998). In vivo formation of a Schiff base of aminoguanidine with pyridoxal phosphate. *Biochemical Pharmacology*, Vol.55, No.10, pp. 1667-1671, ISSN 0006-2952

Thornalley, P.J. (2003). Use of aminoguanidine (Pimagedine) to prevent the formation of advanced glycation endproducts. *Archives of Biochemistry and Biophysics*, Vol.419, No.1, pp. 31-40, ISSN 0003-9861

Thornalley, P.J.; Yurek-George, A. & Argirov, O.K. (2000). Kinetics and mechanism of the reaction of aminoguanidine with the [alpha]-oxoaldehydes glyoxal, methylglyoxal, and 3-deoxyglucosone under physiological conditions. *Biochemical Pharmacology*, Vol.60, No.1, pp. 55-65, ISSN 0006-2952

Tupcová, P. (1996). *Influence of sugars on activity of aminotransferases in vitro*, Diploma thesis, Charles University in Prague, Faculty of Pharmacy, Hradec Králové

Ulrich, P. & Cerami, A. (2001). Protein glycation, diabetes, and aging. *Recent Progress in Hormone Research*, Vol.56, pp. 1-22, ISSN 0079-9963

Vasan, S.; Zhang, X.; Zhang, X.; Kapurniotu, A.; Bernhagen, J.; Teichberg, S.; Basgen, J.; Wagle, D.; Shih, D.; Terlecky, I.; Bucala, R.; Cerami, A.; Egan, J. & Ulrich, P. (1996). An agent cleaving glucose-derived protein crosslinks in vitro and in vivo. Nature, Vol.382, No.6588, pp.275-278, ISSN 0028-0836

Willemsen, S.; Hartog, J.W.; Hummel, Y.M.; Posma, J.L.; van Wijk, L.M.; van Veldhuisen, D.J. & Voors, A.A. (2010). Effects of alagebrium, an advanced glycation end-product breaker, in patients with chronic heart failure: study design and baseline characteristics of the BENEFICIAL trial. *European Journal of Heart Failure*. Vol.12, No.3, pp. 294-300, ISSN 1388-9842

Wolff, S.P. & Dean, R.T. (1987). Glucose autoxidation and protein modification. The potential role of 'autoxidative glycosylation' in diabetes. *The Biochemical Journal*, Vol.245, No.1, pp. 243-250, ISSN 0264-6021

Wolff, S.P.; Jiang, Z.Y. & Hunt, J.V. (1991). Protein glycation and oxidative stress in diabetes mellitus and ageing. *Free Radical Biology & Medicine*, Vol.10, No.5, pp. 339-352, ISSN 0891-5849

Wu, C.H. & Yen, G.C. (2005). Inhibitory effect of naturally occurring flavonoids on the formation of advanced glycation end products. *Journal of Agricultural and Food Chemistry*, Vol.53, No.8, pp. 3167–3173, ISSN 0021-8561

Yagi, T.; Kagamiyama, H.; Nozaki, M. & Soda, K. (1985). Glutamate-aspartate transaminase from microorganisms. *Methods in Enzymology*, Vol.113, pp. 83-89, ISSN 0076-6879

Yamagishi, S.; Nakamura, K.; Matsui, T.; Ueda, S.; Fukami, K. & Okuda, S. (2008). Agents that block advanced glycation end product (AGE)-RAGE (receptor for AGEs)-oxidative stress system: a novel therapeutic strategy for diabetic vascular complications. *Expert Opinion on Investigational Drugs*, Vol.17, No.7, pp. 983-996, ISSN 1354-3784

Yan, H. & Harding, J.J. (1997). Glycation-induced inactivation and loss of antigenicity of catalase and superoxide dismutase. *The Biochemical Journal*, Vol.328, No.Pt 2, pp. 599-605, ISSN 0264-6021

Yan, H. & Harding, J.J. (2006). Carnosine inhibits modifications and decreased molecular chaperone activity of lens alpha-crystallin induced by ribose and fructose 6-phosphate. *Molecular Vision*, Vol.12, pp. 205-14, ISSN 1090-0535

Yegin, A., Özben, T. & Yegin, H. (1995). Glycation of lipoproteins and accelerated atherosclerosis in non-insulin-dependent diabetes mellitus. *International Journal of Clinical & Laboratory Research*, Vol.25, No.3, pp. 157-161, ISSN 0940-5437

Zeng, J.; Dunlop, R.A.; Rodgers, K.J. & Davies, M.J. (2006). Evidence for inactivation of cysteine proteases by reactive carbonyls via glycation of active site thiols. *The Biochemical Journal*, Vol.398, No.2, pp. 197-206, ISSN 0264-6021

Zhao, W.; Devamanoharan, P.S. & Varma, S.D. (2000). Fructose induced deactivation of antioxidant enzymes: preventive effect of pyruvate. *Free Radical Research*, Vol.33, No.1, pp. 23-30, ISSN 1071-5762

5

Inhibition of Nitric Oxide Synthase Gene Expression: *In vivo* Imaging Approaches of Nitric Oxide with Multimodal Imaging

Rakesh Sharma[1,2]
[1]*Center of Nanomagnetics and Biotechnology, Florida State University, Tallahassee, FL*
[2]*Amity Institute of Nanotechnology, Amity University, NOIDA, U.P.*
India

1. Introduction

Nitric oxide is an uncharged free radical abundant in nanomolar quantities and detected by measuring nitric oxide synthase (NOS). Nitric oxide is a gas highly reactive, short lived free radical generated enzymatically by NOS involved in diverse physiological (neurotransmission, immune system) and pathophysiological (tumor progression) mechanisms. Nitric oxide is a biological mediator for its role as EDRF (endothelial-derived relaxing factor) responsible for the regulation of blood vessel relaxation and blood pressure maintenance [Labet et al. 2009]. In recent years, inhibition of NOS gene expression has become a scientific interest to measure NO in tissues and synthetic NOS/NO inhibitors have revolutionized molecular imaging of tissues. Present chapter presents a journey from major mechanistic concepts to the detection of NO, NOS and their multimodal bioimaging applications with limitations.

- Several genes involve in NOS enzyme for its synthase activity (NOS1, NOS2A, NOS3); oxidoreductase activity (NOS1, NOS2A, NOS3, NQO1); as positive regulators(HSP90AB1 or HSPCB, INS); as negative regulators (DNCL1, GLA, IL10);other AKT1, ARG2, DDAH2, DNCL1, EGFR, GCH1, GCHFR genes
- NO diffuses freely across cell membranes. It is short lived but combines with metallic aromatic 'spin traps' to make stable compounds. NO acts in a paracrine or autocrine manner influencing only cells near its point of synthesis.
- NO is synthesized within cells by a flavocytochrome enzyme NO synthase (NOS). The human contains 3 different NO synthases:
- nNOS (NOS-1) is found in neurons (hence the "n"); iNOS (inducible NOS-2), triggered by inflammatory cytokines found in macrophages; and eNOS (NOS-3): constitutively distributed in the vascular endothelium lining the lumen of blood vessels, lung, and platelets (called constitutive NOS or *c*NOS) [Perrier, et al. 2009]. All types of NOS produce NO from arginine with the aid of molecular oxygen and NADPH as shown in following redox reaction and Figure 1. Inhibition of NOS gene expression in cells may detect NO [Nie et al.2008; Terashima et al. 2010].
- NADH-diphorase stain the NOS expression to detect NO in tissues
- NO is generated in vascular endothelium cells (NO plays role in the regulation of vascular tone), peripheral and central neurons (NO acts as synaptic neuronal

messenger) by three isoforms of NO synthase, endothelial, neuronal, and inducible form [Perrier, et al. 2009, Claudette, et al. 2005]. Detection and in vivo monitoring of NO is very difficult. Spin traps, fluorescent dyes, chemi/bioluminescent sensors detect and image NO by ESR,NMR,PET,US,and optical methods.

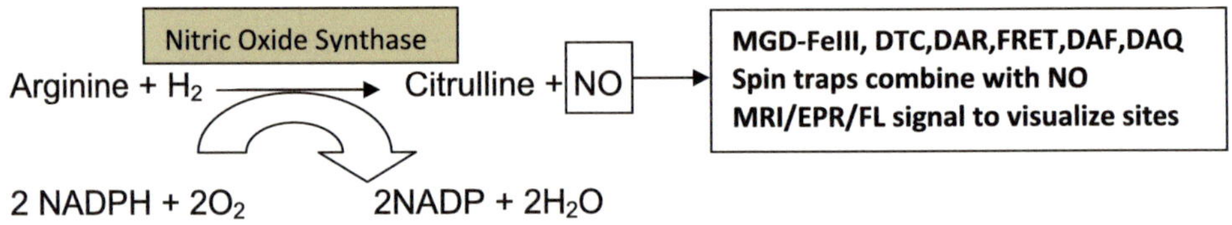

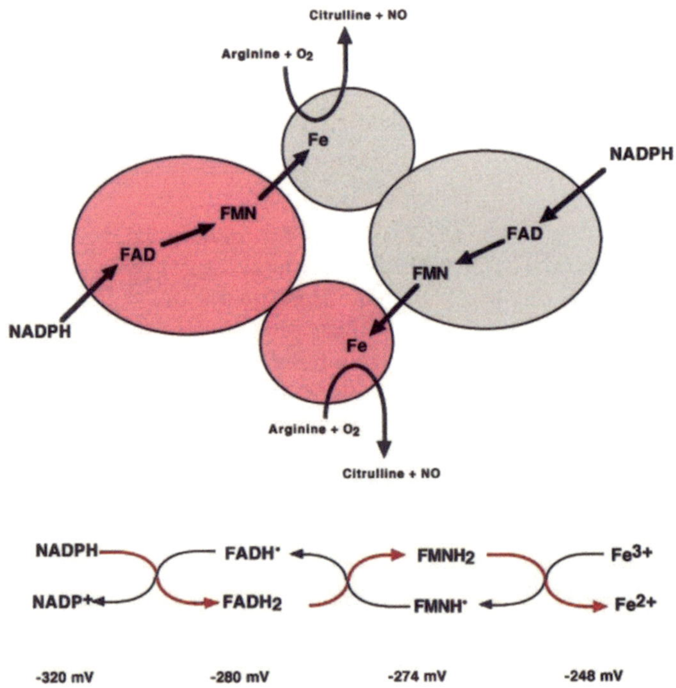

Fig. 1. Redox potentials and direction of electron flow in nNOS enzyme action are shown. The electron flow in the NOS dimer goes via NADPH→FAD→FMN in the reductase of one monomer to the haem iron in the oxygenase domain of a separate monomer. The redox potentials are poised thermodynamically to make this occur. The potentials for the two-electron oxidation of NADPH and the one-electron oxidations of $FADH_2$, $FMNH_2$ and ferric haem are illustrated at the bottom of the Scheme, with the red arrows indicating the direction of electron flow. Note that given the variety of redox couples and the closeness of the $FADH_2$ and $FMNH_2$ potentials the detailed picture is more complex than that illustrated. For example the $FMNH^+$/FMN couple has a redox potential of -49 mV and is likely to donate electrons to the high-potential ferric superoxide species to form the ferryl intermediate. Reduction of the FMN back to $FMNH^+$ would then require an electron from $FMNH^+$. Adapted from Alderton, et al. 2001.

- Mapping the distribution of NO generation or formation of stable spin-trapping species at different locations in tissues or organs by *in vivo* perfusion (**3D visualization of NO**

structural-functional or conformational dynamics as switch OFF and ON) may allow us to understand the real physiological function of NO gas or ionic form under different tissue physiological conditions [Thatte et al. 2009]. Using *in vivo* nitric oxide direct detection by stabilizing NO with suitable spin-trapping reagents is a challenge to estimate the *in vivo* NO concentration by MRI techniques [Hong et al. 2009; Sari-Sarraf, et al. 2009; Liu, et al. 2009; Ny et al. 2008; Fujii, et al, 2007; Vandsburger, et al.2007; Samouilov et al. 2007; Flögel, et al. 2007; Bobko, et al. 2005; Day, et al. 2005; Waller, et al. 2005; Sirmatel, et al. 2007; Itoh, et al. 2004; Berliner, et al. 2004; Haga,et al. 2003; Li, et al. 2003; Hsiao, et al. 2008; Kuppusamy, et al. 2001; Fichtlscherer, et al. 2000; Fujii,et al.1999; Fujii, et al. 2002].

- Other nitric oxide bioimaging technique is flourimetry using fluorescent biomarkers in EPR spectroscopy [Kojima, et al.2001; Fuji, et al.1997; Yoshimura, et al. 1996]. Our immediate focus is to highlight the existing multimodal mechanisms of NO sensitive MRI/EPR signal generation using NOS enzyme expression inhibitors and biosensors to map cellular events.

1.1 Nitric oxide as a second messenger in cellular signaling

Nitric oxide signal transduction through induced-nitric-oxide modifications relies on the system of Cys-based posttranslational modifications. Accordingly, S-nitrosylation of proteins plays an essential role in downstream cascades (Do et al., 1996). Nitric oxide exerts an ubiquitous influence on cellular signaling in large part by means of S-nitrosylation/denitrosylation of protein cysteine residues. S-NO undergo a regulated post-translational protein modification specific to NO-derived effects. These NO-dependent modifications influence protein activity, protein-protein interactions, and protein location. S-nitrosylation thus serves as the prototypical redox-based signal (Janssen-Heininger et al., 2008). S-Nitrosylation has been implicated in transmitting signals downstream of all classes of receptors, including G-protein-coupled receptor (GPCR), receptor tyrosine kinase, tumor necrosis factor, Toll-receptors, and glutaminergic receptors, acting locally within subcellular signaling domains as well conveying signals from the cell surface to intracellular compartments, including the mitochondria and the nucleus (Janssen-Heininger et al., 2008). Cell signaling through S-nitrosylation is useful tool in signaling transduction for:

1. Temporal regulation of response through a rapid and controlled stimulation;
2. The existence of motifs within proteins that provides S-nitrosylation specificity;
3. Colocalization of target proteins with a source of NO;
4. Reversibility of protein S-nitrosylation;
5. Enzymatic control of S-nitrosylation through the action of S-nitrosoglutathione reductase (Janssen-Heininger et al., 2008).

1.2 Why inhibition of NOS expression as bioimaging of NO technique

Mapping the distribution of NO generation at different locations in different tissues and organs by *in vivo* perfusion reveals the physiological function of NO (gas or ionic form) in different tissue physiological conditions [Thatte, et al.2009]. Major nitric oxide imaging techniques utilize mapping NO in tissue using NO specific imaging contrast agents sensitive to fluorescence, magnetic resonance and electron spin resonance. Imaging *in vivo* physical properties of tissue cells such as cell calcium signaling and NO-biomarker is new way by proton magnetic resonance techniques to achieve nanomolar range of nitric oxide mapping without any toxic effects[Thatte, et al.2009, Hong et al.2009]. Fluorescent nitric oxide cheletropic traps are currently available choices in nitric oxide imaging but all of these have pitfalls of causing neurotoxicity [Reif et al.2009]. In this chapter, we display evidence of the NO sensitive fluorescent probes as cell calcium signaling indicator and possibility of NO specific perfusion MRI tool to visualize physiological nanomolar dynamics of NO in living cells and tissues up to the detection limit of 0.1 nM. The cell signaling indicators such as intracellular calcium revealed that ~1 nM of NO was enough to detect apoptosis events such as caspase 3 activation [Li et al.2009].

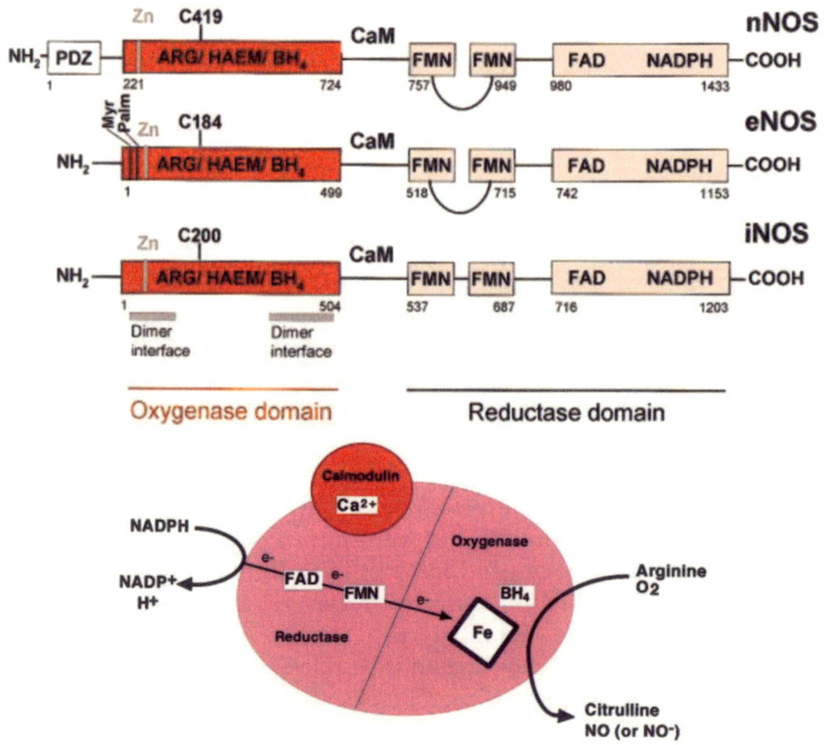

Fig. 2. Domain structure of human nNOS, eNOS and iNOS(on left), Overall reaction catalysed and cofactors of NOS(on right). Adapted from Alderton, et al. 2001.

1.2.1 Feasibility of NOS expression inhibitors as imaging Contrast Agents

NO is a gaseous and highly active neuronal messenger, short lived free radical generated enzymatically by NOS in the brain[Perrier et al.2009]. NOS is a flavocytochrome that is constitutively distributed in the vascular endothelium, brain, lung, and platelets (called constitutive NOS or *c*NOS), but is also found as an inducible form (called inducible NOS or *i*NOS) in many cells and organs, triggered by several factors such as inflammatory cytokines [Claudette et al.2009]. However, NOS is abundant in three isoforms cNOS, iNOS, eNOS as shown in Figure 2. NOS undergo dimerization in presence of BH4 or heme or arginine binding sites. Different roles of flabin, heme, pterin cofactors in NOS are described in detail [Alderton, et al. 2001, Matter et a. 2004]. Authors reviewed NOS activity regulation by calmodulin, phosphorylation, protein inhibitors of NOS(PIN), heat shock protein 90(Hsp 90), myristoylation, palmitoylation, caveolin, at different domain located in NOS or spilce variants: nNOSβ and nNOSγ, nNOSμ, nNOS-2, iNOS/eNOS splice variants. However, cross-activities of all three isoforms of NOS and their dependence on calcium, common locations have put challenge to specificity and gave way to a new way of specific NO/NOS inhibitor compounds. In recent years, NOS gene expression and its regulation have got attention for two reasons:

- NOS gene expression can be stained, imaged for multimodal molecular imaging
- The NOS inhibitors inhibit NOS as potent anti-inflammatory agents in recent years.
- Visualization of NOS inhibition is believed to serve as in vivo or in vitro biomarkers of the proinflammatory status of NOS inhibitors or in vivo proinflammation bioimaging of tissues.

Some examples of NOS ihibitors are given below.

1.2.2 NOS Inhibitors

NOS inhibition occurs at arginine site, tetrahydrobiopterin site, pteridine site, and heme domain by partially selective or highly selective inhibitors [Alderton, et al. 2001, Matter et a. 2004]. Excellent information of binding site interactions of NOS with relationships at different sites is available [Alderton, et al. 2001, Matter et al. 2004]. To understand better the NOS inhibitor structural-activity relationship, X-Ray analysis, 3D QUSAR, comparative molecular field analysis (CoMFA) analysis, GRID/PCA interpretations have enhanced the scope of NOS in pharmacology as shown in Figures 3,4,5 [Matter et al.2004]. There are array of NOS inhibitors described in the literature as drug testing tools. Table 1 shows efficacy of some of these in inhibiting the three human NOS isoforms.. Of these the most widely used have been l-NMMA, l-NNA and its methyl ester prodrug (NG-nitro-l-arginine methyl ester, `l-NAME' and aminoguanidine. However, inhibitors show pitfalls on selectivity to NOS, types of interactions with iNOS and nNOS isoforms. **Selective** NOS inhibitors may be selective in the physiological range (l-arginine concentration etc). Inhibitor agents with 10-50-fold selectivity are useful as `partially selective inhibitors.

inhibitor	IC50 (μM)			Selectivity (fold)		
	iNOS	nNOS	eNOS	iNOS vs nNOS	iNOS vs eNOS	nNOS vs eNOS
L-NNA*	3.1	0.29	0.35	0.09	0.11	1.2
L-NMMA	6.6	4.9	3.5	0.7	0.5	0.7
7-NI*	9.7	8.3	11.8	0.9	1.2	1.4
ARL17477*	0.33	0.07	1.6	0.2	5	23.
Aminoguanidine*	31	170	330	5.5	11	1.9
L-NIL	1.6	37	49	23.	49.	1.3
1400W	0.23.	7.3	1000	32.	4000*.	130*
GW273629	8.0.	630	1000	78.	125*.	1.6*
GW274150	1.4.	145	466	104.	333	3.2

The data shown are for inhibition of the human NOS isoforms in the presence of 30 μM L-arginine at 37 °C over 15 min after a 15 min pre-incubation with inhibitor under turnover. The data serves as progressive inhibitory mechanisms for the NOS assay. Data are from Young et al. 2000.

Table 1. Selectivity of inhibitors of NOSs is compared for different inhibitors

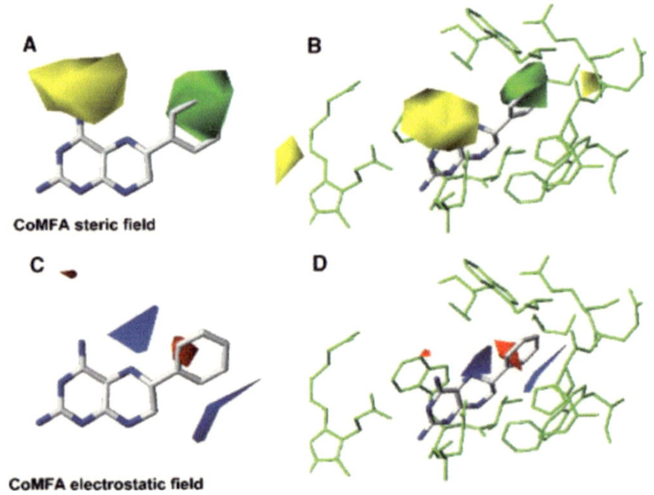

Fig. 3. Contour maps for NOS-I comparative molecular field analysis (CoMFA) analysis with a 4-amino-pteridine inhibitor.A: Steric contour map, green contours indicate sterically favored regions, yellow contours indicate unfavored areas. B: Samethan, A: with NOS-I binding site. C: Electrostatic contour map, blue contours refer to regions, where negatively charged substituents are unfavorable, red contours indicate regions, where negatively charged substituents are favorable. D: Same as C with NOS-I binding site. Reproduced with permission from Matter et al. 2004.

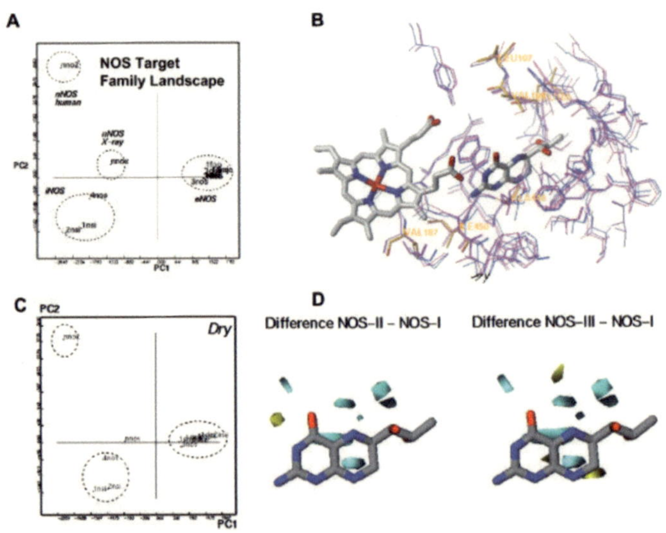

Fig. 4. A: Score plot from GRID/principal componentanalysis (PCA) based on13 conformations and 3NOS isoforms. B: SuperpositionofNOS-III (1nse, blue) toNOS-IX-ray (magenta) andhumanbrainNOS-I (homology, purple).C:Scoreplot fromGRID/CPCA for hydrophobic interactions (GRID dry probe). D:GRID/CPCA differential plots highlighting differences between NOS-I/NOS-II and NOS-I/NOS-III.Favorable interactions toachieveNOS-I isoformselectivities are shownincyancontours, unfavorable interactions are displayed in yellow with respect to H4Bipfromthe NOS-III1nse X-ray structure. Reproduced with permission from reference [Matter et al. 2004]

The mechanistic nature of three human NOS isoforms was expressed in the baculovirus expression system, and cell lysates as the enzyme source. iNOS potency and selectivity underestimates progressive inhibition of iNOS but not eNOS or nNOS; e.g. for 1400W the steady-state values of iNOS IC50 and selectivity have been estimated to be 0.1 μM, 250-fold (versus nNOS) and 5000-fold (versus eNOS) as reported by Young, et al. 2000. Highly selective compounds of over 50- or 100-fold selectivity, inhibit the NOS activity of a single isoform without affecting others. They have potential as selective therapeutic agents without side effects. Recently, 7-nitroindazole (7-NI) was reported as non selective nNOS inhibitor, of isolated NOS enzyme (Table 1). 7-NI showed inhibition of nNOS independent of increases in blood pressure but showed eNOS-dependent celltype specificity (neuronal verus endothelial), intracellular BH4 concentration, or depending on specific cellular transport or metabolism [Handy, et al. 1998]. All three NOS isoforms can be expressed in neurons and both eNOS and iNOS in endothelial cells. Cell-type specificity is clearly a very distinct phenomenon from isoform selectivity. Other pitfall is particular dose. For example, in humans, the non-selective NOS inhibitor l-NMMA causes a five-fold increase in vascular resistance with only a 10% change in blood pressure, because of reflex decreases in cardiac output 185±187 [Ross et al. 1998] Suppression of inhibitor-induced plasma nitrate (mediated predominantly by iNOS) and no effects on blood pressure has led to inhibitors as non selective, e.g. *S*-ethylisothiourea [Raman, et al. 1998]. Selectivity for iNOS versus eNOS distinct enzyme targets is another pitfall. Time-dependent NOS inhibition in assessing efficacy and selectivity of NOS inhibitors were first reviewed by Bryk and Wolff, 1999.

1.2.3 NOS inhibitor interactions with the NOS enzymes

Inhibitors of NOS have been described which interact with the NOS enzymes in a variety of ways: different sites, differing time- and substrate-dependence, and mechanism of inhibition.

L-Arginine site inhibitors identified so far as competitive with the substrate L-arginine binding at the arginine-binding site inhibitors (aminoguanidine, *S*-ethylisothiourea, thiocitrulline) [Raman, et al.1998].

Mechanism-based inhibitors of iNOS require active enzyme and NADPH substrate to permit inhibition to proceed through multiple enzyme covalent modification by a complex formation pathway: EI-EI* complex where E is iNOS, I is the inhibitor, EI is the initial non-covalent complex and EI* is a modified complex, either with a conformational change to tight binding or with covalent changes to the enzyme, inhibitor or both. The interactions of some of these pterin-site inhibitors with NOS reveal unexpected complexity. An example of a mechanism based iNOS inhibition was cited by aminoguanidine [197±199] and by the acetamidine inhibitors N-α-iminoethyl-l-ornithine (l-NIO) and *N*'-iminoethyl-lysine (l-NIL) [200±203], GW273629 (*S*-[2-[(1-iminoethyl)-amino]ethyl]-4,4-dioxo-l-cysteine) and GW274150 (*S*-[2-[(1-iminoethyl)amino]ethyl]-l-homocysteine) (see Table 1). **Heme-binding inhibitors** have been shown to bind with each NOS monomer, one to the haem iron and one to the arginine-binding region (Glu) in competition with CaM. These compounds affect the assembly of iNOS monomers into active dimer inhibiting the dimerization [Sennequier, et al. 1999]. A class of substituted pyrimidine inidazoles have been identified which inhibit dimerization of iNOS during its synthesis and assembly.

1.2.4 Flavoprotein and CaM inhibitors

A range of flavoproteins (e.g. diphenylene iodonium) or CaM (e.g. trifluoperazine) has been shown to inhibit NOS. NOS inhibitors display selectivity of their isoforms as partially selective, highly selective.

Partially and highly isoform-selective NOS inhibitors

Identification of selective inhibitors of iNOS and nNOS "100-fold selectivity for iNOS versus eNOS are reported [Anon, 1999a, 1999b]. Partially-selective nNOS inhibitor amino acids for nNOS versus eNOS and iNOS were reported. For example, *S*-ethyl- and *S*-methyl-1-thiocitrulline, vinyl-l-NIO showed timedependent inhibition of nNOS with significant selectivities versus isolated eNOS and iNOS enzymes [Babu and Griffith, 1998]. Other partially-selective iNOS inhibitors such as acetamidine-containing analogues of arginine, l-NIO and l-NIL have been widely used to probe the effects of iNOS inhibition. For example, some 2-iminohomopiperidines and 2-iminopyrrolidines with high (100±900-fold) selectivity for iNOS versus eNOS, but similar potency was observed on iNOS and nNOS (1±13-fold selectivity), with dual action iNOS-nNOS inhibitors.

Highly-selective iNOS inhibitors

The `highly selective ' iNOS inhibitors versus eNOS are mostly bis-isothioureas. Of these, *S,S*-[1,3-phenylene bis-(1,2-ethanediyl)bis-thiourea (`PBITU') is an l-arginine-competitive, rapidly reversible inhibitor of human iNOS with a K_i of 47 nM, and a selectivity (in K_i terms) of 190-fold versus eNOS showing substrate-binding sites of full-length human iNOS and eNOS in solution or in cells and tissues. Inhibition of human iNOS by 1400W was competitive with l-arginine, NADPH-dependent either an irreversible, or reversible. Mechanism-based inhibitor action was reported as K_d value %7 nM and steady-state selectivity against eNOS and nNOS of 5000 and 250-fold respective effects on vascular leakage [Lazlo, et al.1997]. GW273629 and GW274150 are two novel NOS inhibitors for iNOS versus both eNOS and nNOS. Both are sulphur-substituted acetamidine amino acids acting in competition with l-arginine as NADPH-dependent, whereas the inhibition of human eNOS and nNOS is rapidly reversible. The heme-binding substituted pyrimidine imidazoles inhibit assembly of active dimeric iNOS during its synthesis. It will be interesting to see what the pharmacology and utility of such compounds will be, and whether other compound series are discovered with this NOS inhibition mechanism of action.

1.3 The pharmacological inhibition of inducible nitric-oxide synthase (iNOS) gene expression

Presently, inhibition of NOS gene expression approach is used in testing inhibitors of pharmacological value. In past, inhibition of NOS was ideal assay to measure less stable NO in tissues but less specificity of NOS was discouragement and spin trap agents have emerged as a new approach of detection and measurement of NO in both in vivo and in vitro assays and bioimaging. Some examples of NOS inhibitors are cited in following section.

1. AMP Kinase protein kinase induced inhibition of inducible nitric-oxide synthase (iNOS) Inducible NOS inhibition in endotoxic shock in chronic inflammatory states was observed in several cell types (myocytes, adipocytes, macrophages) and primarily resulted from post-transcriptional regulation of the iNOS protein. Best example is inhibition of inducible nitric oxide synthase by activators of AMP activated protein kinase to explain a new mechanism of insulin sensitizing drug action [Pilon, et al.2004].. Inflammatory cytokines and LPS trigger the iNOS transcription through a complex network of intracellular pathways including NF-κB, Janus kinase/signal transducers and activators of transcription, and mitogen-activated AMPK protein kinase by reducing the transcription of iNOS and mRNA expression [Blanchette et al. 2003]. AMPK switches off protein synthesis either through suppression of the mTOR-p70S6 kinase pathway or by direct activation of eukaryotic elongation factor 2 kinase, resulting in the phosphorylation and inactivation of eukaryotic elongation factor 2 [Horman et al. 2002]. AMPK reduces iNOS protein content by promoting its ubiquitination, required for targeting iNOS through the proteasome proteolysis pathway [Kolodziejski et al.2002].
2. Expression of exogenous Kalirin in pituitary cells dramatically reduces iNOS inhibition of ACTH secretion. Kalirin inhibits iNOS activity by affecting iNOS homodimerization, which is required for iNOS activity. Thus Kalirin may play a neuroprotective role during inflammation of the central nervous system by inhibiting iNOS activity [Ratovitski, et al. 1999].
3. N^5-(Iminoalkyl)- and N^5-(Iminoalkenyl)-ornithines (VNIO) and several L-VNIO analogs showed minor structural changes to produce inhibitors either iNOS-selective or nonselective [Bretscher et al. 2003]. Furthermore, derivatives having a methyl group added to the butenyl moiety of L-VNIO and L-VNIO derivatives display slow-on, slow-off kinetics rather than irreversible inactivation. These results elucidate isoform-selective inhibition by L-VNIO and may provide information useful in rational design of isoform-selective inhibitors. [Bretscher, et al. 2003]
4. Constitutive and inducible isoforms of NOS are inhibited by S-alkyl-L-thiocitrullines with n-alkyl groups of any one carbon. The NOS inhibition is reversible, stereoselective, and competitive with L-arginine[Narayanan, et al. 1995].
5. Autoinhibition of endothelial NOS was reported by presence of an electron transfer control element in the NOS. [Nishida, et al. 1999]. Investigators examined the role of the insert in its native protein context by deleting the insert from both wild-type eNOS and from chimeras obtained by swapping the reductase domains of the three NOS isoforms. The Ca^{2+} concentrations required to activate the enzymes decrease significantly when the insert is deleted, consistent with suppression of autoinhibition. Furthermore, removal of the insert greatly enhances the maximal activity of wild-type eNOS, the least active of the three isoforms. Despite the correlation between reductase and overall enzymatic activity for the wild-type and chimeric NOS proteins, the loop-free eNOS still requires CaM to synthesize zNO. However, the reductive activity of the CaM-free, loop-deleted eNOS is enhanced significantly over that of CaM-free wild-type eNOS and approaches the same level as that of CaMbound wild-type eNOS. Thus, the inhibitory effect of the loop on both the eNOS reductase and zNO-synthesizing activities may have an origin distinct from the loop's

inhibitory effects on the binding of CaM and the concomitant activation of the reductase and zNO-synthesizing activities.

6. A mechanism was reported as Ca^{2+} triggers cross-talk signal transduction between CaM kinase and NO and CaM-K IIa phosphorylating nNOS on Ser847, which in turn decreases the gaseous second messenger NO in neuronal cells by Calcium/Calmodulin-dependent Protein Kinase IIa in NG108-15 neuronal Cells. [Komeima, et al. 2000].
7. Nitric oxide (NO) is moderately produced under control of iNOS gene expression. In recent years, scientists solved the problem of visualizing very unstable NO by mapping iNOS inhibition using gene expression array, *in vivo* nitric oxide direct detection (by stabilizing NO with suitable spin-trapping reagents), bioluminesence and MRI techniques to estimate *in vivo* NO concentration [Hong et al. 2009]. New approaches are in the direction of multimodal bioimaging iNOS gene expression and NO bioimaging as biomarkers. Success depends on NOS gene sensitive MR relaxation signal in cells and specificity to inflammation [Nie et al.2008; Terashima et al. 2010].

2. NO inhibitors in imaging

Major nitric oxide imaging techniques utilize mapping NO in tissue using NO specific imaging contrast agents sensitive to fluorescence, magnetic resonance and electron spin resonance. Recently, focus is diverted towards imaging *in vivo* physical properties of tissue cells such as cell calcium signaling and NO-biomarker based proton magnetic resonance techniques to achieve nanomolar range of nitric oxide mapping without any toxic effect. Fluorescent nitric oxide cheletropic traps are currently available choices in nitric oxide imaging but all of these have pitfalls of causing neurotoxicity [Reif, et al.2009]. In following sections, we put evidence of the NO sensitive fluorescent probes as cell calcium signaling indicator in tissues and possibility of NO specific perfusion MRI tool to visualize physiological nanomolar dynamics of NO in living cells up to the detection limit of 0.1 nM. The cell signaling indicators such as intracellular calcium revealed that ~1 nM of NO was enough to detect apoptosis events such as caspase-3 activation [Green, 1998, Nagata, 1997, Kwon et al.2009]. Furthermore, the possibility of superparamagnetic iron oxide nanoparticle bound complexes serve as MRI imaging contrast agents based on dephasing contrast. The iron oxide chelates are still in active evaluation phase to test their toxicity. The nanomolar range of basal endothelial NO appears to be fundamental to vascular homeostasis, hypoxia, apoptosis and inflammation.

- different contrast mechanisms and contrast characteristics of known nitrosyl-iron complexes display possibility of potential multimodal MRI and EPR probes specific to NO with examples of dithiacarbamates and $Fe(MGD)_2$ complexes in different applications.
- fMRI detects the interaction of paramagnetic species with NO in blood but possibility is still controversial. We review different applications of NO bioimaging.

2.1 Dithiacarbamates, LPS and MGD complexes for bioimaging of NO

In following sections we highlight the existing mechanisms and description to NO sensitive signal generation. Existing mechanisms of NO sensitive signal generation are explored by

MRI and fluorescence arising out from paramagnetic metals, dithiacarbamates or lipopolysaccharides complexes. The mechanisms depend on three approaches:

- Use of paramagnetic metals (SPIO) in MRI and dithiacarbamates (DTC) or lipopolysaccharides (LPS) complexes sensitive to MRI and fluorescence effects;
- Use of cytochrome proteins sensitive to EPR effect;
- Use of NO synthase inhibitors to measure the reduced NO concentrations for fluorometry and blood oxygen sensitivity to reduced NO concentrations.

The following sections are focused on dithiacarbamates in fluorometry and less known imaging contrast agents in MRI to image nitric oxide in tissues.

The first evidence of dithiacarbamates (DTC) reported them as electron Fe(II)-chelate spin trap agents. Examples are N-methyl-D-glucamine dithiocarbamate (MGD), $(MGD)_2$-Fe(II)-NO and NO-Fe-DTC metal complexes as multimodal imaging agents. These were initially verified for EPR with possibility of visualizing the radical distribution by MR images [Kubrina, et al.1992]. The $(MGD)_2$-Fe(II)-NO complex enhanced the contrast in the vascular structures such as hepatic vein and inferior vena cava. The paramagnetic NO-Fe-DTC metal complex is also a potential MRI signal enhancer and acts as contrast agent. These contrast agents showed the magnetic relaxation changes of neighboring protons to visualize the NO generated in living animal tumors [Jordan et al. 2000]. Other contrast enhancement effect showed an impact of short NO exposure to hemoglobin during MRI signal recording as source of *in vitro* MRI and *in vivo* functional MRI (fMRI) [Di Salle, et al. 1997]. fMRI signal intensity of venous blood in T1-, T2-, and T2*-weighted images proportionately changed with NO real-time generation in brain. Later, different approaches of blood hemoglobin and NO interaction were attempted to monitor fMRI signal sensitive to NO: mainly metHb and NO-Hb enhanced the MRI signal intensity. These observations suggested a blood flow-independent effect and less utility [Di Salle, et al. 1997]. Still it is hope that NO sensitive fMRI techniques can detect slow epithelial intracellular processes such as metabolic integrity, vascular tonicity, stress, shear and inflammatory effects at early stages of the disease processs, allowing precise monitoring of onset in intact biological systems at cellular level. Other approaches are also emerging to use NO biosensors for multimodal imaging. Currently, use of fMRI as a non-invasive NO sensitive technique has emerged as potential and remarkable tool to detect apoptosis *in vivo*. The NO sensitive fMRI techniques can detect slow epithelial intracellular processes such as metabolic integrity, vascular tonicity, stress, shear and inflammatory effects at early stages of the process, allowing the onset in intact biological systems, providing a useful tool for monitoring at cellular level.

2.2 The source of intracellular NO and metabolic integrity- feasibility of MRI

The nitric oxide is released from the L-arginine in tissue along with molecular oxygen in the oxidative L-arginine degradation reaction of L-arginine pathway catalyzed by either of any three different NO synthase (NOS) isoenzymes. NO controls the intracellular redox state in tissue and protects the metabolic integrity in two ways [Kuppusamy, et al. 1994]. First, anions and cations in intracellular space or cellular redox state prevent apoptosis for example, NO in hepatocytes, neurons, glial cells and fibroblasts controls the release of mitochondrial apoptogenic factors and induces apoptosis by activation of caspases [Hortelano, et al.2005]. Second, peroxynitrites accumulate as product of nitric oxide and superoxide anion. Peroxynitrites promote apoptosis through the indirect activation of caspases [Kuppusamy, et al. 1994, Hortelano, et al.2005]. However, all these approaches are still invasive to evaluate apoptosis.

Overproduction of nitric oxide and imaging the NO accumulation due to neurotoxicity was reported in neurological disorders or neurodegeneration. The reason of neurotoxicity was reported due to NO reactive oxidative properties in tissue [Caramia, et al. 1998]. This approach was reported not to solve the purpose of *in vivo* functional monitoring but NO properties indicated the state of tissue and follow-up of dynamic status of neurodegenerative factors in brain [Caramia, et al. 1998]. In other application of NO bioimaging to detect apoptosis, T2 weighted maps from control vs GSNO treated and z.VAD treated animals exhibited hyperintense areas perhaps due to toxicity of NO on the T2 maps while z-VAD treated animals showed small lesion areas due to reduced NO toxicity. The response of z-VAD injection was presumed as reduced toxicity due to caspases and NO-dependent apoptosis. These authors explained that GSNO increased T2 intensity 25% while z-VAD reduced this MRI signal [Komarov, et al. 1995]. These classic reports indicated that T2 hyperintensities on MRI positively offer the possibility of in vivo evaluation of cell death at different locations in whole tissues undergoing apoptosis. The apoptosis is also a major mechanism in brain neurodegeneration and post-myocardial infarction heart [Caramia, et al. 1998]. However, this approach seems as potential tool of functional imaging in near future to monitor the therapeutic intervention of new drugs to reduce neurodegeneration[Kuppusamy, et al. 1994, Hortelano, et al.2005, Foster, et al. 1998]. Currently, non-invasive NO sensitive techniques are big hope as potential and remarkable tools to detect apoptosis *in vivo*.

Initial studies on NO with iron-dithiocarbamate complexes had succeeded in direct detection of NO in mice by whole body electron paramagnetic resonance spectroscopy (EPR) at the L-band [Fujii, et al. 1997, Komarov, et al. 1995] with new possibilities by Magnetic Resonance Spectroscopy [Reif, et al. 2009, Li, et al. 2009]. Several authors demonstrated the feasibility of EPR imaging in visualizing free radical distributions *in vivo* at low resolution where the intrinsic line width of the radical is large, such as the spin-trapped NO [Yoshimura, et al. 1996, Kubrina, et al. 1992, Foster, et al. 1998]. This drawback of low resolution caught attention for feasible MRI contrast agents, such as stable (*N*-methyl-D-glucamine)$_2$-Fe(II)-NO complex to generate better resolution. The complex has a much longer *in vivo* half-life than most (stable) nitric oxide derived compounds. Recently N-methyl glucamine iron complexes have shown greater affinity with NO to make (MGD)$_2$-Fe(II)-NO complex useful for in vivo NO measurements by EPR [Caramia, et al. 1998]. In recent years, the art of MRI combined with NO spin trapping mechanism was evaluated as feasible method of mapping the distribution of NO spin-trap complex in animals with possibility in clinical use. In following section, our design of ultrafast MRI protocol is described using NO/NOS sensitive biosensor complexes for use at 21-Tesla MRI microimager.

2.3 Approach of stable NO sensitive lipopolysaccharides (LPS) as feasible imaging complexes

The LPS serves as encaged bag holding contrast agent. The LPS based NO imaging approach has following presumptions:

- spin-trapped NO is stable in intracellular tissue environment;
- NO-LPS contrast enhancement properties are MRI visible;
- NO-LPS complex is stable in tissues and organs; and
- simultaneous visualization and mapping of NO free radicals are possible by MRI.

The routine method of LPS injected animals after 6 hours of $(MGD)_2$-Fe(II) injection, usually serve as experimental model of subsequent MRI detection and visualization of NO generated in animals. It was confirmed by using a specific NOS inhibitor *N*-monomethyl L-arginine (L-NMMA) [Fujii, et al. 2007]. The observations indicated that the suppressed NO levels were proportional to NO sensitive LPS concentration independent of $(MGD)_2$-Fe(II)-NO contrast agent [Fujii, et al. 2007]. Next issue of feasibility and stability of $(MGD)_2$-Fe(II)-NO imaging contrast agent was major breakthrough reported in model systems to generate EPR and MRI signals. It is established that the NO complex $(MGD)_2$-Fe(II)-NO is very stable in model aqueous media. The NO complex $(MGD)_2$-Fe(II)-NO complex, if injected, is stable in tissues and organs as confirmed by L-band EPR measurements as reported elsewhere [Lai, et al. 1994, Mulsch, et al. 1999]. These authors confirmed the assignment as $(MGD)_2$-Fe(II)-NO from the hyperfine coupling J constant of this signal (see Figure 5). However, LPS complex accumulates in liver, brain, heart, kidney [Yoshimura, et al. 1996]. NOS inhibitor, *N*-monomethyl L-arginine (L-NMMA, 2 mM) confirmed that the NO complex is not bio-reduced or biodegraded *in vivo* based on the EPR spectrum of $(MGD)_2$-Fe(II)-NO complex. The stability of $(MGD)_2$-Fe(II)-NO complex was at least 12 hours. In other report, the reduction/ decomposition of the NO complex occurred in the presence of 1 mM ascorbic acid or glutathione inhibitor with a half-life of 40 and 48 min, respectively [Perrier, et al. 2009]. All these evidences indicate the NO complex as stable, non-biodegradable or decomposed form of $(MGD)_2$-Fe(II)-NO complex as strong proton relaxation enhancer with paramagnetic properties [Perrier, et al. 2009, Fujii, et al. 2007]. All these evidences also indicated the NO complex as stable, non-biodegradable or decomposed form of $(MGD)_2$-Fe(II)-NO complex as strong proton relaxation enhancer with paramagnetic properties.

Biosensor#		T1 (600/15) msec		r1 L.mmol^{-1}.sec^{-1}	r2 L.mmol^{-1}.sec^{-1}
DNIC (240 μmol/L)		931 ±3 7			
DNIC-cysteine				0.23 ± 0.06	0.33 ± 0.03
DNIC-GSH(230 μM)		850 ± 24		0.11 ± 0.02	0.19 ± 0.03
DNIC-BSA(160 μM)		247 ± 13		0.71 ± 0.09	1.37 ± 0.30
MNIC-MGD				0.97 ± 0.09	1.37 ± 0.03
MGD		55.5	52.4		
$(MGD)_2$-Fe(II)-NO	T1	13.3*	13.9**		
(100 nmol/g)	T2	8.3*	8.7**		
(MGD)2-Fe(II)		24.9	25		

The strong magnetic moment of the unpaired electron promotes both spin lattice and spin-spin relaxation of the surrounding water protons, resulting with decrease in their spin-lattice (T1) and spin-spin (T2) relaxation times or enhancement in signal intensity on T1 or T2 weighted MR images. Relaxation constants of $(MGD)_2$-Fe(II)-NO, $(MGD)_2$-Fe(II) at 500 MHz* and 300 MHz** is shown indicating that relaxivity is not magnetic field dependent over this range.

Table 2. Relaxation constants of biosensor complexes are shown for NO detection in aqueous solutions.

In other study, NO complex was prepared from NO gas and $(MGD)_2$-Fe(II) in glass capillaries. The strong magnetic moment of the unpaired electrons promote both spin lattice and spin-spin relaxation of the surrounding water protons and show a decrease in their spin-lattice (T1) and spin-spin (T2) relaxation constants or show enhanced signal intensity in T1 or T2 weighted MR images as shown in Table 2 and Figure 5 [Hortelano, et al. 2005]. Major points were:

- The relaxation constants of $(MGD)_2$-Fe(II)-NO in aqueous media were characteristic and complex concentration dependent;
- T1 and T2 relaxation constants of liver were measured 1.17(1/mM * sec) and 1.32 (1/mM * sec) respectively at 500 MHz at increasing concentration of NO;
- T1 relaxation constant of $(MGD)_2$-Fe(II), was reported 0.044 (1/mM * sec) at both 20 MHz and 85 MHz magnetic fields. The T1 relaxation constant was field independent;
- Distinct increase in relaxivity after complexing NO with $(MGD)_2$-Fe(II). These unique properties of complex suggested its feasibility to visualize the regional distribution in tissue *in vivo* where the NO was trapped.

In previous study, $(MGD)_2$-Fe(II)-NO and LPS complexes were used as in vivo MR contrast agents [Fujii, et al. 2007]. First, intraperitoneal injection of 2 ml of 9.1 mM $(MGD)_2$-Fe(II)-NO complex (150 nmol/g tissue) showed relative difference on T1-weighted MR images of rat liver site before and after injection of NO complex as shown in Figure 5.

- The images indicated the NO complex very effective "intrinsic contrast agent," to enhance the contrast in several other organs.

The Second, LPS injected animals after 6 hours of $(MGD)_2$-Fe(II) injection, showed subsequent MRI detection and reproducible visualization of NO generated in liver of animals as shown in Figure 5 and Fig. 6a (indicated by the arrow) as plots against time with the intensity of the reference water sample. Figure 6b showed that the image intensity in the liver increased with the NO generation, without any intensity change in reference sample in the presence of the specific NOS inhibitor *N*-monomethyl L-arginine (L-NMMA).

Interestingly, other face of $(MGD)_2$-Fe(II)-NO complex is 'spin trap'. Spin traps are chemical compounds used to detect unstable free radicals such as hydroxyl (*OH) and superoxide (O_2^{*-}) in biological systems. Spin trapping compounds (spin traps) selectively react with NO and trap the nitric oxide. The spin-trap stock $(MGD)_2$-Fe(II) complex solution (mixture of deoxygenated saline MGD (100 mM) and $FeSO_4$ (20 mM) solutions in nitrogen) makes $(MGD)_2$-Fe(II)-NO complex by mixing $(MGD)_2$-Fe(II) with either pure NO gas or *S*-nitroso-*N*-acetyl DL-penicillamine (SNAP). The pure NO gas is prepared by passing NO gas through a KOH solution. Major concern was the coordination of NO to the Fe complex altered the solubility of N0-Fe(II)$(MGD)_2$ < 1 mM. The in vivo half life of N0-Fe(II)$(MGD)_2$ was 41 minutes and only 40% trapped NO was estimated by detection of Fe(II)$(MGD)_{2\ \text{induced}}$ inhibited NOS and decrease in NO formation and perfusion in the tissue [Fichtlscherer, et al. 2000].

$$2NO + \mathbf{Fe(III)}(MGD)_2 \xrightarrow{H_2O\ (^*OH)} NO\text{-}\mathbf{Fe(II)}(MGD)_2 + NO_2^- \tag{1}$$

$$NO + \mathbf{Fe(III)}(MGD)_2 \xrightarrow{\text{Reducing Equivalent}} NO\text{-}\mathbf{Fe(II)}(MGD)_2\ (40\%)$$

MR contrast due to trapped NO can be expressed in Eq 2, 3:

$$1/T1[NO\text{-}Fe(II)(MGD)_2] = 1/T1_{(observed)} - 1/T^* = r1_i[NO\text{-}Fe(II)(MGD)_2] \times \text{concentration} \quad (2)$$

$$1/T_{obs} = 1/T^* + r1_i[NO\text{-}Fe(II)(MGD)_2];\ i = 1, 2... \quad (3)$$

where $T1_{(observed)}$ is long longitudinal relaxation time constant and T1* is short longitudinal relaxation time constant in absence of $[NO\text{-}Fe(II)(MGD)_2]$, $r1_i$ is longitudinal relaxivity dpends on I = 1, 2, 3 the number of carbons in branch of MGD chain, $[NO\text{-}Fe(II)(MGD)_2]$, concentration and interaction with NO specific bioactive receptor, enzyme, antibody, hormone molecules in tissue.

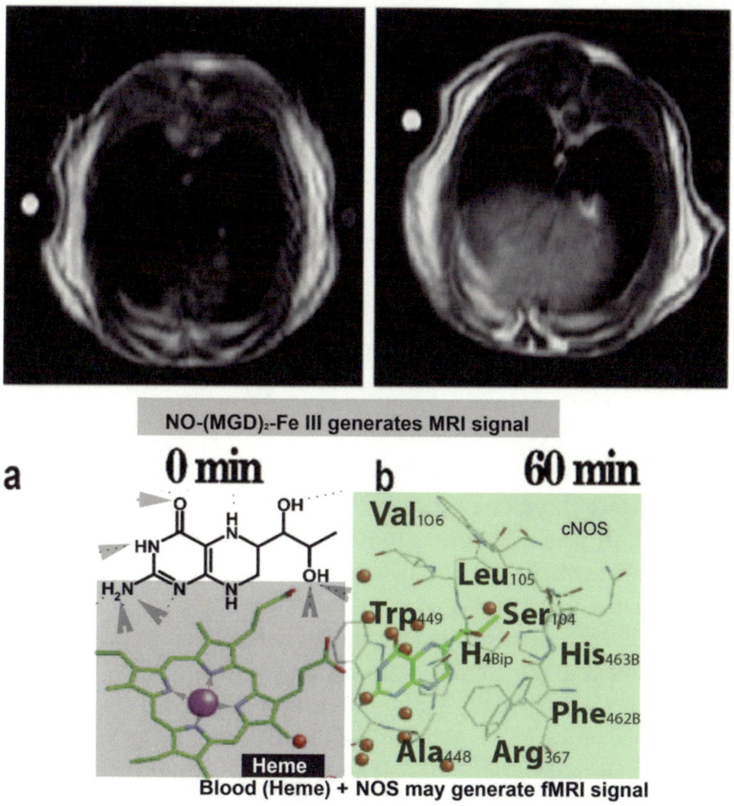

Fig. 5. Imaging of preformed $(MGD)_2$-Fe(II)-NO complex. Transverse T1-weighted MR images in the axial plane of the liver of Wistar rats. Two ml of 9.5 mM $(MGD)_2$-Fe(II)-NO complex was injected i.p. in the lower abdomen of 250 g rats (n = 3). **a:** Control, before injection; **b:** 60 min after injection of $(MGD)_2$-Fe(II)-NO complex. A sketch at bottom shows the interaction of No with heme as basis of MRI signal. Reproduced with permission for reference Fujii, et al. 2007.

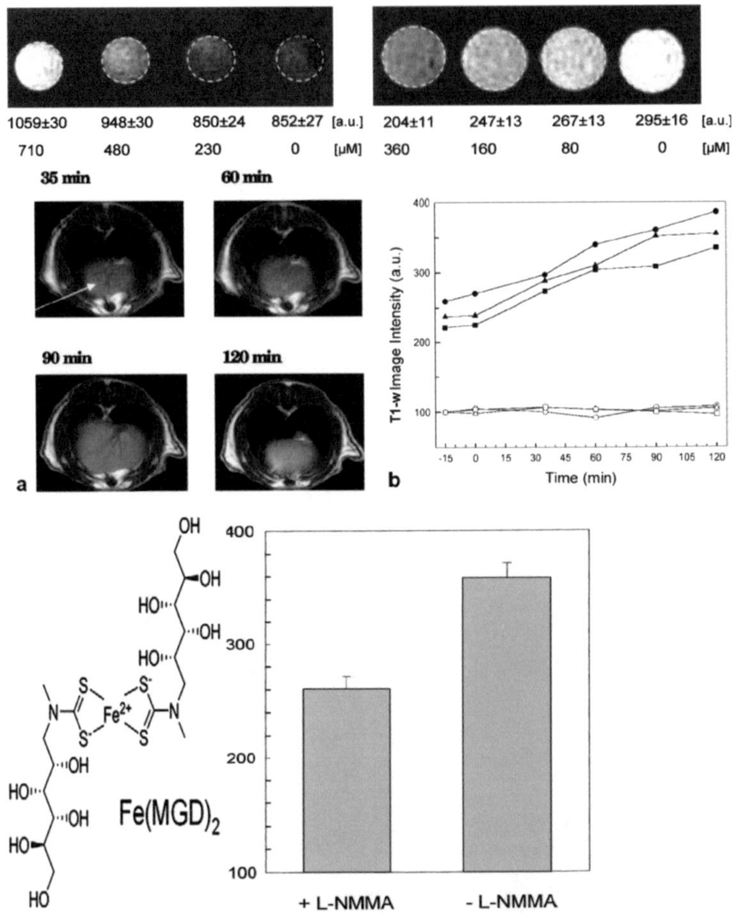

Fig. 6. Imaging of NO in six phantoms and LPS-treated rats. **a:** Transverse T1-weighted MR images focussed on a selected region of the liver in LPS-doped rats. The MR images were measured at the times indicated. Six hours after LPS injection, the NO spin-trap [3 ml of $(MGD)_2$-Fe(II), MGD: 100 mM, Fe: 20 mM] was administered i.p. **b:** Plot of MR image intensity with time. Signal intensities were averaged over the selected region indicated by the arrow in a (7*7 mm^2) for three different animals (filled symbols), normalized to intensity of a reference (water in a tube placed next to the animal, open symbols). At zero time, the spin-trap, (MGD)2-Fe(II), was added. **c:** Comparison of the signal intensity in a selected slice region of the liver in the presence and absence of the NOS inhibitor, L-NMMA. The image intensity in the liver was first normalized to the external standard noted above. The averaged intensity in the selected regions (7 x 7 mm^2) was averaged (n = 3). L-NMMA (50 mg/kg in saline) was injected i.p. 3 h after LPS injection (Reproduced from reference Fujii et al. 2007).

2.4 MRI measurement protocol design

We developed a MRI protocol to investigate the regional distribution of $(MGD)_2$-Fe(II)-NO in various organs. For whole animal imaging, protocols for relaxivity of $(MGD)_2$-Fe(II)-NO were designed on a Bruker 500 MHz, to achieve millimolar relaxivity. For the animal experiments, anesthesia was administered as sodium pentobarbital (30 mg/kg, i.p.) and ketamine (20 mg/kg, intra-muscular injection) before injection of contrast agent in animals. For imaging, animals were anesthetized by intubating with 14 gauge, 2 inch intravenous catheter (Abbocath Lab, IL) on nose with 30% oxygen/70% nitrogen mixture containing 5% isoflurane/air mixture to continuous supply through nose during MRI session. Animals were kept in vertical direction to the side of MRI gantry and 30 mm diameter RF insert covering kidneys in the center of magnet. During six hours after administration of LPS to Wistar rats (150–200 g) they were injected intraparitoneally with 3 ml of $(MGD)_2$-Fe(II) [MGD:100 mM, Fe:20 mM] at different intervals. MR images of the anesthetized rats (n = 5) were obtained after 1 hr spin trap injection with a Bruker Biospin 500 MHz scanner (PARAVISION 3.2).

Different imaging techniques are available in our lab on 500 MHz and 900 MHz scanners. In vivo NO induction was carried out according to previously reported methods with slight modifications [Lai, et al. 1994, Yoshimura, et al. 1996, Green, et al. 1998].

- Axial T2-weighted spin echo sequence with fat suppression (TR 2000 ms, TE 100 ms, flip angle 30; EC 1/1 15.6 kHz) was used for more detailed T2-weighted information in detecting solid tissues.
- Axial T1-weighted images were acquired at TR/TE/flip angle = 750 ms/ 4.18 ms/25°, FOV/matrix size/spatial resolution = 2.6 x 1.4 cm^2/ 256 x 256 ×75 µm, and the inversion time (TI approximately 250 ms) set to null the normal the background tissue.
- Gradient echo sequence in-phase and opposed-phase(TR 180 ms, TE 2.3 ms/4.5 ms, flip angle 30 were done as a dual echo sequence to show renal lesions hypointense and parenchyma isointense with hyperintense protein rich cyst. Opposed phased T1-weighted gradient echo sequences were sensitive to fat.
- Axial T1-weighted gradient echo sequence for dynamic perfusion/ diffusion weighted imaging (TR 130 ms, TE 1.0 ms, flip angle 30) was done after 30 ml intravenous gadolinium contrast injection for acquiring pre-contrast and post-contrast images in arterial phase to distinguish perfusion in the tissues. Parameters for T1-weighted spin-echo images were TR 1500 msec, TE 10 msec, 2 NEX, 1-mm slice thickness, 0.5-mm slice gap, field of view 12 x 3 x 12 cm^3, and matrix, 256 x 256.

The original bird-cage coil with 2.6 cm diameter Rf insert was used for MR imaging. We established in lab MR image analysis by shareware software NIH Image J, a PC version supplied by Scion Corporation. The analysis of averaged image intensities from selected regions of interest 15 x 3 x 15, and matrix , 256 x 256 served as best for measurement of signal enhancement. An external reference (saline solution in a sealed test tube) placed near the animal serves to normalize intensities from both control and test groups. Percentage enhancements can be calculated from the normalized data [35]. Functional dependence of the MR imaging signal intensity S on the intrinsic properties (proton spin density $N(H)$, T1, T2) has been derived based on the assumption of a monoexponential dependence on T1 and T2 with a spin-echo sequence: $S(TE,TR) = N(H)\exp(-TE/T2)\ [1 - \exp[-(TR/\ TE)/T1]]$. The term on the right reflects the T1 dependence, and the exponential expression on the left reflects the T2 effects. TE << T2 generates a T1-weighted. T1 image is obtained by means of least-squares fitting (Marquardt-Levenberg algorithm) of the mean intensity values versus

TR. T2 weighted image is obtained by TR >> T2. Mean intensity values are fitted versus the TE (Marquardt Levenberg algorithm).

T1 weitghted MR contrast due to trapped NO as NO-Fe(II)X spin trap in tissue can be expressed in Eq 4, 5:

$$1/T1[\text{NO-Fe(II)X}] = 1/T1_{(observed)} - 1/T^* = r1_i[\text{NO-Fe(II)X}] \times \text{concentration} \quad (4)$$

$$1/T_{obs} = 1/T^* + r1_i[\text{NO-Fe(II)X}];\ i = 1, 2... \quad (5)$$

where $T1_{(observed)}$ is long longitudinal relaxation time constant and T1* is short longitudinal relaxation time constant in absence of [NO-Fe(II)X],

$r1_i$ is longitudinal relaxivity dpends on i= 1, 2, 3 the number of carbons in branch of X contrast agent, its concentration and interaction with NO specific bioactive receptor, enzyme, antibody, hormone molecules in tissue.

2.5 Approaches to NO evaluation by Magnetic Resonance Imaging (MRI) techniques

Two major developments reported in context with NO. First, a search of NO specific stable multimodal MR/EPR sensitive spin trap complex; second, calibrating imaging characteristics of spin trap complex on NMR/MRI/PEDRI techniques. NO is perfused in gaseous phase or remains in ionic phase by redox state in the tissues. Both forms of NO are captured or trapped by $Fe(II)\text{-}(DTC)_2$ complexes and show MRI signal change due to reactive nitrosylation (reaction of intracellular NO with $Fe(II)(DTC)_2$ to form NO-$Fe(II)(DTC)_2$ complex) shown in Eq 6. Second-order rate constant of the reaction of NO with $Fe(II)(ProDTC)_2$ was reported to be $(1.1 \pm 0.3)\times 10^8\ M^{-1}\ s^{-1}$[Caramia, et al. 1998].

$$Fe(II)(DTC)_2 \rightarrow Fe(II)(ProDTC)_2 + NO \rightarrow NO\text{-}Fe(II)(DTC)_2 \quad (6)$$

Recently, attempts have been made to overcome the limitations of EPR-based imaging with low resolution and large sample size by employing new approaches of nuclear magnetic resonance (NMR) and magnetic resonance imaging (MRI) techniques. Several paramagnetic metal complexes have been widely used as contrast agents in magnetic resonance imaging (MRI) due to their ability to enhance the NMR relaxation of neighboring protons [von Bohlen, et al. 2003]. Paramagnetic NO-Fe-DTC complex has been utilized to visualize NO generated in living rats with septic shock [Fichtlscherer, et al. 2000, Fujii, et al. 1999]. Paramagnetic nitrosyl iron complex showed contrast properties to enhance the signal intensity of SNP-perfused rat liver in MRI [Fichtlscherer, et al. 2000, Fujii, et al. 1999]. Thus, a paramagnetic nitrosyl iron complex may be potentially useful as a microscopic localizer MRI contrast agent specific for NO accumulation in living organisms. Other proton-electron-double-resonance-imaging (PEDRI) technique was based on the enhancement of proton NMR signal intensity in the presence of radicals through the Overhauser effect Mulsch et al. 1999,Foster, et al.1998]. Rat livers were exposed to SNP as an NO donor exhibited intense PEDRI images [Rast, et al. 2001, Jordan, et al.2004]. PEDRI imaging is a potential tool to study NO in large biological specimens. Still imaging application of MR and EPR spectroscopy using nitric oxide synthase inhibitors, oxygenation or perfusion are in infancy [von Bohlen, et al. 2003, Zhou, et al. 2009, Nagano, et al. 2002, Anbar, 2000]

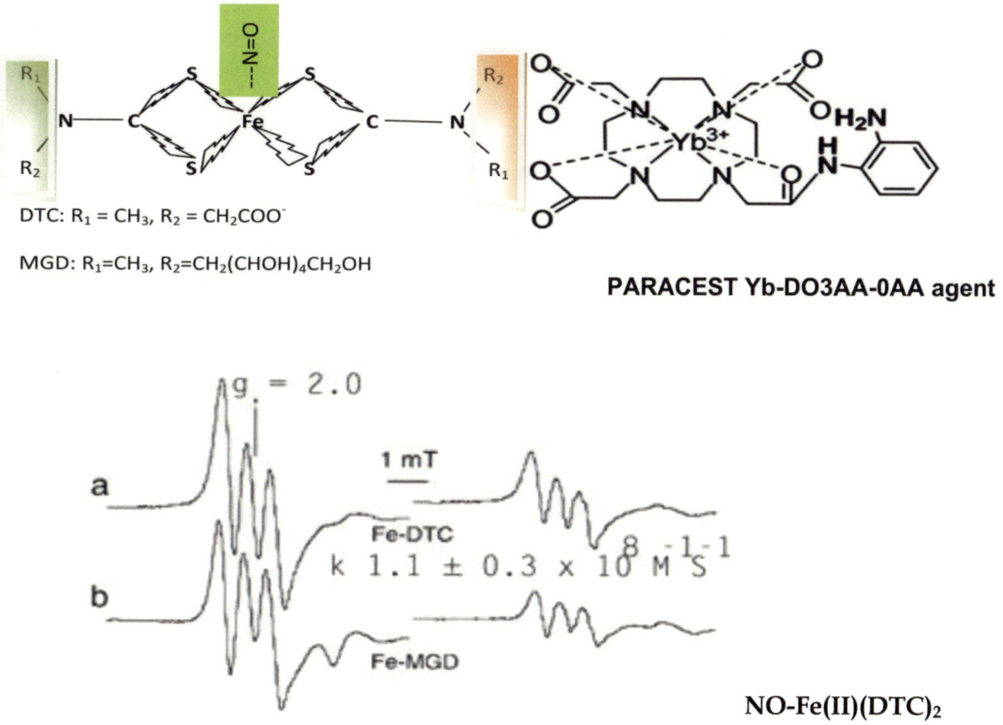

Fig. 7. NO sensitive biomarker imaging contrast agents. Example of NO-Fe(II)(DTC)$_2$ is cited to illustrate the signal change during changes in cerebral blood flow and possibility of functional MRI [Caramia, et al. 1998].

Two decades ago, NO activation-induced signal changes in functional MRI (fMRI) was observed [Di Salle, et al. 1997]. Hb–NO was found to produce marked stimulator dose and concentration-dependent NO sensitive fMRI signal intensity changes (increased after aqueous NO solution, nitrite, or dithionite and nitrite added to blood while it decreased on the addition of ascorbic acid) examined by T_1-, T_2-, and T_2*-weighted MRI in venous erythrocytes with possibility of physiological MRI evaluation associated with NO change [Di Salle, et al. 1997]. Transient formation of two paramagnetic species (met Hb and NO-Hb) enhanced the signal intensity, while ascorbate reduced the signal intensity in fMRI experiment on healthy volunteers during standard tasks. Further, infusion of NO precursor, L-arginine increased the cerebral blood volume (measured by MRI) in nonischemic spontaneously hypertensive rats [Foster, et al. 1998]. In other study, administration of an NO donor, isosorbide dinitrate (ISDN) increased both tumor blood flow and partial oxygen pressure in mice implanted with liver tumor in the thigh [Jordan, et al. 2000]. These findings indicate blood flow-independent change in MRI signal produced by different additives or stimulators. These observations open a new perspective on the monitoring of NO and the in vivo speculation of NO effects to test new drugs by using magnetic resonance techniques. Fuji et al. 1999 first time reported *in vivo* MRI of NO using Fe(MGD)$_2$ as NO–Fe(MGD)$_2$ spin-trapping technique in a septic shock rat model [Fujii, et al. 1999]. (Fig. 6A). The reduced signal after administration of an iNOS inhibitor confirmed the MRI visibility of NO–

$Fe(MGD)_2$ complex especially in inflammation which has high levels of NO. Now many MRI visible nitrosyl-iron complexes are known for MRI mapping of NO using a conventional 1.5 T MR scanner. The principle is that after exposure to NO in tissues $Fe(MGD)_2$ imaging agents form nitrosyl-iron complexes, which shorten the T_1 and T_2 relaxation time in a concentration-dependent manner [Hong, et al. 2009]. Mainly, four major techniques are reported in quest of developing new NO sensitive MRI contrast agents. First, unpaired electrons of the nitrosyl–iron complexes can enter into dynamic nuclear polarization with water protons, a technique called proton-electron-double-resonance imaging (PEDRI). Second, superparamagnetic iron oxide (SPIO) or ferumoxides dephase T_2-weighted MRI signal dependent on SPIO: NO donor ratios indicating inverse relation of contrast enhancement by SPIO with increasing levels of NO in tissue. The MRI signal change was due to reduction of ferric to ferrous iron with result of decrease in paramagnetic relaxation of water protons [Kojima, et al.2001]. Third, technical development emerged as irreversible paramagnetic chemical exchange saturation transfer (PARACEST) based MRI contrast agents in imaging [Liu, et al. 2007, Zhang, et al. 2003, Wu, et al. 2008]. A PARACEST MRI agent, Yb(III)-(1,4,7,10-tetraazacyclododecane-1,4,7-triacetic acid)-orthoaminoanilide (Yb-DO_3A-OAA), was developed for NO detection (Fig. 6B) [9]. Fourth, NO and O_2 combine by irreversible covalent bond that causes a quick disappearance of the PARACEST effect captured in the MR images. PARACEST MRI is fast data acquisition (in few minutes) vs SPIO based imaging is slow (in few hours). PARACEST effect has poor sensitivity (only detects millimolar concentrations of NO).

2.6 Approach of real-time NO synthase inhibition as blood oxygenation dependent MRI signal

Nitric oxide synthase inhibition (NOSi) in humans by blood oxygenation level-dependent (BOLD) MRI as reduced response to NOSi was reported using BOLD MRI to compare changes in R_2* to direct measures of renal medullary oxygen levels and blood flow using N-nitro-L-arginine methyl ester invasive probes (OxyLite/OxyFlo) to examine for the first time the effect of NOSi on intrarenal oxygenation in humans [Li, et al. 2009, Di Salle, et al. 1997]. Authors showed that NOSi decreased medullary pO_2 and blood flow in a dose-dependent manner, and BOLD MRI showed an increase in medullary R_2* consistent with the invasive pO_2 measurements.

Other alternate approach of positron emission tomography (PET) was proposed in recent study using nitric oxide imaging probe prepared from [^{18}F] 6-(2-fluoropropyl)-4-methyl-pyridin-2-amine by substitution at position-6 of substituted 2-amino-4-methylpyridine with favorable properties as a PET tracer to image iNOS activation/expression with PET of animal lungs [Zhou, et al. 2009].

3. Major developments in NO bioimaging probes

3.1 Magnetic resonance relaxation enhancement mechanism

In 1999, the first in vivo MRI of NO was reported using the spintrapping technique in a septic shock rat model [Fujii, et al. 1999]. NO was trapped by $Fe(MGD)_2$ in vivo, which can be visualized by MRI. The NO–$Fe(MGD)_2$ complex displayed significantly enhanced contrast in the vascular structure, such as the hepatic vein and inferior vena cava, which can be reduced after administration of an iNOS inhibitor.Initially, NO mapping was reported by whole body EPR imaging at poor resolution [Claudette, et al. 2005,Yoshimura, et al. 1996]. Recently, MRI strategies combined with in vivo spin-trapping map "NO distributions" within tissues and organs showed much higher spatial resolution using $(MGD)_2$-Fe(II)-NO as a NMR contrast agent in vivo [Fichtlscherer, et al. 2000]. The MR images also provided follow-up of NO generation kinetics at different sites within the organ at much higher spatial resolution than with EPR [Claudette, et al. 2005,Yoshimura, et al. 1996]. NO makes stable complex such as $(MGD)_2$-Fe(II)-NO. In vivo, hemoglobin is normally the natural NO spin-trap such that NO tends to bind with hemoglobin or oxidize the hemoglobin, followed by conversion to nitrosyl-hemoglobin or methemoglobin, both of these complexes are paramagnetic. Authors believe that during brain stimulation, NO is generated in some regions and it is quite possible that (paramagnetic) nitrosyl-hemoglobin and methemoglobin are formed in these active regions in brain. The MRI signal intensity enhancement in these regions results from changes in blood flow rich with these complexes. However, for blood flow–independent effects in MRI, the paramagnetic relaxation from spin-trapped NO might provide a new fMRI contribution in future. To our knowledge, there have been no reports successfully imaging or detecting free radicals such as NO in vivo by NMR or fMRI. Further, both short lifetime and rapid diffusion of NO in time preclude any effective relaxation enhancement mechanisms. NMR spin trapping serves to overcome these problems such as NO short lifetime and its fast diffusion in tissues to some extent. In other classic study on rats, spin relaxation constants of dinitrosyl-iron bound albumin or GSH and mononitrosyl-iron dithiocarbamate MGD complexes showed concentration dependence by MRI technique [Fujii, et al.1999].

3.2 Emission enhancement mechanism

Present state of art in NO detection was developed over last 2 decades using fluorescein cheletropic compounds visualized by emission enhancement and photoinduced electron transfer mechanisms. Fluorescence properties of fluorescein derivatives are controlled by the photoinduced electron transfer (PET) process from the benzoic acid moiety to the fluorophore, the xanthene ring. PET is a widely accepted mechanism for fluorescence quenching in which electron transfer from the PET donor to the excited fluorophore diminishes the fluorescence of the fluorophore. A brief survey of these compounds and their fluorescence mechanisms will catch the attention of readers with the newer developments of NO detection in tissues. EPR (also known as electron spin resonance or ESR) explores the magnetic moment associated with the unpaired electron(s) in free radicals and paramagnetic metal ions [Mader, 1998, Swartz, 2007]. It has been applied for the detection of free radicals and paramagnetic metal ions in various biological conditions [Berliner, et al. 2004]. When an unpaired electron in a paramagnetic molecule is subjected to an external magnetic field, the energy level of the electron splits into two quantum states with different energy levels. Typically, the magnetic field is scanned with a fixed microwave frequency of approximately 9.5 GHz (usually called the X-band). Specimens less than about 100 μL are measurable in an Xband spectrometer. However, larger biological samples (e.g., tissues, organs, and live animals) cannot be measured with a conventional X-band spectrometer due to the high dielectric loss of water at such frequencies and the small size of the EPR cavity resonator.

EPR spectroscopy at lower frequencies, the L-band (0.4–1.6 GHz) or S-band(1.6–4 GHz), can be used for in vivo EPR imaging. EPR imaging is generally considered to be the most effective technique for noninvasive observation of the spatial distribution of free radicals [Kleschyov, et al. 2007]. Direct EPR detection of endogenous free radicals in biological samples requires that the radical of interest is present at a concentration higher than the practical detection limit (0.1–0.01 μM) and has a relatively long lifetime. Although such requirement can be met at cryogenic temperature, in vivo detection is almost impossible since most radicals do not have a sufficiently long lifetime. Therefore, spin-trapping techniques have been developed in which the spin traps can react with unstable free radicals to form a relatively stable adduct [Berliner, et al. 2004, Davie, et al. 2004]. The formation of the long lived adduct results in accumulation of steady-state free radicals which can be detected readily by EPR spectroscopy.

In principle, the fluorescence properties of fluorescein derivatives are controlled by the photoinduced electron transfer (PET) process from the benzoic acid moiety to the fluorophore, the xanthene ring. PET is a widely accepted alternate mechanism for fluorescence quenching in which electron transfer from the PET donor to the excited fluorophore diminishes the fluorescence of the fluorophore. The fluorescein structure has two parts confirmed by X-ray analysis, i.e., the benzoic acid moiety as the PET donor and the xanthene ring shown as the fluorophore in previous study [Liu, et al. 2007]. The study showed that only small alterations in absorbance were observed among fluorescein and its derivatives. The dihedral angle (ϕ) between the benzoic acid moiety and the xanthene ring was almost 90°. It also suggested that there is little ground-state interaction measured by HOMO energy levels between these two parts.

The relationship of HOMO energy levels and (ϕ) angle of fluorescent compounds predict the fluorescence mechanism of benzoic acid moiety in the compounds. If the HOMO energy level of the benzoic acid moiety is high enough for electron transfer to the excited xanthene ring, the ϕ value will be small. In other words, fluorescein derivatives with a high ϕ value must have benzoic moieties with low HOMO energy levels. The HOMO energy levels of 3-aminobenzoic acid, 3-benzamidobenzoic acid, 3,4-diaminobenzenecarboxylic acid (DAB-COOH), and benzotriazole-5-carboxylic acid (BT-COOH), which are the benzoic moieties of aminofluorescein, benzamidofluorescein, DAF-2, and DAF-2T, respectively were estimated by semi-empirical (PM3) calculations. DAB-COOH and aminobenzoic acid, which are the benzoic acid moieties of weakly fluorescent fluorescein derivatives, have higher HOMO levels than BT-COOH and amidobenzoic acid. These results were consistent with the PET mechanism.

Further, to confirm this mechanism 9-[2-(3-carboxy)-naphthyl]-6-hydroxy-3H-xanthen-3-one (NX) and 9-[2-(3-carboxy)anthryl]-6-hydroxy-3H-xanthen-3-one (AX) were synthesized [Nagano, et al. 2002]. It is well-known that expanding the size of aromatic conjugates makes their HOMO levels higher, similar with by introducing electron-donating substituents. Thus, naphthoic acid and anthracene-2-carboxylic acid would have higher HOMO levels than benzoic acid, and therefore, the fluorescence properties of NX and AX would differ from those of fluorescein.. The absorbance maximum (Ex_{max}) and emission maximum (Em_{max}) of NX and AX were closer and not much altered among these fluorescein derivatives. However, the ϕ values were greatly altered; NX is highly fluorescent, whereas

AX is almost nonfluorescent [Nagano, et al. 2002]. Thus, a small change in the size of conjugated aromatics, namely, from naphthalene to anthracene, causes a great alteration of the fluorescence properties. Authors compared the HOMO levels of the benzoic acid moieties with xanthine ring and naphthoic acid. Napthoic acid moiety is present in highly fluorescent fluorescein and NX. The HOMO levels of benzoic acid moiety were lower than that of the xanthene ring, while the HOMO level of anthracenecarboxylic acid moiety in fluorescent AX, was higher than that of the xanthene. These results were consistent with the idea that a PET process controls the fluorescence properties of fluorescein derivatives and that these properties can be predicted from the HOMO level of the benzoic acid moiety, with a threshold at around -8.9 eV. This in turn provides a basis for developing novel fluorescence probes with fluorescein-derived structure. Still much remains to justify the fluorescent probes in NO bioimaging.

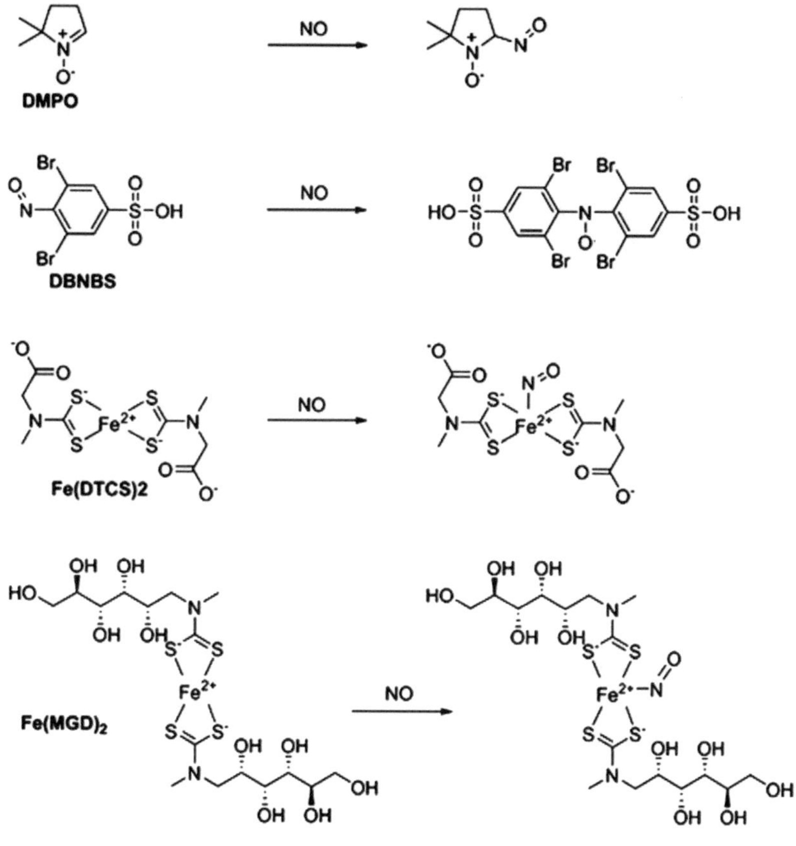

Fig. 8. Representative spin-trapping agents for EPR imaging of NO.

$Fe(MGD)_2$ or $Fe(DETC)_2$ have also been tested for EPR imaging to map the spatial distribution of NO generation in the ischemic myocardium [Kuppusamy, et al. 1996]. Subsequently, $Fe(MGD)_2$ was used in both EPR and MR imaging in rats with sepsis shock [Fujii, et al.1999]. It was reported that $Fe(MGD)_2$ can reduce nitrite to NO at physiological pH, which may cause some detection Inaccuracy as shown in Figure 8, although the rate constant is quite slow (1–5 M^{-2} s^{-1}) [Tsuchiya, et al. 2000]. $Fe(DETC)_2$ has been used for imaging vascular NO production in rabbits [Kleschyov, et al. 2000], animal brain injury [Ziaja, et al. 2007], vascular disease [Khoo, et al. 2004], and even in plants [Xu, et al. 2004, Vanin, et al. 2007]. Heme proteins, such as myoglobin, cytochrome c, and catalase, have been used as NO-trapping agents in various biological specimens[Kleschyov, et al. 2000]. Although the EPR signals of heme–NO species can be observed in tissues exposed to elevated levels of NO, it is almost impossible to discriminate between the individual nitrosylated heme proteins due to overlapping of the signal. Generally, the dominant heme-NO EPR signal belongs to the major heme protein expressed in the tissue (e.g., myoglobin in the heart [Konorev, et al. 1996]). The most frequently used heme protein in EPR research is hemoglobin (Hb) [Kosaka, et al. 1994], which can interact with NO to generate several paramagnetic Hb–NO derivatives detectable by EPR spectroscopy [Picknova, et al. 2005]. Different preparations of deoxy-Hb have been used for NO trapping in isolated cells and organelles [Kozlov, et al. 1996].

3.3 NO imaging and fluorescent dyes

Several fluorescence indicators for direct detection of NO have been generated, which might be useful for in vitro or in vivo NO-imaging. These substances interact with NO to form a fluorescent complex. Several fluorescent dyes have been developed, which can be used for NO bioimaging. However, several of these promising fluorescent dyes have been found not to be very specific such as 2′,7′-dichlorofluorescein (DCF) or to be cytotoxic such as 2,3-diaminonaphthalene (DAN). Of specific mention, iron (II) N-(dithiocarboxyl) sarcosine($Fe(DTCS)_2$; Fig. 5) trapping agents within 1 h after injection of the spin trap generated stable signal which indicated that the formation of the steady-state free radical had reached equilibrium. The same trapping agent was also tested for NO imaging in lipopolysaccharide-treated mice [Quaresima, et al. 1996]. Subsequently, EPR measurements and EPR-CT (CT denotes computed tomography) imaging with $Fe(DTCS)_2$ were carried out in live mice [Yokoyama, et al. 1997]. The use of $Fe(DTCS)_2$ resulted in a clear EPR-CT image showing high intensity areas in the ventral regions, while other NO-bound iron complexes such as Nmethyl-D-glucamine dithiocarbamate (Fe $(MGD)_2$; Fig. 5) or N,Ndiethyl-dithiocarbamate ($Fe(DETC)_2$) gave much lower signal-to-noise ratios. Several compounds such as 5,5-dimethyl-1-pyrroline N-oxide (DMPO; Fig. 5), α-phenyl-N-tert-butylnitrone(PBN), or 3,5-dibromo-4-nitrosobenzene sulfonate (DBNBS), NO cheletopic traps (NOCTs), aci anions of nitroalkanes, iron (II)–dithiocarbamate(DTC) complex and(DAFs) are fluorescein derivatives. DAFs themselves scarcely fluoresce while DAF-Ts are strong fluorescent compounds. However, the mechanisms accounting for the diminution of fluorescence in DAFs and fluorescence enhancement in DAF-Ts remain unclear.Other fluorescent $Fe(DTCS)_2$ and Cobalt based dyes have been tested in some cell culture systems, but these substances have not been tested in in vivo or in vitro acute brain slice preparations or tissues [von Bohlen, et al. 2003, Zhou, et al. 2009]. Until now, several other fluorescent dyes have been promising candidates for the evaluation of the temporal and spatial aspects of NO generation and distribution even under microscopic observation, including diaminofluoresceins (DAFC), diaminorhodamines (DAR), diaminoanthraquinone (DAQ), fluorescent resonance energy transfer (FRET) and the "fluorescent nitric oxide cheletropic traps" as shown in Table 3 [von Bohlen, et al. 2003, Zhou, et al. 2009].

In following section, we will summarize the progress to date on multimodality imaging of NO and NOS expression and address future research directions and obstacles for NO/NOS imaging using biosensors as outlined in Figure 9. For interested readers, basic methods of NO detection and imaging are given in Appendix in then end of this chapter.

- **Fluorescence imaging of NO**:
  2,7-dichlorofluorescin was introduced for NO detection. When it is oxidized by NO, a fluorescent compound (dichlorofluorescein) is formed. Besides a relatively poor detection sensitivity (~ 10 μM of NO), 2,7-dichlorofluorescein also has very low specificity for NO. Later several iron (II) complexes, fluorescently labeled cytochrome c (a hemoprotein that binds NO selectively), cobalt complexes and fluorescent NO cheletropic traps (FNOCTs) were introduced. The iron(II) complex of quinoline pendant cyclam has fluorescence emission at 460 nm, which can be quenched effectively by NO from NO releasing agents. 2,2,6,6-tetramethylpiperidine-N-oxyl (TEMPO) labeled with acridine and Fe(II)-(N-(dithiocarboxy) sarcosine)$_2$ complex (Fe(DTCS)$_2$) (Fig.5), was developed. The addition of an NO-releasing reagent causes a decrease in the fluorescence signal due to the irreversible binding of NO to the Fe(II).
- DAFCs are the most successful and extensively investigated agents for NO imaging. In vivo imaging of NO in whole living vertebrate with the DAF compounds was reported, where diaminofluorophore 4-amino-5-methylamino-2′-7′-difluorofluorescein diacetate (DAF-FM-DA) was used to detect NO production sites in the zebrafish Danio rerio. The specificity of the fluorescent signal was confirmed by a decrease in animals exposed to a NO scavenger or a NOS inhibitor, as well as an increase in the presence of a NO donor. Local changes in NO production in response to stressful conditions could also be imaged by this agent, suggesting that DAF-FM-DA can be used to monitor changes in NO production which may have future applications in drug screening and molecular pharmacology. Best example of NO imaging in vivo with the DAF compound in zebrafish is quite transparent. The major limitation of the DAF-2 compound is that it is not specific to NO because it can also react with dehydroascorbic acid (DHA) and ascorbic acid (AA) to generate new compounds that have fluorescence emission profiles similar to that of DAF-2T.
- Similar to DAF and DAR, rhodamine fluorophore was developed and it was termed "the DAR". To image NO in living cells, DARs should be membrane permeable. Therefore, an ethyl group was introduced into the carboxyl group of DAR-1 (4,5-diamino-N,N,N′,N′-tetra ethyl rhodamine), anticipating that the ethyl ester would be hydrolyzed by cytosolic esterases after permeation through the cell membrane [Kojima et al. 2001].
- New class of DAQ compounds of DAFC family called diaminocyanines (DACs) were prepared with a tricarbocyanine as the NIR fluorophore and o-phenylenediamine as the NO sensitive fluorescence modulator. It was reported that these compounds can react faster with NO than the DAFs and the reaction efficiency was at least 50 times higher than that of DAF-2. Although it was demonstrated that NO can be imaged in isolated rat kidneys, in vivo imaging of NO using these compounds has not been reported.
- A new class of difluoroboradiaza-s-indacene-based agents have been widely used in the detection of NO in cells or tissues [Huang, et al. 2002a]. One of these compounds, 1,3,5,7-tetramethyl-8-(4′-aminophenyl-N-(2″-amino)-phenzyl)-difluoroboradiaza-s-indacene (TMAPABODIPY), was shown to have high photostability, high sensitivity (detection limit is 5 nMof NO), and no pH dependency over a wide pH range (see

Figure 9). Subsequently, a similar compound, 1,3,5,7-tetramethyl-2,6-dicarbethoxy-8-(3',4'diaminophenyldifluoroboradiaza-s-indacene (TMDCDABODIPY), was applied for real-time imaging of NO in cells [Huang, et al. 2002b].

	MGD	DAF	DAR	DAQ
Emission (color)	~520 nm (green)	~515 nm(green)	>580 nm(red)	460 nm(blue)
Excitation	~460 nm	~495 nm	~550 nm	~520 nm
Reacts with ROS	Generates	No	--	No
Tested in neurotoxic tissues	No	Yes	No	Yes
Neurotoxic	--	No	--	No

Table 3. Properties of fluorescent NO-indicators

- In copper-based fluorescent compounds copper (II) is coordinated with certain fluorophores, the base-catalyzed reaction with NO leads to the reduction of the metal center and the intramolecular nitrosylation of a secondary amine. Based on this mechanism, a series of fluophore coordinated copper (II) complexes were designed for NO imaging [Lim, 2007]. Reduction of the Cu (II) core by NO to Cu(I) with nitrosation of the fluorescent ligand is accompanied by bright fluorescence emission. Any nitrogen/oxygen reactive species or ascorbic acid do not cause any significant fluorescence alteration in Cu(II)-based agents. Cu(II) complexes are quite successful in NO imaging in cell culture but they have not been tested in vivo likely due to the suboptimal emission wavelength, the potential toxicity, and the poor in vivo stability of the Cu(II) complex. Recently, several benzimidazole derivatives have been tested in vivo for NO imaging [Quyang, et al. 2008]. The fluorescent 5'-chloro-2-(2'-hydroxyphenyl)-1H-naphtho[2,3-d]imidazol and 4-methoxy-2-(1H-naphtho [2,3-d]imidazol-2-yl)phenol can coordinate with Cu(II) to form a nonfluorescent coordination compound, which can detect NO production in inflammation with a turn-on fluorescence. NO production was imaged successfully in both activated murine macrophages and an acute severe hepatic injury (ASHI) mouse model. Although these agents represent the first successful attempt of in vivo fluorescence imaging of NO, the actual imaging was performed ex vivo on frozen tissue slices due to the suboptimal fluorescence emission wavelength as shown in Figure 10. Much further optimization of these Cu(II)-based agents will be needed in the future, in particular fluorescence emission in the NIR range and reduction of the toxicity.
- Imaging NO/NOS with FRET: FRET agents react with NO covalently a irreversible sensors for NO. Further, some of the organic dyes can accumulate in subcellular membranes and emit fluorescence signals in an NO-independent manner which can cause significant background signal and interfere with the detection of NO. To overcome these limitations, a NO-selective sensor with the heme domain of soluble guanylate cyclase (sGC) was developed [Barker, et al. 1999]. The heme domain of the sGC was labeled with a fluorescent dye and domain changes in the fluorescence intensity of the dye based on the sGC heme domain's characteristic binding of NO. It was found that these sensors have fast, linear, and reversible responses to NO unaffected by other radical species. However, the detection limit is only about 10 μM of NO which makes less useful.

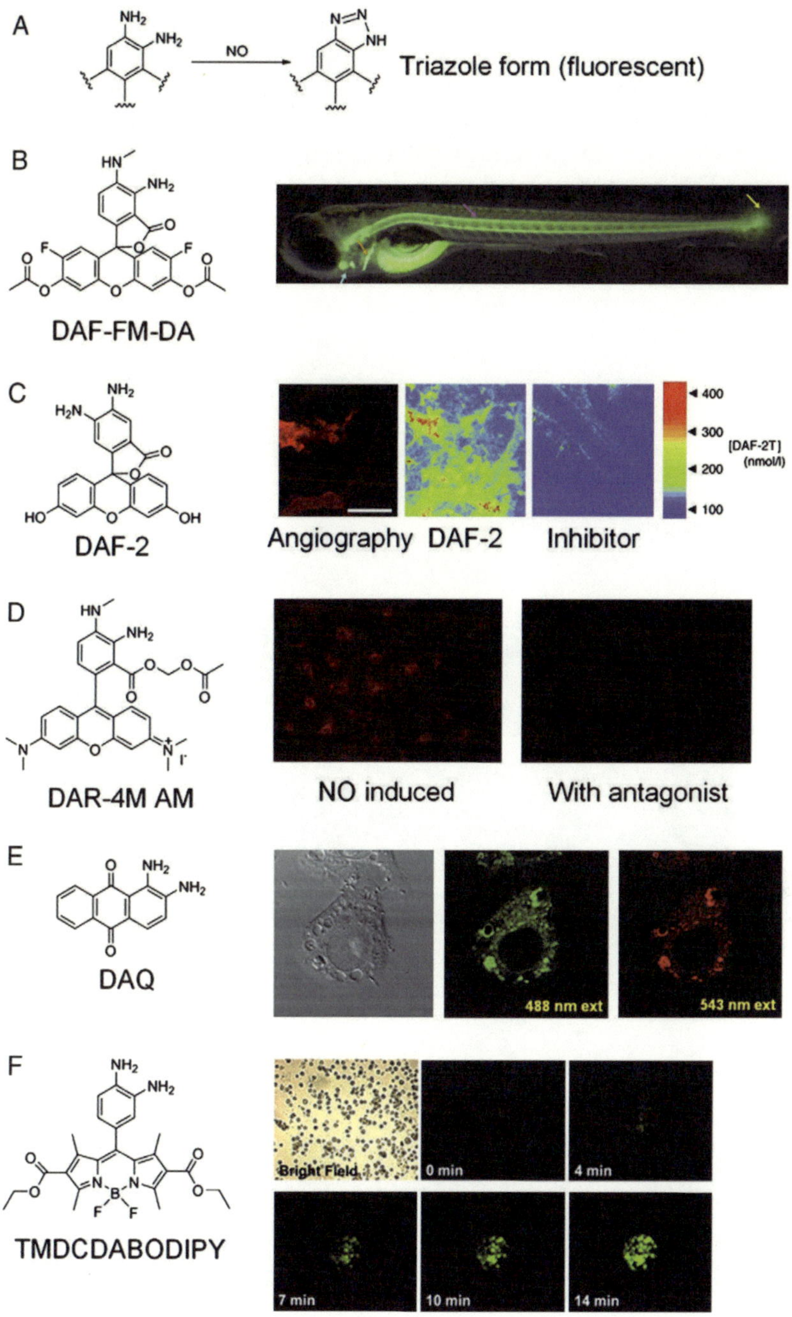

Fig. 9. Imaging NO with diamino-aromatic fluorescent compounds. (A) The general reaction that leads to NO detection by fluorescence with this class of agents.
(B) Fluorescence image of live zebrafish larvae loaded with DAF-FM-DA. Blue arrow, heart; orange arrow, the cleithrum; purple arrow, the notochord; yellow arrow, the caudal fin.
(C) Distribution of NO in B16F10 tumors grown in the cranial window in mouse. Left: microangiography using tetramethylrhodamine-dextran; middle: representative

microfluorography captured 1 h after the loading of DAF-2 in tumors; right: The animals were treated with NOS inhibitor. Color bar shows calibration of the fluorescence intensity with known concentrations of DAF- 2T. (D) Left: fluorescent image of endothelial cells with DAR-4M AM at 10 min after the stimulation. Right: an antagonist of the stimuli strongly abolished the fluorescence intensity. (E) Confocal laser fluorescence microscopy image of cells loaded with DAQ and stimulated to produce NO. Left: bright-field; middle: 488 nm excitation; right: 543 nm excitation. (F) Bright-field and fluorescence images of activated PC12 cells loaded with TMDCDABODIPY in the presence of arginine. Adapted from [Huang, et al. 2007a, 2007b].

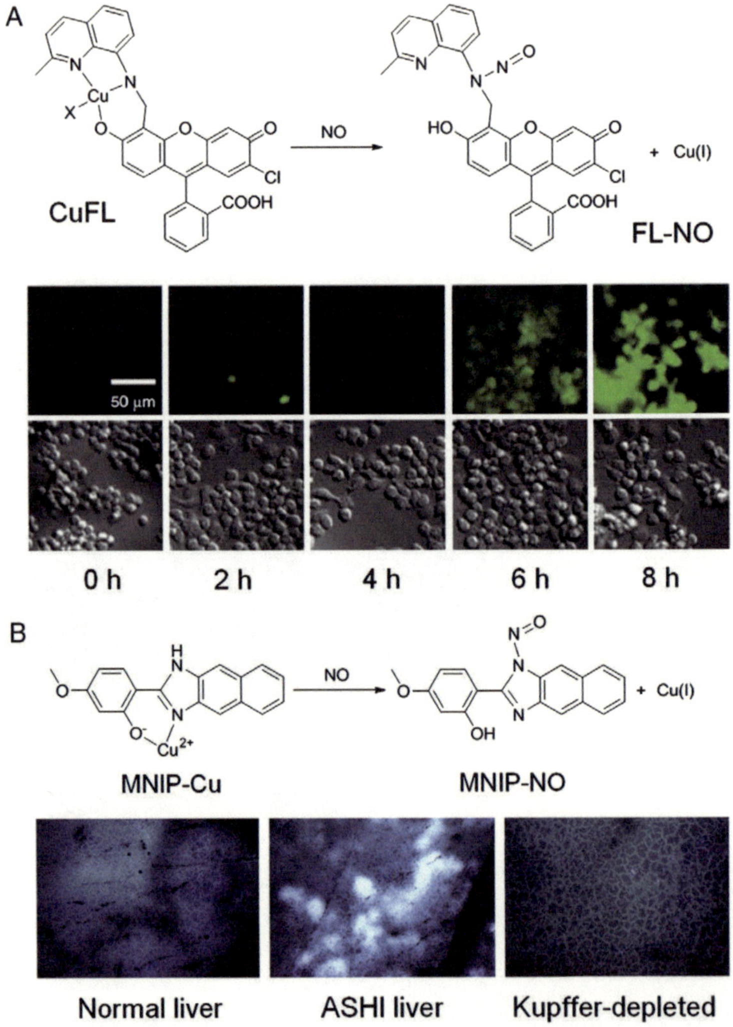

Fig. 10. Copper-based fluorescence imaging of NO. (A) CuFL detection of NO produced by iNOS in macrophage cells after different time of incubation. Top: fluorescence images; Bottom, bright-field images. (B) Fluorescence images of excised liver slices harvested from live animals after intravenous injection MNIP-Cu. Left: normal mouse liver; center: ASHI liver; right: Kupffer cell-depleted ASHI liver. Adapted from reference [Quyang, et al. 2008].

Subsequently, a genetically encoded fluorescent indicator for NO was developed based on FRET, which can reversibly detects NO with high sensitivity (detection limit of 0.1 nM) and visualizes the nanomolar dynamics of NO in single living cells on NO induction [Jares-Erijman, et al. 2003]. However, this approach requires genetic encoding and it generates the biologically active molecule cyclic guanosine monophosphate (cGMP), which can induce other cellular responses. The cGMP molecules can bind to both the NO-associated and the No free probe, resulting in an increase in fluorescence. Therefore, the signal generated is a reflection of the intracellular concentration of cGMP rather than that of NO per se. Recently, the use of confocal based spectral imaging with NO-sensitive FRET reporters enabled NO imaging in the vasculature of intact, isolated perfused mouse lung[St Croix, et al. 2008]. Using calibration spectra, it is possible to unambiguously separate the cross-talk between overlapping donor and acceptor emissions. Overall, FRET is a very powerful technique in cell-based experiments but with only very limited potential applications in animal studies as shown in Figure 11.

3.4 NO bioimaging by bioluminescence and chemiluminescence imaging

The major application of chemiluminescence in NO imaging/ detection lies in gas phase or liquid phase measurements[Taha, et al. 2003]. The major chemiluminescence activators for NO imaging include ozone, luminol, and lucigenin. Gas phase chemiluminescence detection of NO is generally based on the reaction of NO with ozone to produce excited-state nitrogen dioxide. When it returns to the normal state, light is emitted at about 600 nm wavelength. Such gas phase detection has been widely used for measuring NO in exhaled breath to detect inflammatory diseases of the lung. Gas phase chemiluminescence detection has very high specificity, as well as extreme sensitivity, which can reach parts per billion (ppb). However, drawbacks include inconvenient instrumentation, high cost, and timeconsuming detection procedure. To overcome these problems, an optimized chemiluminescence detection system for NO was developed. For liquid phase chemiluminescence imaging of NO, currently the luminol/H_2O_2 detection system is the most useful [Robinson, et al. 1999]. Alkaline luminol can react with NO to generate luminescence. The presence of H_2O_2 as an activator can enhance the sensitivity by about 20-fold. Two approaches have been used for luminol/H_2O_2-based imaging of NO. In the first approach, the sample is directly added to the luminol/H_2O_2 mixture. This strategy can detect very low levels of NO, which is quite useful for visualizing the production of NO from cells or organs, but it lacks NO specificity since luminol can react with many free radical species. In the second strategy, a selective dialysis membrane is placed between the sample and the luminol/H_2O_2 mixture. This procedure can offer better specificity yet it suffers from poor sensitivity, slow response time, and undesirable NO detection limit (μM level). Besides ozone and luminol, lucigenin is can also be used for NO detection. Lucigenin has been widely used as a chemiluminescent substrate to monitor vascular superoxide formation. In particular, lucigenin at a concentration of 5 μM could be used as an indirect probe to estimate basal vascular NO release. Generally, chemiluminescence is a very sensitive method for NO imaging/detection. However, the disadvantages such as low specificity and time-consuming procedure make chemiluminescence not very useful as an independent technique for NO imaging/detection.

Several other imaging techniques have been investigated for NO detection. As an important component of molecular imaging, ultrasound is not inherently suitable for direct NO imaging, although it can be used for evaluating the endothelial function or vasodilation caused by NO [Heiss, et al. 2008]. It is worth noting that ultrasound at different frequencies can stimulate NO production in different contexts such as in endothelial cells, osteoblasts, and vascular dilation. A few other techniques such as time-resolved photoion and photoelectron imaging, Hb spectral alteration imaging, laser induced fluorescence detection (LIF), and calcium amplification imaging have also been studied for NO detection. In particular, LIF imaging was able to detect the NO release in a single cell, which makes it a very powerful tool for studying the kinetics of NO release by neuronal cells during neurotransmission. PET imaging of anesthetized dogs with inhaled radiolabeled ^{13}NO has been performed to study the in vivo kinetics of NO [McCarthy, et al. 1996]. In future, molecular imaging may provide more accurate information on NO in intact living systems. EPR-CT [Yokoyama, et al. 1997], EPR-MRI [Berliner, et al. 2004], and EPR-chemiluminescence imaging [Nagano, 2002] have all been attempted and much further research/optimization will be needed before any potential clinical investigation can be in place.

3.5 Multimodal bioimaging of NOS gene expression

Initially NOS was considered a sole enzyme to produce NO. Nitric oxide synthase (NOS) is also under extensive investigation [Crowell et al. 2003]. NO is biosynthesized as three distinct mammalian NOS isoforms shown in section 1. Neuronal NOS (nNOS or NOS I) and endothelial NOS (eNOS or NOS III) are constitutively expressed in neuronal and endothelial cells, respectively and both are referred to as cNOS. Inducible NOS (iNOS or NOS II) is expressed in cells involved in inflammation, such as macrophages and microglias when stimulated by cytokines and/or endotoxins. Although cNOS and iNOS do not have a significant difference in activity (both function through NO), cNOS expression can be stimulated by Ca^{2+} whereas iNOS activity is Ca^{2+}-independent. Generally, the NO levels produced by cNOS in stimulated endothelial and neuronal cells are much lower (in nanomolar [nM] range) than those generated by iNOS in macrophages (micromolar [μM] range). Aside from NOS synthesis, recent studies have pointed out that nitrite reduction can also serve as a possible source of biologically relevant NO [Gladwin, et al. 2005]. Excessive production of NO is believed to be responsible for various pathophysiologies, while very low levels of NO are needed to maintain normal physiological conditions. Since the production of NO is closely associated with the expression of NOS, blocking NOS expression is a more convenient approach to inhibit the function of NO than intervention through NO itself.

Molecular imaging refers to the characterization of NO/NOS and measurement of biological processes at the molecular level [Massoud, et al. 2003]. It takes advantage of traditional diagnostic imaging techniques and introduces molecular probes to measure the expression of indicative molecular markers at different stages of diseases. Molecular imaging modalities include molecular magnetic resonance imaging (MRI), magnetic resonance spectroscopy (MRS), optical imaging, targeted ultrasound, single photon emission computed tomography (SPECT), and positron emission tomography (PET). Many hybrid systems that combine two or more of these modalities are already commercially available and certain others are under active development. Molecular imaging can give whole body readout in an intact system at the cost of less workload, inexpansive drug development trial, and provide more statistically relevant results since longitudinal studies can be performed in the same subjects [Cai, et al. 2008a, 2008b]. In the clinical setting, molecular imaging can significantly aid in early lesion detection, patient stratification, and treatment monitoring, which can allow for much earlier diagnosis, earlier treatment, better prognosis, and individualized patient management. To date, the multimodal imaging modalities for visualization of NO include optical imaging (fluorescence and chemiluminescence), electron paramagnetic resonance (EPR) imaging, and MRI. Meanwhile, NOS has been mainly investigated using optical and PET imaging. In following section, we describe recently developed multimodal techniques in their infancy.

3.5.1 Optical imaging of NOS expression

Imaging enzyme expression and function is quite challenging. Compared to the vast number of literature reports on NO imaging, much less effort has been directed toward NOS imaging. Briefly, the major modalities for NOS imaging include optical imaging, PET, and some other techniques such as electron microscopy. Historically, the study of NOS physiology was constrained by the lack of suitable probes to detect NOS in living cells or animals. In one pioneering study, a fluorescent iNOS inhibitor, pyrimidine imidazole FITC (PIF), was developed and investigated for microscopy of iNOS in living cells [Panda, et al.2005]. It is an isoform-selective molecule with high affinity to iNOS and the binding is essentially irreversible, which allowed for iNOS imaging in many cell types as well as freshly obtained human lung epithelium. This study indicated that fluorescent probes (e.g., PIF) can be valuable for studying iNOS cell biology and understanding the pathophysiology of diseases that involve dysfunctional iNOS expression. In another report, immunofluorescence staining was performed to evaluate nNOS expression on left inferior alveolar nerve (IAN) injury in ferrets [Davies, et al. 2004]. A possible translocation of nNOS protein from the cell body to the site of nerve injury was proposed. Lastly, some ruthenium(II) and rhenium(I)-diimine wires were reported to be capable of binding to iNOS in vitro to generate NIR fluorescence (710 nm emission), which may potentially be useful as a fluorescent imaging probe for iNOS [Dunn, et al. 2005]. The inflammatory modulating effects of iNOS on laser therapy were studied using bioluminescence imaging (BLI) [Moriyama, et al. 2005]. In this study, the efficiency of different wavelengths laser therapy at different wavelengths in modulating iNOS expression was determined using transgenic animals with a luciferase gene driven by the iNOS gene. On induction of inflammation followed by laser treatment, imaging of iNOS expression was performed at various times by measuring the bioluminescence signal. When compared with the nonirradiated animals, a significant increase in the bioluminescence signal was observed after laser irradiation which demonstrated that iNOS expression can be detected by noninvasive BLI. Comparing the two strategies for NOS imaging with optical techniques, many fluorescently labeled NOS inhibitors have charges which makes them not cell membrane permeable. NOS-controlled luciferase expression appears to be a preferable choice, yet much further optimization will be needed in future studies. Generally speaking, the limited penetration of light in tissue renders optical imaging only suitable for a limited number of scenarios (e.g., superficial regions-of-interest and endoscopy-accessible tissues). Another drawback of optical imaging is that it cannot give accurate quantitative analysis of the imaging result. Combination of quantum dot optical and radionuclide-based imaging techniques (e.g., PET) may significantly help in this aspect [Cai, et al.2007a, 2007b].

3.5.2 PET imaging of NOS expression

Small animal PET scanners have very good spatial resolution (1 mm) and they are also becoming increasingly widely available. PET does not have a tissue penetration issue, it is highly quantitative, and it can be used for both diagnosis and treatment monitoring. To date, PET imaging of NOS expression has been focused on the clarification of nNOS function in neurological diseases [Pomper, et al. 2000, Zhang, et al. 1997]. PET imaging of nNOS was initially investigated with a ^{11}C-labeled nNOS inhibitor, S-methyl-L-thiocitrulline (MTICU) (FIG 11, 12) [Zhang, et al. 1997]. Biodistribution of ^{11}C-MTICU gave a brain-to-blood ratio of 1:2 at 30 min postinjection. Uptake in the cerebellum (high nNOS expression) was 20% higher than that of the cortex or the brain stem (low nNOS expression). Blockage using 1 mg/kg of MTICU was able to reduce the uptake in the cerebellum and the cortex, but not in the brain stem. Study demonstrated the potential of ^{11}CMTICU as a potentially useful tracer in determining nNOS levels in vivo. In another report, two nNOS-selective PET tracers, AR-R 17443 [N-(4-(2-(phenylmethyl) (methyl)-amino)ethyl)phenyl)-2-thiophenecarboximidamide)] and AR-R 18512 [(N(2-methyl-1,2,3,4-tetrahydroisoquinoline-7-yl)-2 thiophenecarboximidamide)] were synthesized and evaluated in rodents and primates [Pomper et al. 2000]. Another strategy has been employed for iNOS imaging using antisense oligonucleotide with hybridization properties for iNOS mRNA using RT-PCR. The oligonucleotide was then labeled with ^{18}F (FIG 12) [de Varies, et al. 2004]. Cellular uptake or efflux showed no selectivity for iNOS expressing cells. Thus PET imaging of NOS expression has not been very successful which is likely due to several reasons. First, the expression of NOS is in the cytoplasm. Although the radiolabeled NOS inhibitors usually have high affinity for NOS, they may not permeate through cell membrane efficiently which dramatically affects the tracer uptake. Second, the specificity of these NOS inhibitors may not be high enough for imaging applications; many of them can undergo nonspecific adsorption to other proteins such as albumin. Third, the stability of those tracers in vivo is a major concern. Uptake in the tumor or other targeted organs may not truly reflect the distribution of the intact tracer. Lastly, these NOS inhibitors are small molecules. Although replacing the $^{12}CH_3$ with $^{11}CH_3$ will not change the chemical structure, in many cases this cannot be achieved. Recently, a novel NOS inhibitor design strategy, termed the "anchored plasticity approach," may give rise to novel NOS inhibitors for PET imaging applications in the future [Garcin et al. 2008].

3.5.3 Imaging NOS expression with other techniques

Besides optical and PET imaging, other techniques have been explored for NOS-related research, mainly electron microscopy and immunohistochemistry [Heinrich, et al. 2005, Lajoix, et al. 2006]. Electron microscopy was applied to visualize the subcellular localization of NOS in different cell organelles using gold particles conjugated with anti-NOS antibodies, which gave exquisite distribution information of NOS in cells. Immunohistochemistry also uses anti-NOS antibodies, easier to perform than electron microscopy yet the resolution is not as high. Both approaches are invasive and not suitable for in vivo imaging; however, the experimental findings are typically quite reliable and thus they can serve as ex vivo validation for in vivo imaging studies.

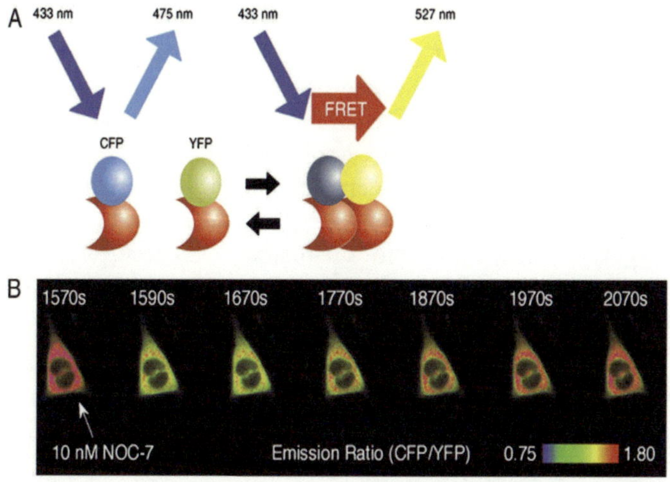

Fig. 11. (on left) Imaging NO with FRET. (A) Basic principles of FRET between CFP and YFP. CFP: cyan fluorescent protein; YFP: yellow fluorescent protein. (B) Pseudocolor images of the CFP/YFP emission ratio of the cell before (1570 s) and after addition of NOC-7, an NO-donating reagent that releases 2 mol of NO per mole. Adapted from reference [Hong et al. 2009]

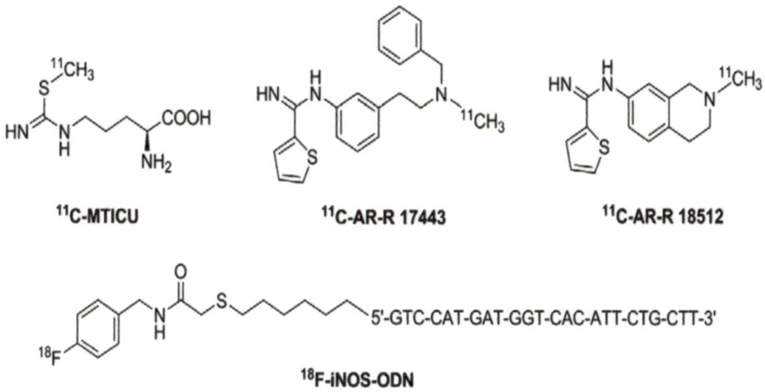

Fig. 12. (on right) Several NOS inhibiters and an antisense oligonucleotide for iNOS mRNA have been radiolabeled for PET imaging of NOS expression.Adapted from reference [Hong, et al. 2009]

4. Biological applications of NO imaging

Still development and applications of fluorescent agents in cell imaging and NO sensitive MRI contrast agents for in vivo physiological fMRI are in infancy. MRI imaging of NO based on nitrite concentration can detect physiological changes such as cerebral blood volume change in nonischemic spontaneously hypertensive rats [Caramia et al. 1998], increase in tumor blood flow and partial oxygen pressure after NO donor administration [Jordan et al. 2000], NO production alteration in humans after a 30-min exposure to a 1.5 T magnetic field [Sirmatel, et al. 2007]. Interestingly, high magnetic strength can cause NO concentration fluctuations during a MRI scanning of NO-related diseases.

The major advantages of MRI include high resolution, good soft tissue contrast, wide availability, sufficiently large field-of-view, and the ability to provide both functional and anatomical information. Nitric oxide bioimaging has emerged as technique to visualize the intracellular metabolism based on redox reactions, mitochondrial oxidation and nitric oxide synthase enzyme NOSi [Anbar, 2000, Zhao, et al. 2009]. Recently, nitric oxide was invented as active mechanism to play role in vascular physiology in lungs, kidney, heart, brain, sex organs. Endothelial cell, smooth muscle cells, oxygenation and perfusion, hypoxia, ischemia are the possible targets of nitric oxide biotransformation. These physiological changes are likely to be visualized by nitric oxide bioimaging in real time manner and reviewed in the following sections.

4.1 Cardiovascular and endothelial cell injury

DAF-FM is useful tool to visualize the temporal and spatial distribution of intracellular NO. Endogenous ATP plays a central role in HTS-induced NOS in endothelial cell injury. Endothelial cNOS, a calmodulin-dependent enzyme is critical for vascular homeostasis and makes detectable basal level of NO production at low extracellular Ca^{2+}. cNOS expression can be stimulated by Ca^{2+} while iNOS activity is Ca^{2+}-independent. Actin microfilaments in primary artery endothelial cells (PAEC) regulates L-arginine transport and this regulation can affect NO production by PAEC, DAR-4M may be useful for bioimaging of samples that have strong autofluorescence [Lai, et al. 1994, Kojima, et al. 2001]

4.2 Apoptosis in cells

Apoptosis plays a key role in many pathological circumstances, such as neurodegenerative diseases. In these processes, the involvement of nitric oxide (NO) has been well established, and the ability of NO to exert cellular damage due to its reactive oxidative properties is perhaps the primary neurotoxic mechanism. Emerging evidences indicate that apoptosis contributes to neuronal death to explain neurodegeneration [Fujii et al. 1997]. The major biochemical change in activation of the cystein protease (caspase-3) was reported to execute apoptosis in the central nervous system disorders such as stroke, spinal cord trauma, head injury and Alzheimer's disease [Green,1998, Nagata,1997]. Although numerous techniques have been described to evaluate apoptosis, these approaches involve invasive techniques and cannot provide detailed information about apoptosis *in vivo* [Anbar, 2000].

4.3 Renal and cardiovascular imaging

A new approach of irreversible responsive PARAmagnetic Chemical Exchange Saturation Transfer (PARACEST) for MRI contrast agents constituted a new type of agent for molecular imaging. A novel PARACEST MRI contrast agent, Yb(III)-(1,4,7,10-tetraazacyclododecane-1,4,7-triacetic acid)-orthoaminoanilide (Yb-DO3A-oAA), was developed to image nitric oxide (NO). The agent exhibited two CEST effects at -11 ppm and +8 ppm chemical shifts, which were assigned to chemical exchange from amide and amine functional groups, respectively. This responsive PARACEST MRI contrast agent indicated the change in the presence of NO and $O_{(2)}$, which caused an irreversible disappearance of both PARACEST effects from MR images. This report highlighted the advantages of irreversible MRI contrast agents and demonstrated that large changes in PARACEST can be used to in vivo biomedical applications in molecular imaging [Liu, et al. 2007]. In other study, the clinical detection of evolving acute tubular necrosis (ATN) and differentiating it from other causes of renal failure was performed. Herein, sodium magnetic resonance imaging (^{23}Na MRI) was applied to study the early alteration in renal sodium distribution in rat kidneys 6 h after the induction of ATN combined with inhibition of nitric oxide and prostaglandin synthesis. Hence, ^{23}Na MRI non-invasively quantified changes in the corticomedullary sodium gradient in the ATN kidney during morphologic tubular injury [Maril, et al. 2006]. Other newer application of ^{19}F NMR spin-trapping technique for in vivo *NO detection was employed to elucidate the significance of *NO availability in animal models of hypertension. In vivo *NO-induced conversion of the hydroxylamine from fluorinated nitrosyl nitroxide (HNN) to the hydroxylamine of the iminonitroxide (HIN) was evidenced in hypertensive ISIAH and OXYS rat strains [Bobko, et al. 2005]. The NMR detected positive levels of nitrite/nitrate for *in vivo* evaluation of *NO production and provided the basis for in vivo *NO imaging [Kojima, et al. 2001, Kojima, et al. 2001].

The cardiovascular ischemia in tissues generates the nitric oxide (NO). It is an enzyme-independent mechanism of NO generation and more pathogenesis. So, the NO formation in mice after cardiopulmonary arrest could be imaged. Real-time measurement of NO generation was performed by detection of naturally generated NO-heme complexes in tissues using L-band electron paramagnetic resonance (EPR) spectroscopy. Measurements of NO generation were imaged on the intact animal at the levels of the head, thorax, and abdomen within 3 hours [Ueno, et al. 2002]. Very recently, nitric oxide was recognized as biomarker of vascular dysfunction in renal, clitoris and penile sex organs to detect the physiology [Ockalli, et al. 1999, Kakiailatu, 2000, Burnett, 2002, Barouch, et al. 2002, Hillebrandt, et al. 2009, Ciplak, et al. 2009, Gragasin, et al. 2004, Richards, et al. 2003]. The NO complexes were found to have maximum levels in lung, heart, and liver by three-dimensional spatial mapping of the NO complex in the intact animal subjected to cardiopulmonary arrest. The images also confirmed the maximum NO formation in the lungs, heart, and liver [Kuppusamy, et al. 2001].

4.4 Calcium-imaging and NO-imaging

The combined real-time bioimaging of calcium and nitric oxide with continuous electrophysiological intracellular recording has gained enormous interest. For intracellular monitoring by calcium imaging in cells, Fura-2 is a calcium-sensitive fluorescent probe. Using Fura-2, intracellular free calcium ion concentration ($[Ca^{2+}]_i$) of single neurons was measured [Richards, et al. 2003]. Calcium-imaging with Fura-2 was used to monitor and modulate the intracellular Ca^{2+} evoked by NMDA and AMPA in cultured hippocampal pyramidal cells [Keelan, et al. 1999] Calcium and NO play role in excitotoxic mitochondrial depolarization in the hippocampus [Berkels, et al. 2000]. The combination of NO bioimaging with calcium imaging may distinguish the temporal rise in NO and calcium to suggest that NO is being produced as a consequence of Ca^{2+} enhancement and Ca^{2+} induced NO Synthase activity). Recently, imaging intracellular calcium imaging was achieved together with NO imaging by using two monochromators DAF-2 and Fura-2 AM as excitation light source. The DAF-2 excited at 485 nm and Fura-2 AM excited at 340 nm [Berkels, et al. 2000]. The emission maximum of fura-2 is about 510 nm different while the emission maximum of DAF-2 is about 515 nm. By using dichroic or multichroic mirrors calcium and NO levels can be measured simultaneously [Pittner, et al. 2003]. Other example of monochromator complex is DAQ and DAR-4M AM. These are calcium sensitive non-neurotoxic dyes with unique fluorescent patterns of DAQ (excitation ~520 nm; emission ≥580 nm) and DAR-4M AM (excitation ~ 550 nm; emission ≥ 580 nm). These can also combine with the calcium-sensitive dye Fura-2 as potential tools in fluorescence microscopy.

4.5 Kupffer cell NO bioimaging

A new mechanism was reported in nonparenchymal cells based on Bcl-2/adenovirus EIB 19-kDa interacting protein 3 (BNIP3), a cell death-related member of the Bcl-2 family, was upregulated in vitro and in vivo by nitric oxide (NO) downregulation of BNIP3 gene expression as one mechanism of hepatocyte cell death and liver damage. Authors proposed that inflammatory stresses can lead to the modulation of BNIP3 [Metukuri, et al. 2009].

4.6 Combining electrophysiology and NO-imaging

The calcium-imaging with electrophysiological recording was excitement in the hippocampus imaging art because of calcium accumulation in hippocampus pyramidal cells during synaptic activation [Regehr, et al. 1989]. Now a day, calcium-imaging combined with electrophysiological stimulation and recording has emerged as present state of art of calcium-signaling and intracellular free calcium ion concentration $[Ca^{2+}]_i$ of single neurone. Moreover, little is understood how neuronal signals stimulate NO production. Potential applications of fluorescent NO indicators in bioimaging are to monitor NO in response to pharmacological manipulation using NO-donors, NOS substrate, and NOS or NOS isoform inhibitors as enhancers of NO production in vivo and in vitro [Moore, et al. 1997, Southan, et al. 1996]. Other emerging techniques of bioimaging by NO inhibitor are targeting the differential cofactor requirements of the various isoforms; pharmacological agents targeting the differential substrate requirements of cell expression of various isoforms of the NOS [Salerno, et al. 2002]. However, this approach suffers due to lack a marked isoform selectivity or direct affecting the arginine transport into cells, like N^G-methyl-L-arginine (L-NMA). Other example of NOS inhibitors such as 7-nitroindazole (7-NI), might have unexpected side-effects in neuronal tissues, since 7-NI interferes with monoamine oxidase [Castagnoli, et al. 1999, Desvignes, et al. 1999]. However, selective NOS-inhibitors together with fluorescent NO-imaging appear as useful tools for the investigation of the biological functions of the different NOS-isoforms in neuronal tissues [Chen, et al. 2001, Brown, et al. 1999]. The combined NO-imaging with in vitro electrophysiology methods visualize the NO generation and NO distribution with localization of NO induced by neurotransmitter stimulations, e.g., NO formation after neuronal stimulation with NMDA, in normal synaptic transmission, in paired-pulse facilitation, in long-term depression or LTP. Thus, in a single preparation, the spatial distribution of NO after induction of hippocampal LTP has been evaluated [Schuppe, et al. 2002]. Main unsolved issues of precise time-point of NO release and selective NOS isoform-inhibitors in neuronal activity still remain to investigate before the combined fluorescent NO-imaging and electrophysiology method becomes an acceptable bioimaging tool. In near future, the researchers will thrive to combine electrophysiology with simultaneous calcium-imaging and NO-imaging to get better answers of the role and interaction of calcium and NO in physiological and pathological neuronal processes with continuous monitoring the calcium and NO by dichroic or multichroic mirrors in synaptic plasticity.

4.7 Alveolar cells

Human alveolar macrophages and alveolar epithelium is the site of nitric oxide synthase-2 (NOS-2) messenger ribonucleic acid (mRNA) and protein expression in alveolar wall in lung of patients with pulmonary malignancies. In the process of NO production, pro-inflammatory cytokines interleukin (IL)-1β, tumour necrosis factor (TNF)-α, interferon (IFN)-γ or lipopolysaccharide (LPS) play significant role to induce the effect of human surfactant protein-A (SP-A) on IFN-γ-mediated NOS-2 mRNA expression [Pechkovsky, et al. 2002].

4.8 Endothelial cells

Kojima et al. 2001, reported NO bioimaging in cultured bovine aortic endothelial cells. The cultured bovine aortic endothelial cells were incubated with DAF-FM DA for 1 h for dye loading. After stimulation with bradykinin, the fluorescence intensity in the cells increased and that in the cytosol increased more than that in the nucleus. NO is produced in the cytosol, where NOS exists [von Bohlan, et al. 2002]. This observation implies that little of the produced NO diffuses into the nucleus or if it does diffuse into the nucleus little of it is oxidized there. The augmentation of the fluorescence intensity was suppressed by an NOS inhibitor. In conclusion, DAF-FM is a useful tool for visualizing the temporal and spatial distribution of intracellular NO. Kimura et al.2001, examined the effects of acute glucose overload on the cellular productivity of NO in bovine aortic endothelial cells (BAEC) using DAF-2. ATP induced an increase in NO production, assessed by DAF-2, mainly due to Ca^{2+} entry as shown in Figure 13. In contrast, the ATP-induced increase in DAF-2 fluorescence was impaired by glucose overload. The results indicate that glucose overload impairs NO production via the O_2^--mediated attenuation of Ca^{2+} entry. In addition, Kimura et al. 2000, examined in detail the mechanism by which mechanical stress induces NOS in endothelium. Hypotonic stress (HTS) induced ATP release, which evoked Ca^{2+} transients in BAEC. HTS also induced NOS, assessed by DAF-2 fluorescence. The results obtained indicate that endogenous ATP plays a central role in HTS-induced NOS in BAEC. Broillet et al. 2001, reported that Ca^{2+}, Mg^{2+}, or incident light promoted the reaction of DAF-2 and NO. However, Nagano et al. 2002, probed that the enhancement of fluorescence intensity was caused not by increasing the reaction rate between DAF-2 and NO but by promotion of NO-releasing rate of NO donors. Endothelial NOS, a Ca^{2+}/calmodulin-dependent enzyme, is critical for vascular homeostasis [Suzuki, et al. 2002]. To determine the signaling pathway for endogenous eNOS, Lin et al. 2000, directly measured NO production in bovine pulmonary artery endothelial cells (PAEC). Endothelial cells grown in a monolayer produce very low levels of NO which are below the detection range of traditional assays. Therefore, they employed DAF-2 DA to measure NO production in real-time. PAEC had detectable basal NO production which increased slightly after thapsigargin treatment in the presence of low extracellular Ca^{2+}. In contrast, a large increase in NO production was detected in the presence of 4 mM extracellular Ca^{2+}, suggesting that capacitative Ca^{2+} entry regulates wild-type eNOS activity. Zharikov et al. 2001, investigated possible involvement of the actin cytoskeleton in the regulation of the L-arginine/NO pathway in pulmonary artery endothelial cells (PAEC). DAF-2 DA was used for detection of NO in PAEC. They conclude that the state of actin microfilaments in PAEC regulates L-arginine transport and that this regulation can affect NO production by PAEC. Berkels et al. 2000, presented a new method to measure intracellular Ca^{2+} and NO simultaneously in endothelial cells. The method makes it possible to follow intracellular Ca^{2+} and NO distributions online and is sensitive enough to monitor changes of NO formed by the constitutive endothelial NOS. Franz et al. 2000, reported DAR-4M AM applied to imaging of NO in bovine aortic endothelial cells using aminotroponiminates[Franz et al. 2000]. DARs are more photostable than DAFs. The fluorescence in cells was observed after stimulation with bradykinin (see Figure 14), which raises the intracellular Ca^{2+} level and thereby activates NOS, and this increase was suppressed by addition of an NOS inhibitor. DAR-4M should be useful for bioimaging of samples that have strong autofluorescence in the case of 490 nm excitation in which the intracellular pH may fall below 6. Hiratsuka et al. 2009, reported in vivo visualization of nitric oxide and intracellular inflammatory interactions among platelets, leukocytes, and endothelium following hemorrhagic shock and reperfusion.

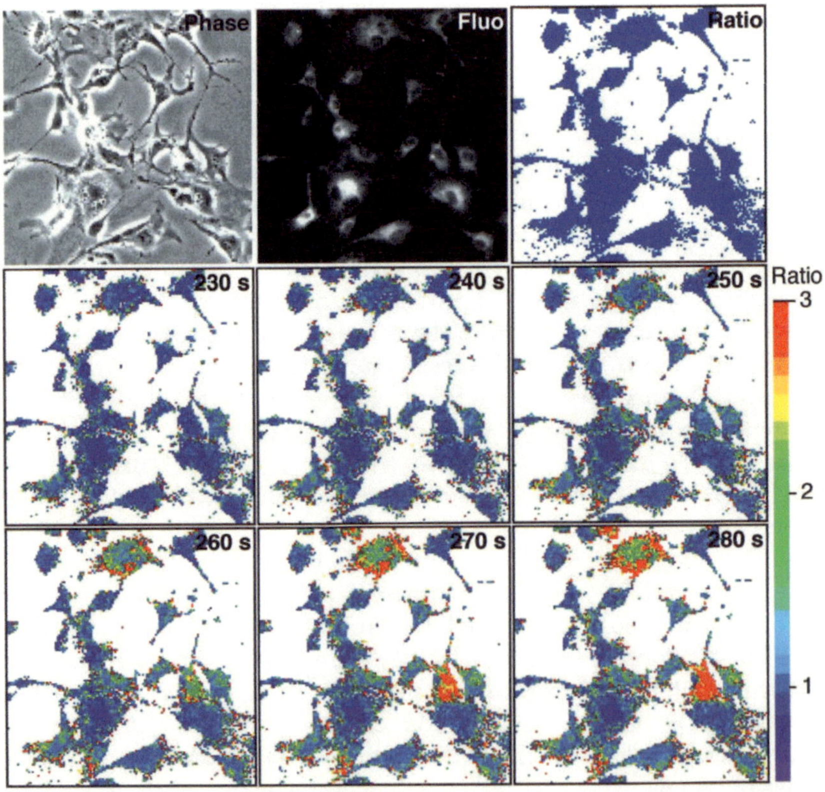

Fig. 13. Bright-field and fluorescence images of cultured bovine aortic endothelial cells loaded with DAR-4M AM. The upper images are the bright-field, the fluorescence, and the fluorescence ratio images at the start of measurement, respectively. The middle and the lower images are fluorescence ratio images at the indicated times after the start of measurement. The fluorescence ratio images indicate the ratio of the intensity to the initial intensity at the start. Adopted from reference Kimura, et al. 2001

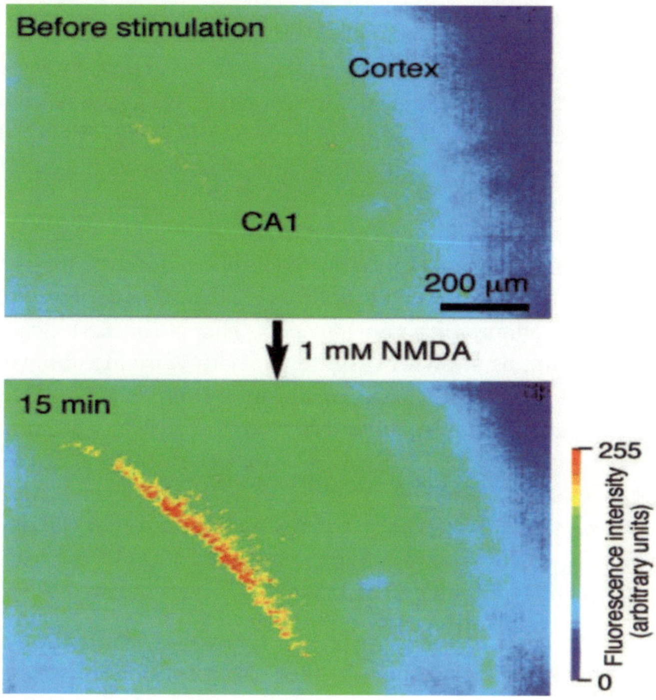

Fig. 14. Fluorescence images of a rat brain slice loaded with DAF-2 DA. The fluorescence intensity is shown in pseudocolor. (Reproduced from reference Nagano, et al. 2002)

4.9 Smooth muscle cells

DAF-2 DA was applied to the imaging of NO in cultured rat aortic smooth muscle cells by using a fluorescence microscope equipped with fluorescence filters for fluorescein chromophores. Confocal laser scanning microscopy indicated that fluorescence was emitted from the whole cell body, including the nucleus. This implies that DAF-2 regenerated intracellularly by esterase is distributed throughout the whole cell. There was no indication of any cell damage caused by loading the dye. The results show that the fluorescence intensity in endotoxin- and cytokine-activated cells increased owing to DAF-2 T production from DAF-2 by reaction with NO. The fluorescence did not increase in inactivated cells, which did not produce NO. The probe is very useful for bioimaging of NO in cultured rat aortic smooth muscle cells. Itoh et al. detected DAF-FM T using reversed-phase high-performance liquid chromatography with a fluorescence detector [Itoh et al. 2000]. This sensitive method enabled them to detect the spontaneous and substance P-induced NO release from isolated porcine coronary arteries, both of which were dependent entirely on the NOS activity in vascular endothelial cells. Furthermore, they obtained fluorescence images of cultured smooth muscle cells of the rat urinary bladder after loading with DAF-FM DA. In the cells pretreated with cytokines, the fluorescence intensity increased with time after DAF-FM loading. In recent studies on imaging of nitric oxide in nitrergic neuromuscular neurotransmission suggested the new insight of molecular mechanisms in animal organs [Thatte, et al. 2009, Perrier, et al. 2009, Di Cesare Mannelli, et al. 2009].

4.10 Brain and neuronal behavior

Reif et al. 2009, reported the influence of functional variant of neuronal nitric oxide synthase on impulsive behaviors in humans. The biological functions of NO in the neuronal system remain controversial. Using DAF-2 DA for direct detection of NO, Nagano et al. 2002, examined both acute rat brain slices and organotypic culture of brain slices to ascertain NO production sites. The fluorescence intensity sensitive to Ca^{++} in the CA1 region of the hippocampus was augmented, especially after stimulation with NMDA, in acute brain slices. This NO production in the CA1 region was also confirmed in cultured hippocampus. Glutamate and NO were important mediators, since antagonists of both N-methyl-D-aspartate (NMDA) and NOS inhibitors were effective in attenuating neurotoxicity. This is the first direct evidence of NO production in the CA1 region as shown in Figure 4. There were also fluorescence cells in the cerebral cortex after stimulation with NMDA. Imaging techniques using DAF-2 DA should be very useful for the clarification of neuronal NO functions. Calcium sensitive fluorescent probe can monitor intracellular calcium by FURA-2. Combined calcium and NO together can be imaged by excitation light source (DAF-2) excited at 485 nm, and Fura-2 AM excited at 340 nm using dichroic or multichroic mirrors.DAQ, DAR-4M AM are other choices. NOS-II may be involved in Alzheimer's disease as b-amyloid induces NOS-II expression in astrocytes Astrocytes which which can be potentiated by cytokines [Vodovotz et al. 1996].

DAF-FM DA was also applied to imaging of NO generated in rat hippocampus slices by exposure to an aglycemic medium [Reif et al. 2009]. NO production was observed mainly in the CA1 area and was dependent on the concentration of O_2. During exposure to an anoxic-aglycemic medium, NO was hardly produced while marked elevation of intracellular Ca^{2+} was observed. Production of NO increased sharply as soon as the perfusate was changed to the normal medium. These results suggest that NOS is activated after reperfusion rather than during ischemia. ROS and NO are important participants in signal transduction that could provide the cellular basis for activity-dependent regulation of neuronal excitability [Yermolaieva, et al. 2000]. In young rat cortical brain slices and undifferentiated PC 12 cells, paired application of depolarization/agonist stimulation and oxidation induces long-lasting potency of subsequent Ca^{2+} signaling that is reversed by hypoxia. This potency critically depends on NO production and involves cellular ROS utilization. Using the fluorescent dye DAF-2, NO production was measured in PC 12 cells during depolarization or histamine application [Yermolaieva, et al. 2000]. Application of 100 μM histamine induced a significant increase in NO production. The crucial role for NO in oxidative potentiation of Ca^{2+} signaling triggered by pairing of electrical/agonist stimulation and oxidation demonstrated here provides an insight into the cellular mechanisms involved in neuronal developmental plasticity. NO could influence its effectors by stimulating the cGMP-dependent signaling pathway and/or by acting as a weak radical. Perrier et al. 2009, reported the effect of uncoupling endothelial nitric oxide synthase on calcium homeostasis in aged porcine endothelial cells in heart. In other report, Sergeant et al. 2009, reported the spontaneous Ca^{2+} waves in rabbit corpus cavernosum were modulated by nitric oxide. Role of cGMP was crucial in regulation of calcium dependent neuroactive energy changes.

4.11 Ion channels

Most voltage-gated Na^+ channels are almost completely inactivated at depolarized membrane potentials, but in some cells a residual Na^+ current is seen that is resistant to inactivation. The biological signaling mechanisms that regulate the persistence of Na^+ channels are not well understood. Ahern et al. 2000, showed that in nerve terminals and ventricular myocytes, NO reduces the inactivation of Na^+ current. This effect was independent of cGMP and blocked by *N*-ethylmaleimide. Thus, ROS act directly on the channel or a closely associated protein. Application of ionomycin to raise the intracellular Ca^{2+} concentration in myocytes activated NOS. The NO produced in response to ionomycin was detected with DAF-2 DA. They concluded that NO is a potential endogenous regulator of persistent Na^+ current under physiological and pathophysiological conditions. Choi et al. 2000, reported that the NMDA receptor (NMDAR)-associated ion channel is modulated not only by exogenous NO, but also by endogenous NO. They examined inhibition of NMDAR responses by endogenous NO to determine the underlying molecular mechanism. For this purpose, HEK 293 cells transfected with nNOS were transiently transfected with wild-type NR1/NR2A or mutant NR1/NR2A (C399A) receptors. NMDA responses were monitored by digital Ca^{2+} imaging with fura-2. Production of endogenous NO by these HEK-nNOS cells was measured with DAF-2. It was concluded that endogenous *S*-nitrosylation may regulate ion channel activity.

4.12 Fertilization and developmental biology

Kuo et al. 2000, demonstrated that NO is necessary and sufficient for egg activation at fertilization. The early steps that lead to the rise in Ca^{2+} and egg activation at fertilization are unknown but are of great interest, particularly with the advent of in vitro fertilization techniques for treating male infertility and whole-animal cloning by nuclear transfer. They showed that active NOS is present at high concentration in sperm after activation by the acrosome reaction. An increase in nitrosation within eggs is evident seconds after insemination and precedes the Ca^{2+} pulse of fertilization. They used DAF-2 DA for detection of NO. It was concluded that NOS- and NO-related bioactivity satisfy the primary criteria of an egg activator, i.e. they are present in an appropriate place, active at an appropriate time, and necessary and sufficient for successful fertilization.

4.13 Inflammation

NO plays important roles in inflammatory processes. Lopez-Figueroa et al. 2000, examined whether changes in NOS mRNA expression lead to similar temporal and anatomical changes in NO production in an experimental model of CNS inflammation. iNOS mRNA expression was analyzed 2, 4, 6, and 24 h after intracerebroventricular (icv) injection of interleukin-1β or the vehicle. Increased expression of iNOS mRNA was observed surrounding the microinjection site and meninges. Using DAF-2 DA for the direct detection of NO production, they observed a significant increase in NO production after 4 and 6 h. They conclude that increase in iNOS mRNA following icv administration of IL-1β leads to increases in NO production. They proposed that the DAF-2 DA method can be used as a potential marker in the diagnosis of CNS inflammation. MRI of NO is very useful because NO is available at high concentrations in millimolar levels in immune-related diseases (e.g., inflammation).Terashima et al. 2010, reported in vivo detection of iNOS expression in murine carotid lesions by biuoluminescence may provide a valuable approach for monitoring vascular gene expression and inflammation in small animal models. The phasic nature of inflammation is also associated with a decrease in NOS activity. NOS inhibitors are effective at alleviating pain and have disease modifying effects. Induction of NOS-II is seen in models of septic shock. NOS-II inhibitors (dexamethasone) prevent hypotension and decrease in PO_2.

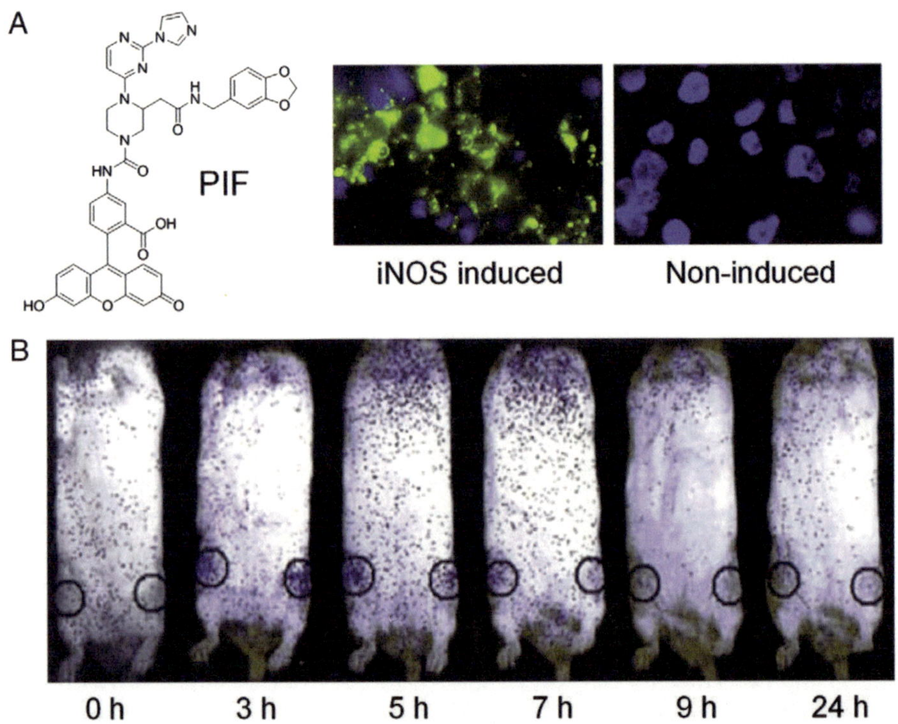

Fig. 15. Optical imaging of iNOS expression. (A) Fluorescence imaging of iNOS in living cells. Left: cells were induced for iNOS expression; Right: cells were not induced for iNOS expression. (B) Bioluminescence signal from the knee joints (circle) indicates iNOS expression following inflammation induction. Adapted from reference [Panda, et al. 2000].

4.14 Adrenal zona glomerulosa

Adrenal zona glomerulosa (ZG) cells do not contain NOS. Hanke et al. 2000, conferred endothelial NOS activity upon adrenal ZG cells through transduction with a recombinant adenovirus encoding the endothelial NOS gene (AdeNOS) to determine the effect of endogenous NO on aldosterone synthesis. AdeNOS-transduced cells exhibited DAF-2 DA fluorescence, which was blocked by pretreatment with an NOS inhibitor. In conclusion, adenovirus-mediated gene transfer of eNOS in ZG cells results in the expression of active endothelial NOS enzyme, and this endogenous NO production by ZG cells decreases aldosterone synthesis.

4.15 Bone marrow stromal cells

Gorbunov et al. 2000, reported that using the reverse transcription-polymerase chain reaction and immunofluorescence analysis of murine bone marrow stromal cells after γ-

irradiation doses of 2-50 Gy stimulated the expression of iNOS. The activation of iNOS was accompanied by an increase in the fluorescence of DAF-2 and accumulation of 3-nitrotyrosine within cellular proteins in a dose-dependent manner.

4.16 Mitochondria

NO has been implicated in the modulation of mitochondrial respiration, membrane potential, and subsequently apoptosis. Although the presence of mitochondrial NOS was reported while there is no direct evidence in vivo of the presence of NO within mitochondria. Using DAF-2 DA, Lopez-Figueroa et al. observed NO production in PC 12 and COS-1 cells by conventional and confocal fluorescence microscopy [Lopez-Figueroa, et al. 2000]. The subcellular distribution of NO production is consistent with the presence of a mitochondrial NOS.

4.17 Retina

In the retina, NO functions in network coupling, light adaptation, neurotransmitter receptor function, and synaptic release. Neuronal NOS is present in the retina of every vertebrate species so far investigated. However, although nNOS can be found in every retinal cell type, little is known about the production of NO in specific cells or about the diffusion of NO within the retina. Blute et al. 2000, used DAF-2 to image real-time NO production in turtle retina in response to stimulation with NMDA. In response to NMDA, NO was produced in somata in the ganglion cell and inner nuclear layers, in synaptic boutons and processes in the inner plexiform layer, in processes in the outer plexiform layer, and in photoreceptor inner segments. They concluded that NO may function at specific synapses, modulate gene expression, or coordinate events throughout the cell. Matsuo et al. 2000, reported that basal NO production is enhanced by hydraulic pressure in cultured human trabecular cell. He measured the intracellular NO level in real-time using DAF-2.

4.18 Drosophila

Wingrove and O'Farrell 1999, reported that *Drosophila* utilizes components of the NO/cGMP signaling pathway to respond to hypoxia. Hypoxic exposure rapidly induced exploratory behavior in larvae and arrested the cell cycle. These behavioral and cellular responses were diminished by an inhibitor of NOS and by a polymorphism of cGMP-dependent protein kinase. Regions of the larvae that specialize in responding to hypoxia should express significant levels of the activities involved. They used DAF-2 DA to define possible foci of function. DAF-2 DA stained the pouch of tissue surrounding the anterior spiracles as well as neuronal-like processes within the pouch. This staining near the openings of the tracheal system is interesting because the location is consistent with a possible role in governing the opening of the spiracles to increase access to oxygen. DAF-2 DA is useful for clarification of response mechanisms to hypoxia.

4.19 Plants

Leaves and callus of *Kalanchoe daigremontiana* and *Taxus brevifolia* were used to investigate NO-induced apoptosis in plant cells [Huang, et al. 2000]. NO production was visualized in cells and tissues with DAF-2 DA. The NO burst preceded a significant increase in nuclear

DNA fragmentation and cell death. L-NMMA significantly decreased NO production and apoptosis in both species. Pedroso et al. 2000a, 2000b concluded that NO is involved in DNA damage leading to cell death and proposed a potential role of NO as a signal molecule in these plants. Foissner et al.2000, used DAF-2 DA, in conjunction with confocal laser scanning microscopy, for in vivo real-time imaging of an elicitor-induced NO burst in tobacco. A growing body of evidence suggests that NO, an important signaling and defense molecule in mammals, plays a key role in activating disease resistance in plants, acting as a signaling molecule and possibly also as a direct antimicrobial agent. The results revealed additional similarities between plant and animal host responses to infection.

5. NO/NOS bioimaging: Future prospects

The existing techniques based on quenching spin trap or paramagnetic heavy metals pose a concern of contrast agent toxicity with limitations of event capturing to generate NO image of isolated cells by EPR or in vivo body image by MGD enhanced MRI. However, unique role of short lived NO molecule in the cell is peculiar to give insight of rapid ionic regulatory mechanisms such as calcium, sodium dependent signaling events including dietary effects, antimicrobial therapy on inflammation, angiogenesis and hypoxia reported in last three years [Manuel, et al. 2002, Palombo, et al. 2009]. To date, the main imaging modalities for visualization of NO include fluorescence and bio/chemiluminescence optical imaging, electron paramagnetic resonance (EPR) imaging, and MRI. cNOS has been mainly investigated using optical and PET imaging. Paper has presented a progress to date on multimodality imaging of NO and addressed the future research directions and obstacles of NO imaging. Other major developments are expected in design of robust imaging hardware with improved image generating capability and highly sensitive to the rapid changes of NO/NOS molecular EPR signal and MRI frequency with time. In this direction, active research efforts suggest the development and availability of less toxic ion sensitive contrast agents attached with EPR and MRI visible molecules as labels sensitive to fluorometric and magnetic resonance techniques respectively. It is attributed that cytochromes are best candidates as fluorochromic agents while lipopolysaccharides and carbamates with paramagnetic agents are potential as MRI contrast agents. More likely, fast imaging techniques such as parallel phase array, SENSE, FISP MRI imaging, xenon based PARACEST techniques may be available to capture rapid NO changes or distribution associated with metabolic short lived events such as oxygen sensitivity, sodium sensitivity in the intact tissue. In future, advances in ionic ratiomatric techniques will be available to capture the short living NO molecular distribution in cells as real-time events associated with motion. Still major issues to resolve are:

- lack of NO sensitive EPR or MRI contrast agent to generate quicker and enough MRI or EPR signal visible as image;
- the EPR and MRI signals are not true representative of NO concentrations;
- the images generated from MRI visible paramagnetic ions or EPR visible spin trap agents may represent several physical and molecular events sensitive to attached NO quenching carbamates or polysaccharides for MRI and cytochrome proteins for EPR.

The biggest disadvantage of MRI is its low molecular sensitivity. The need for a high dose of imaging agents (typically more than 100 mM injection) undoubtedly raises major concerns about the potential toxicity caused by such studies. Lastly, relatively long data acquisition time is typically needed to generate a detectable MR signal in MR imaging. Since the lifetime of NO and/or related species is usually short, the accuracy of MRI measurement of NO concentration is questionable. After these issues are resolved, it will need sequence of investigations to calibrate linearity of NO concentration and image signal intensity, image reproducibility, image component analysis, signal measurement accuracy and precision of NO sensitive imaging techniques as shown in Table 4.

Target	Modality	Resolution	Sensitivity	Penetration	Clinical potential
NO	Optical	++++	++++	+	Low
NO	EPR	++	++++	+++	Medium
NO	MRI	+++	+	++++	Medium
NO	PET	++	++++	+++	Medium
cNOS	Optical	++++	+++	++	Low
cNOS	PET	+++	++++	++++	Medium-High

Table 4. Comparison of molecular imaging of NO and cNOS expression

5.1 Real-time NO imaging

Nitric oxide (NO) is a small uncharged free radical abundant in nanomolar quantities that is involved in diverse physiological and pathophysiological mechanisms.

- When generated in vascular endothelial cells, NO plays a key role in vascular tone regulation and EPR/optical/MRI visible signals visualize the vascular physiology (induced fitting of L-arginine specific NOS active site) as prosposed by author to inhibit NOS and low NO production shown in Figure 5.

In this direction, short lived NO specific amplifier-coupled fluorescent indicator was reported to visualize physiological nanomolar dynamics of NO in living cells (detection limit of 0.1 nM)[Ueno, et al. 2002, Burnett, et al. 2002]. An amplifier-coupled fluorescent indicator visualizes NO in single living cells. Its domain structures are sGCα, sGCβ, CGY, sGCα-CGY, and sGCβ-CGY. This genetically encoded high-sensitive indicator revealed that ≈1 nM of NO was enough to relax blood vessels [Ueno, et al. 2002, Burnett, et al. 2002, Barouch, et al. 2002, Hillebrand et al. 2009, Ciplak, et al. 2009, Gragasin, et al. 2004]. The nanomolar range of basal endothelial NO thus revealed appeared to be fundamental to vascular homeostasis. A fuorescent probe, 1,3,5,7-tetramethyl-2,6-dicarbethoxy-8-(3,4-diaminophenyl)-diXuoroboradiaza-s-indacence (TMDCDABODIPY), produced real-time image NO of PC12 cells, Sf9 cells and human vascular endothelial cells in the presence of L-arginine with inverted fuorescence microscope [Manuel, et al. 2002]. NO production in the cells was successfully captured and imaged with improved temporal and spatial resolution. The bioimaging applications of TMDCDABODIPY as a fuorescent imaging probe for NO in living cells was compared with the other fuorescent imaging probes for NO. The TMDCDABODIPY had the advantages of shorter reaction time, higher photostability and minimal background. The study has indicated that TMDCDABODIPY-based fuorescence microscopy may be a promising approach for NO-related pathological and physiological studies [Hiratsuka, et al. 2009]. A recent review highlighted the possibilities if nitric oxide bioimaging becomes reality as clinical imaging tool [Hong, et al. 2009]. Current state-of-the-art multimodality imaging may detect NO and NOS enzyme expression [Matter et al. 2004, Zhang, et al. 2011]. Optical (fluorescence, chemiluminescence, and bioluminescence), electron paramagnetic resonance (EPR), magnetic resonance (MR), and positron emission tomography (PET) noninvasive imaging techniques may reveal the biodistribution of NO or

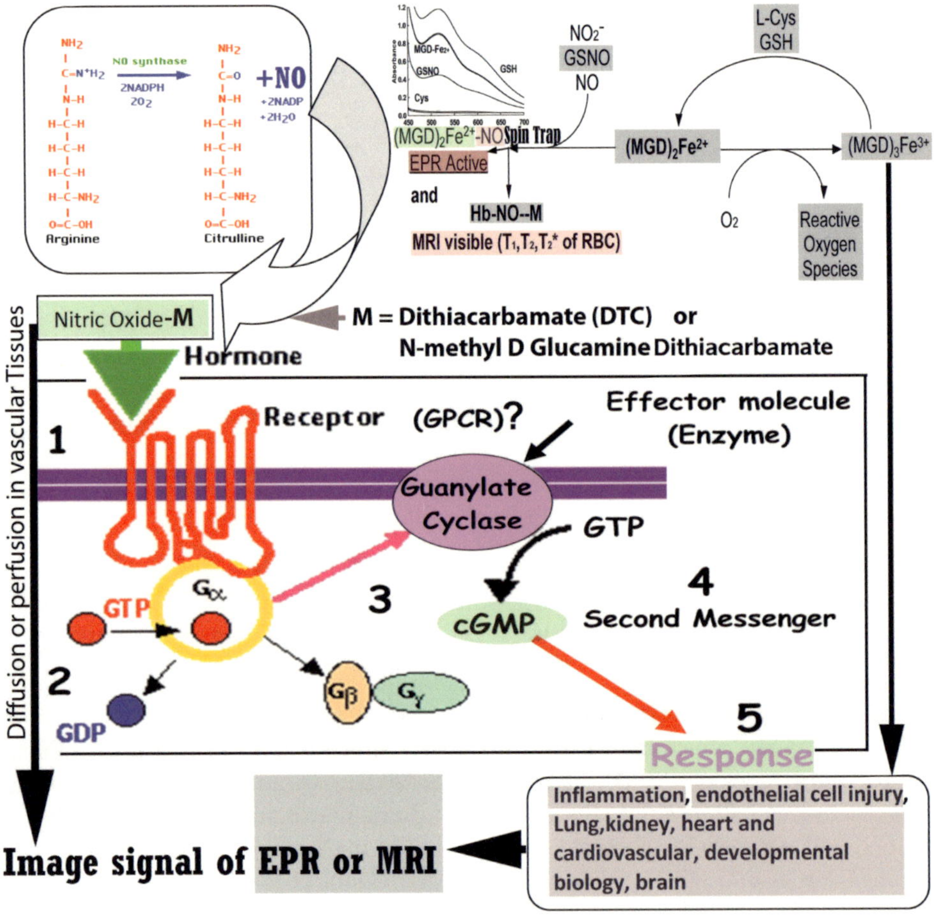

Fig. 15. (on top) Formation and release of NO in tissues is shown. No is trapped by $Fe^{2+}(MGD)_2$ to make MRI/EPR visible trap NO-(MGD)2-Fe^{3+}. (in middle row)The figure represents how different levels of Nitric oxide (1) activate cytosolic guanylate cyclase (2) and thus elevate intracellular levels of cyclic GMP (3 and 4) as secondary messenger. The cyclic GMP stimulates response (5) in different tissues and nitric oxide can be detected as biomarker of tissue pathology.(at bottom) The relation of released NO across the membrane is shown with different biological and physiological changes to generate EPR and MRI signal visible by imaging techniques. Notice the NO regulation by enzyme guanylate cyclase over intracellular responses (cell injury, inflammation etc.) sensitive to spin trap MRI/EPR image signals. A proposal of multimodal nitric oxide imaging is shown for development of multimodal spin trap applicable in imaging.

NOS in living subjects with high fidelity which will greatly facilitate scientists and clinicians in the development of new drugs and patient management[Oliveira, et al. 2010, Payne, et al. 2010]. Lastly, novel NO/NOS imaging agents with optimal in vivo stability and suitable pharmacokinetics such as pteridines, imidazoles, quinolines for clinical translation will benefit in patient management [Payne, et al. 2010, Bonnefous, et al. 2009].

Imaging of NOS expression mainly relies on labeling high affinity, preferably also high specificity, NOS inhibitors with optimal membrane permeability and pharmacokinetics. Of all the fluorescence imaging agents for NO, the DAFC compounds are the most widely used. The requirement of oxygen restricts their applicability in cancer-related NO imaging to a certain extent, although some have been tested for NO imaging during tumor angiogenesis [Kashiwagi, et al. 2005]. Copper-based fluorescent compounds do not require oxygen for NO imaging, yet the stability and toxicity of those compounds are questionable. FRET-based NO imaging is a clever approach, yet it is too technically challenging to be widely available and useful. With spin-trapping technology, EPR imaging of NO can be quite accurate, yet it is not applicable for clinical studies due to many reasons. MRI can have high resolution and large field-of-view for potential clinical imaging studies, yet it is usually quite slow in data acquisition and the intrinsic low sensitivity makes molecular MRI unlikely to succeed in the clinic. NOS imaging is also important in elucidating NO-related physiological and pathological processes. NOS imaging has not been well studied to date. To the best of our knowledge, no clinical studies of NO or NOS imaging have been reported. Whether NO/NOS imaging can help and improve patient management remains to be demonstrated in the future. The deregulation of iNOS in melanoma has been correlated directly with poor survival [Madunapantula, et al. 2008] and NOS expression in patients with neurological disorder has been studied [Broholm, et al. 2004]. Given the fact that NO and NOS plays diverse roles in many diseases such as inflammation, neurodegeneration, cancer, and vascular malfunctions, further understanding of the disease mechanisms and clarification of the temporal/ spatial NO/NOS expression are certainly very important in patient management. Whether imaging NO or NOS is more relevant depends primarily on the disease. eNOS and nNOS typically only produces low levels of NO; therefore, imaging NO is expected to be more difficult than imaging the NOSs themselves. On the other hand, iNOS can produce high concentrations of NO in immunological processes and/or cancer; hence imaging NO may be more desirable in these scenarios. Granted that NO imaging is more relevant since it directly affects the physiological and pathological processes, imaging of NO may not always be feasible and imaging NOS expression (as an alternative) can provide valuable insights. The notion that nitrite reduction can also serve as a possible source of biologically relevant NO tilted the balance toward NO imaging [Bryan, 2006]. Nitrite reduction to NO can occur via several routes involving enzymes, proteins, vitamins, or even simple protons [Galdwin, 2004, Lundberg, et al. 2005]. Nitrite is now under active investigations in physiology, pathophysiology, and therapeutics. The ideas that nitrite can be a "storage" form of NO and the possible importance of dietary nitrite are two vibrant research areas today. Similar to the important role of L-arginine for optimal NOS activity, nitrite may emerge as an essential nutrient. This pathway may serve as a backup system for NO generation in conditions such as hypoxia, in which the NOS/L-arginine system is compromised. We envision that imaging NO and NOS in the same system will undoubtedly give a more thorough picture in understanding the functions of NO and NOS. For potential clinical translation of NO/NOS imaging probes, PET tracers have the best chances of success. Each imaging modality has its advantages and disadvantages in terms of sensitivity, spatial resolution, temporal resolution, probe availability, and cost [Massoud, et al. 2003]. With the development of hybrid imaging systems such as PET/CT, SPECT/CT, and PET/MRI Catana et al. 2006; Sharma, 2002, 2008, 2011;Townsend, et al. 2002], a multimodality approach may offer synergistic advantages over any single modality alone in the future.

6. Conclusion

Nitric oxide (NO) *in vivo* image visualizes the distribution of NO using the ''spin-trapping'' technique. Available NO visualizing chemical probes combined with the spin traps and fluorescent dyes are promising. Physiological mechanisms of NO production by NOS and role of NOS gene expression altered by NOS inhibitors is illustrated. **Scope:** Different multimodal contrast mechanisms are introduced using nitrosyl-iron complexes as MRI-PET signal intensity enhancers and fluorescent or bioluminescent NO visualizing cheletropic trap agents. NO synthase enzyme is the source of NO formation in presence of competitive inhibitor N-monomethyl-L-arginine in the tissues.

- **General Significance:** NO makes stable paramagnetic with N-methyl-D-glucamine dithiocarbamate (MGD), as $(MGD)_2$-Fe(II)-NO metal complex *in vivo* by spin trapping and may serve as effective EPR and/or MRI contrast agent than other less stable nitrogen containing radicals, such as bound nitric oxides. MRI spin-trapping induces the magnetic resonance relaxation changes of neighboring protons and visualizes spatial distributions of NO free radicals in pathologic organs and tissues but not confirmed. Other paramagnetic NO-Fe-dithiacarbamate metal complexes serve as contrast agents in EPR bioimaging. Recently, real time physiological in vivo monitoring using tissue physical properties such as fluorescence and magnetic resonance in tissues is revisted to visualize nanomolar range of nitric oxide mapping.
- **Major conclusions**: 1. More applications of NO bioimaging and fluorescent probes are emerging in imaging apoptosis, vascular imaging, smooth muscle cells, neurodegeneration, calcium channel imaging and endothelial stress. 2. The status of NO production by NOS in cells gives insight of inflammation, fertilization, mitochondrial oxidation injury, bone structure and adrenal gland and retina. 3. The real time NO imaging application is emerging in animals and plants.
- **Major highlights:**
    - Nitric oxide bioimaging is emerging as rapid noninvasive multimodal molecular imaging techniques by fluorescent, MRI, PET, optical bioimaging for wider applications;
    - Significant progress is made to explain the mechanism of multimodal NO spin trap contrast agent enhanced MRI and fluorescent bioimaging;
    - Both MRI and fluorescent contrast agents may serve as potential MRI/fluorescent multimodal spin traps as well as contrast agents in clinical or bioimaging;
    - NO sensitive MRI of vital soft tissues is major achievement to visualize very short lived NO in real time manner.
    - Development of NOS inhibitors or NO production suppressors and in vivo NO/NOS evaluation is a booming industry in diagnostics of stroke, sepsis, neurodegeneration, inflammation and new drug discovery.

Spin trap contrast agents may enable the simultaneous EPR-MRI and fluorescent imaging of NO distribution in the organs and tissues in muscle vascular system, retina, adrenal gland, brain, bone marrow to monitor intracellular metabolic events such as apoptosis, inflammation, nerve activation and calcium ions etc. To our knowledge, there are no reports of successfully imaging or detecting free radicals such as NO in vivo clinical imaging. Due to short lifetime and rapid diffusion of NO in tissue, it is a challenge to devise any effective relaxation enhancement mechanism(s) for noninvasive clinical imaging. NMR spin trapping may serve to overcome these problems of NO short lifetime and diffusion to some extent.

7. Appendix 1: Protocols on NOS and NO* radical detection

Source: Cai H., Dikalov S, Griendling K.K., Harrison, D.G.(2007) Detection of reactive oxygen species and nitric oxide in vascular cells and tissues. Chapter 20 In: Methods in Molecular Medicine, Vascular Biology Protocols. Editor: Sreejayan N. and Ron J., Humana Press Inc. Totowa, NJ. Pp 293-311.

Electronic configuration in NO* with 11 valence electrons:
$(K^2K^2)(2s\sigma^b)^2(2s\sigma^*)^2(2p\pi^b)^4(2p\sigma^b)^2(2p\pi^*)^1$

Detection of Nitric Oxide Radical

It has been challenging to detect nitric oxide radical (NO•) directly from biological samples. NO• "production" is measured by NO• synthase activity using the L-arginine conversion assay or NO• metabolites nitrite and nitrate using the Griess reagent. New methods are NO•-selective electrode and ESR represent specific and quantitative assays for detection of functional NO•.

Materials needed:

1. NO•-Specific Microelectrode
   Nafion and o-PD coated carbon electrode can directly detect NO• in the low micromolar range.
2. ESR with NO•-Specific Spin Traps
   Dithiocarbamate (DTC), N-methylglucamine dithiocarbamate (MGD) detect extracellular NO*, and diethyldithiocarbamate (DETC) trap and detect the NO* in cellular lipid membrane.
   iron-MGD and iron-DETC.
3. ESR measurement:
   - Endothelial or vascular smooth muscle cells.
   - Modified Krebs/HEPES buffer (see Subheading 2.1.1.2.).
   - Cyclic hydroxylamine CPH or 1-hydroxy-3-methoxycarbonyl-2,2,5,5-tetramethylpyrrolidine
   - (CMH) stock solution (10 mM) (Alexis Biochemicals, San Diego, CA,
   - USA) in modified Kreb's/HEPES buffer containing metal chelator, 25–50mM
   - deferoximine, and 3┘ 5mM DETC. This stock solution should be de-oxygenated by
   - nitrogen gas continuously to maintain low background oxidation of the spin traps.
   - Lysis buffer containing protease inhibitors: 50mM Tris–HCl buffer, pH 7.4,
   - containing 0.1mM ethylenediamine tetraacetic acid (EDTA), 0.1mM ethylene

- glycol tetraacetic acid (EGTA), 1mM phenylmethyl sulfonyl fluoride, 2mM bestatin, 1mM pepstatin, and 2mM leupeptin.
  - NADPH (0.2 mM).
  - Xanthine (0.1 mM).

Detection of Nitric Oxide Radical

NO•-Specific Microelectrode

- Carbon fiber electrodes (100μm length and 30μm outer diameter; Word Precision
- Instruments, Sarasota, FL, USA).
- o-PD solution (in 0.1M PBS with 100μM ascorbic acid).
- Nafion (5% in aliphatic alcohols; Sigma-Aldrich).
- Modified Kreb's/HEPES buffer.
- Axopatch 200B amplifier (Axon Instruments, Union City, CA, USA).
- Silver/silver chloride reference electrode.

Iron-DETC for Trapping of NO•

- Culture endothelial cells.
- Saline (0.9% NaCl).
- $Fe^{2+}(DETC)_2$: FeSO4·7H2O, 4.45 mg/10 ml for 1.6 mmol/l stock and DETC, 7.21 mg/10 ml for 3.2 mmol/l stock.
- PBS.
- Modified Kreb's/HEPES buffer.
- Ferrous sulfate (4 mM).
- N-methyl-d-glucamine dithiocarbamate MGD (20 mM).

Methods:

Detection of Nitric Oxide Radical

NO•-Specific Microelectrode

- Coat bare carbon fiber electrodes (100μm length and 30μm outer diameter; Word
- Precision Instruments) with nafion and o-PD. Coat with freshly made o-PD solution at constant potential (+0.9V vs. Ag/AgCl reference electrode) for 45 min. Dip in nafion solution for 3 s and dry for 5 min at 85°C. The nafion-coating cycle should be repeated 10–15 times.
- Culture endothelial cells on 35-mm dishes or prepare fresh tissue samples in freshly made modified Kreb's/HEPES buffer.
- Place the electrode tip at the surface of an individual cell, endocardium, or lumen of blood vessels, and then withdraw precisely 5μm.
- Record NO•-dependent oxidation currents (voltage clamp mode, hold at 0.65 V, approximately the voltage for peak NO• oxidation) immediately after addition of agonists using an Axopatch 200B amplifier (Axon Instruments). A silver/silver chloride reference electrode is used. Use pCLAMP 7.0 program (Axon Instruments) for delivery of voltage protocols and data acquisition and analysis.
- Calculate NO• concentrations from a standard curve obtained using dilutions of de-oxygenated, saturated NO* gas solutions.

Iron-DETC Protocol for Intracellular Trapping of NO•

- Culture endothelial cells on 100-mm Petri dishes or prepare vascular segments (6–12, 2-mm vessel segments).
- Bubble freshly prepared saline (0.9% NaCl) with nitrogen gas to remove oxygen.
- Aspirate media and rinse cells with warm PBS once, add 1.5 ml modified Kreb's/HEPES buffer with or without desired agonists, then mix Fe^{2+}(DETC)2, and immediately add to culture dish (500µl of each solution, final volume 2.0 ml).
- Incubate in cell culture incubator for desired period for cumulative trapping of NO•.
- Aspirate buffer, gently collect cells into a 1-ml insulin syringe, snap freeze in liquid nitrogen, then transfer sample column into a finger dewer, and capture Fe2+(DETC)2–NO• signal using ESR at the following settings:
- Bruker EMX: Field sweep, 160 G; microwave frequency, 9.39 GHz; microwave power, 10MW; modulation amplitude, 3 G; conversion time, 2621 ms; time constant, 328 ms; modulation amplitude, 3 G; receiver gain, 1×10^4; and four scans.
- Miniscope 200: Biofield, 3267; field sweep, 100 G; microwave frequency, 9.78 GHz; microwave power, 40 mW; modulation amplitude, 10 G; 4096 points resolution; and receiver gain, 900.

Iron-MGD Protocol for Extracellular Trapping of NO•

- Culture endothelial cells on 100-mm Petri dishes or prepare vascular segments (6–12, 2-mm vessel segments).
- Bubble freshly prepared saline (0.9% NaCl) with nitrogen gas to remove oxygen, and then make stock solutions of FeSO4 $\cdot 7H_2O$, 4 mM, and MGD, 20 mM.
- Prepare stock solutions of Fe^{2+}MGD by mixing FeSO4 and MGD at the ratio of 1:5 or 1:10 (final Fe concentration: 0.5 mM).
- Aspirate media and rinse cells with warm PBS once, add 1.5 ml modified Kreb's/HEPES buffer with or without desired agonists, then mix $Fe^{2+}(MGD)_2$, and immediately add to culture dish.
- Incubate in cell culture incubator for desired period for cumulative trapping of NO•.
- Collect 1 ml of post-incubation supernatant into a 1-ml insulin syringe and snap freeze in liquid nitrogen, then transfer sample column into a finger dewer, and capture Fe^{2+}MGD–NO• signal using ESR settings as described above for $Fe^{2+}(DETC)_2$.

8. Acknowledgements

Author acknowledge the manuscript preparation, corrections, expert suggestions by Dr Soonjo Kwon, Utah State University, Logan, UT and Professor Ching J Chen at Florida State University, Tallahassee for extending expert modifications and comments in this manuscript. Prof Avdhesh Sharma, JNV University, corrected manuscript.

9. References

Ahern, G.P., Hsu, S.F., Klyachko,V.A., Jackson, M.B. (2000) Induction of persistent sodium current by exogenous and endogenous nitric oxide. J Biol Chem, Vol 275, No 37, pp 28810-5.

Alderton, W.K., Cooper, C.E., Knowles, R.G.(2001) Nitric oxide synthases: structure, function and inhibition. Biochem J, Vol 357, 593-615.

Anbar, M.(2000) Detection of cancerous lesions by measuring nitric oxide concentrations in tissue. US Patent 6,035,225. In: Omnicorder Technologies, Stoneybrook, N.Y: U.S.A.

Anon (1999a) Nitric oxide synthase inhibitors with cardiovascular therapeutic potential. Expert Opin.Ther. Pat. Vol 95, pp 537-547

Anon (1999b) The inhibition of isoforms of nitric oxide synthase : non-cardiovascular aspects. Expert Opin. Ther. Pat. Vol 95, pp 549-556

Babu, B. R. and Griffith, O. W. (1998) N5-(1-imino-3-butenyl)-L-ornithine. A neuronal isoform selective mechanism-based inactivator of nitric oxide synthase. J. Biol. Chem. Vol 273, pp 8882-8889.

Barker, S. L., Zhao, Y., Marletta, M. A., Kopelman, R. (1999) Cellular applications of a sensitive and selective fiber-optic nitric oxide biosensor based on a dye-labeled heme domain of soluble guanylate cyclase. Anal. Chem. Vol 71, pp 2071–2075.

Barouch, L.A., Harrison, R.W., Skaf, M.W. , Rosas, G.O., Cappola, T.P., Kobeissi, Z.A., Hobai, I.A., et al. (2002) Nitric oxide regulates the heart by spatial confinement of nitric oxide synthase isoforms Nature Vol 416, pp 337-339.

Berkels, R., Dachs, R., Roesen, R., Klaus, W. (2000) Simultaneous measurement of intracellular Ca(2+) and nitric oxide: a new method. Cell Calcium, Vol 27, No 5, pp. 281-286.Berliner, L.J., Fujii, H. (2004) In vivo spin trapping of nitric oxide. Antioxid Redox Signal. Vol 6, No 3, pp 649-56.

Berliner, L. J.; Fujii, H.(2004) In vivo spin trapping of nitric oxide. Antioxid. Redox Signal. Vol 6, pp 649- 656.

Blanchette, J., Jaramillo, M., Olivier, M. (2003) Signalling events involved in interferon-gamma-inducible macrophage nitric oxide generation.Immunology Vol 108, pp 513–522.

Bobko, A.A., Sergeeva, S.V., Bagryanskaya, E.G., Markel, A.L., Khramtsov, V.V., Reznikov, V.A., Kolosova, N.G. (2005) 19F NMR measurements of NO production in hypertensive ISIAH and OXYS rats. Biochem Biophys Res Commun. Vol 330, No 2, pp 367-70.

Bretscher L.E., Li, H., Poulos, T.L., Griffth, O.W. (2003) Structural Characterization and Kinetics of Nitric-oxide Synthase Inhibition by Novel *N*5-(Iminoalkyl)- and *N*5-(Iminoalkenyl)-ornithines. J. Biol. Chem. Vol. 278, No. 47, Issue of November 21, pp. 46789–46797.

Broholm, H., Andersen, B.,Wanscher, B., Frederiksen, J. L., Rubin, I., Pakkenberg,B., Larsson, H. B., Lauritzen, M. (2004) Nitric oxide synthase expression and enzymatic activity in multiple sclerosis. Acta Neurol Scand. Vol 109, pp 261–269.

Broillet, M., Randin, O., Chatton,J. (2001) Photoactivation and calcium sensitivity of the fluorescent NO indicator 4,5-diaminofluorescein (DAF-2): implications for cellular NO imaging. FEBS Lett, Vol 491, No 3, pp 227-32.

Brown, L.A., Key, B.J., Lovick,T.A. (1999) Bio-imaging of nitric oxide-producing neurones in slices of rat brain using 4,5-diaminofluorescein. J Neurosci Methods, Vol 92, No 1-2, pp 101-10.

Bryan, N. S.(2008) Nitrite in nitric oxide biology: cause or consequence? A systems based review. Free Radic. Biol. Med. Vol 41, pp 691–701.

Bryk, R. and Wolff, D. J. (1999) Pharmacological modulation of nitric oxide synthesis by mechanism-based inactivators and related inhibitors. Pharmacol. Ther. Vol *84*, pp 157-178

Bryk, R., Wolff, D.J.(1998) Mechanism of inducible nitric oxide synthase inactivation by aminoguanidine and L-N6-(1-iminoethyl)lysine. Biochemistry.Vol 37, No 14, pp 4844-52.

Burnett, A.L. (2002) Nitric Oxide Regulation of Penile Erection: Biology and Therapeutic Implications. Journal of Andrology,Vol 23, No 5, pp 2002:20-26.

Butle, T.A., Lee, M.R., Eldred, W.D. (2000) Direct imaging of NMDA-stimulated nitric oxide production in the retina. Vis Neurosci, Vol 17, No 4, pp 557-66.

Cai, W., Niu, G., Chen, X.(2008b) Imaging of integrins as biomarkers for tumor angiogenesis. Curr. Pharm. Des. Vol 14, pp 2943–2973.

Cai,W., Chen, K., Li, Z. B., Gambhir, S. S., Chen, X. (2007) Dual-function probe for PET and near-infrared fluorescence imaging of tumor vasculature. J. Nucl. Med. Vol 48, pp 1862–1870.

Cai,W., Chen, X.(2008a) Multimodality molecular imaging of tumor angiogenesis. J. Nucl. Med. Vol 49 (Suppl. 2), pp 113S–128S; 2008.

Cai,W., Hsu, A. R., Li, Z. B., Chen, X. (2007a) Are quantum dots ready for in vivo imaging in human subjects? Nanoscale Res. Lett. Vol 2, pp 265–281.

Caramia, F., Yoshida, T., Hamberg, L.M., Huang, Z., Hunter, G., Wanke, I., Zaharchuk, G., Moskowitz, M.A., Rosen, B.R.(1998) Measurement of changes in cerebral blood volume in spontaneously hypertensive rats following L-arginine infusion using dynamic susceptibility contrast MRI. Magn Reson Med, Vol 39, No 1, pp 160-3.

Castagnoli, K., Palmer, S., Castagnoli, Jr.,N.(1999) Neuroprotection by (R)-deprenyl and 7-nitroindazole in the MPTP C57BL/6 mouse model of neurotoxicity. Neurobiology (Bp), Vol 7, No 2, pp 135-49.

Catana, C.,Wu, Y., Judenhofer,M. S., Qi, J., Pichler, B. J., Cherry, S. R. (2006) Simultaneous acquisition of multislice PET and MR images: initial results with a MR compatible PET scanner. J. Nucl. Med. Vol 47, pp 1968–1976.

Chen, X., Sheng, C., Zheng, X. (2001) Direct nitric oxide imaging in cultured hippocampal neurons with diaminoanthraquinone and confocal microscopy. Cell Biol Int, Vol 25, No 7, pp 593-8.

Choi, Y.B., Tenneti, L., Le, D.A., Ortiz, J., Bai, G., Chen, H.S.V., Lipton, S.A. (2000) Molecular basis of NMDA receptor-coupled ion channel modulation by S-nitrosylation. Nat Neurosci, 2000; 3(1): 15-21

Ciplak, M., Pasche, A., Heim, A., Haeberli, C., Waeber, B., Liaudet, L., Feihl, F., Engelberger, R.(2009) The vasodilatory response of skin microcirculation to local heating is subject to desensitization.Microcirculation. Vol 16, No 3, pp 265-75

Claudette, M., Croix, S., Molly, S., Watkins, S.C., Pitt, B.R.(2005) Fluorescence Resonance Energy Transfer–Based Assays for the Real-Time Detection of Nitric Oxide Signaling. Methods in Enzymology, Vol 396, pp 317-326.

Crowell, J. A., Steele, V. E., Sigman, C. C., Fay, J. R.(2003) Is inducible nitric oxide synthase a target for chemoprevention? Mol. Cancer Ther. Vol 2, pp 815–823.

Davies, M. J.; Hawkins, C. L.(2004) EPR spin trapping of protein radicals. Free Radic. Biol. Med. Vol. 36, pp 1072–1086.

Davies, S. L., Loescher, A. R., Clayton, N. M., Bountra, C., Robinson, P. P.,Boissonade, F. M. (2004) nNOS expression following inferior alveolar nerve injury in the ferret. Brain Res. Vol. 1027, pp 11–17.

Day, R.W., White, K.S., Hedlund, G.L. (2005) Nitric oxide increases the signal intensity of the T1-weighted magnetic resonance image of blood. J Cardiovasc Magn Reson. Vol 7, No 4, pp 667-9.

de Vries, E. F., Vroegh, J., Dijkstra, G., Moshage, H., Elsinga, P. H., Jansen, P. L., Vaalburg, W. (2004) Synthesis and evaluation of a fluorine-18 labeled antisense oligonucleotide as a potential PET tracer for iNOS mRNA expression. Nucl. Med. Biol. Vol 31, pp 605–612.

Desvignes, C., Bert, L., Vinet, L., Denoroy, L., Renaud, B., Lambas-Senas, L. (1999) Evidence that the neuronal nitric oxide synthase inhibitor 7-nitroindazole inhibits monoamine oxidase in the rat: in vivo effects on extracellular striatal dopamine and 3,4-dihydroxyphenylacetic acid. Neurosci Lett, Vol 264, No 1-3, pp 5-8.

Di Cesare, M. L., Nistri, S., Mazzetti, L., Bani, D., Feil, R., Failli, P. (2009) Altered nitric oxide calcium responsiveness of aortic smooth muscle cells in spontaneously hypertensive rats depends on low expression of cyclic guanosine monophosphate-dependent protein kinase type I.J Hypertens. Vol 26, No 6, pp 1258-1267.

Di Salle, F., Barone, P., Hacker, H., Smaltino, F., Marco, I. (1997)Nitric oxide-haemoglobin interaction: a new biochemical hypothesis for signal changes in fMRI. Neuroreport, Vol 8, No 2, pp 461-4.

Dunn, A. R., Belliston-Bittner, W., Winkler, J. R., Getzoff, E. D., Stuehr, D. J., Gray, H. B. (2005) Luminescent ruthenium(II)- and rhenium(I)-diimine wires bind nitric oxide synthase. J. Am. Chem. Soc. Vol 127, pp 5169–5173.

Engelsman, A.F., Krom, B.P., Busscher, H.J., van Dam, G.M., Ploeg, R.J., van der Mei, H.C.(2009) Antimicrobial effects of an NO-releasing poly(ethylene vinylacetate) coating on soft-tissue implants in vitro and in a murine model. Acta Biomater. Vol 5, No 6, pp 1905-1910.

Fichtlscherer, B., Mülsch, A. (2000) MR imaging of nitrosyl-iron complexes: experimental study in rats.Radiology. Vol 216, No 1, pp 225-31.

Flögel, U., Jacoby, C., Gödecke, A., Schrader, J. (2007) In vivo 2D mapping of impaired murine cardiac energetics in NO-induced heart failure. Magn Reson Med. Vol 57, No 1, pp 50-8.

Foissner, I., Wendehenne D, Langebartels C, Durner J. (2000) In vivo imaging of an elicitor-induced nitric oxide burst in tobacco. Plant J, Vol 23, No 6, pp 817-24.

Foster, M.A., Seimenis, I., Lurie, D.J.(1998) The application of PEDRI to the study of free radicals in vivo. Phys Med Biol, Vol 43, No 7, pp 1893-7.

Franz, K.J., Singh, N., Spinler, B., Lippard, S.J.(2000) Aminotroponiminates as ligands for potential metal-based nitric oxide sensors. Inorg Chem, Vol 39, No 18, pp 4081-92.

Freeman, B. A., Lancaster Jr., J. R., Feelisch, M., Lundberg, J. O. (2005) The emerging biology of the nitrite anion. Nat. Chem. Biol. Vol 1,pp 308–314.

Fujii, H., Itoh, K., Pandian, R.P., Sakata, M., Kuppusamy, P., Hirata, H. (2007) Measuring brain tissue oxygenation under oxidative stress by ESR/MR dual imaging system. Magn Reson Med Sci. Vol 6, No 2, pp 83-9.

Fujii, H., Koscielniak, J., Berliner,L.J.(1997) Determination and characterization of nitric oxide generation in mice by in vivo L-Band EPR spectroscopy. Magn Reson Med, Vol 38, No 4, pp 565-8.

Fujii, H., Wan, X. Zhong, J., Berliner, L. J., Yoshikawa, K. (1999) In vivo imaging of spintrapped nitric oxide in rats with septic shock: MRI spin trapping. Magn. Reson. Med. Vol 42, pp 235–239.

Garcin, E. D., Arvai, A. S., Rosenfeld, R. J., Kroeger, M. D., Crane, B. R., Andersson, G., Andrews, G., Hamley, P. J., Mallinder, P. R.et al. (2008) Anchored plasticity opens doors for selective inhibitor design in nitric oxide synthase. Nat. Chem. Biol. Vol 4,700–707.

Gladwin, M. T. (2004) Haldane, hot dogs, halitosis, and hypoxic vasodilation: the emerging biology of the nitrite anion. J. Clin. Invest. Vol 113, pp 19–21.

Gladwin, M. T., Schechter, A. N., Kim-Shapiro, D. B., Patel, R. P., Hogg, N., Shiva, S., Cannon III, R. O., Kelm, M., Wink, D. A., Espey, M. G., Oldfield, E. H., Pluta, R. M.,

Gorbunov, N.V., Pogue-Geile, K.L., Epperly, M.W., Bigbee, W.L., Draviam, R., Day, B.W., Wald, N., Watkins, S.C., Greenberger, J.S.(2000) Activation of the nitric oxide synthase 2 pathway in the response of bone marrow stromal cells to high doses of ionizing radiation. Radiat Res, Vol 154, No 1, pp 73-86.

Gragasin, F.S., Michelakis, E.D., Hogan, A., Moudgil, R., Hashimoto, K., Wu, X., Bonnet, S., Haromy, A., Archer, S.L. (2004) The neurovascular mechanism of clitoral erection: nitric oxide and cGMP-stimulated activation of BKCa channels. FASEB J. 2004;18:1382-1391.

Green, D.R.(1998) Apoptotic pathways: the roads to ruin. Cell, Vol 94, No 6, pp. 695-8.

Haga, K.K., Gregory, L.J., Hicks, C.A., Ward, M.A., Beech, J.S., Bath, P.W., Williams, S.C., O'Neill, M.J. (2003) The neuronal nitric oxide synthase inhibitor, TRIM, as a neuroprotective agent: effects in models of cerebral ischaemia using histological and magnetic resonance imaging techniques. Brain Res. Vol 993, No 1-2, pp 42-53.

Handy, R. L. and Moore, P. K. (1998) A comparison of the effects of L-NAME, 7-NI and L-NIL on carrageenan-induced hindpaw oedema and NOS activity. Br. J.Pharmacol. Vol 123, pp 1119-1126

Heinrich, U. R., Selivanova, O., Feltens, R., Brieger, J., Mann,W. (2005) Endothelial nitric oxide synthase upregulation in the guinea pig organ of Corti after acute noise trauma. Brain Res. Vol 1047, pp 85–96.

Heiss, C., Sievers, R. E., Amabile, N., Momma, T. Y., Chen, Q., Natarajan, S., Yeghiazarians, Y., Springer, M. L. (2008) In vivo measurement of flow-mediated vasodilation in living rats using high-resolution ultrasound. Am. J. Physiol, Heart Circ. Physiol. Vol 294, pp H1086–H1093; 2008.

Hillebrand, U., Lang, D., Telgmann, R.G., Hagedorn, C., Reuter, S., Kliche, K., Stock, C.M., Oberleithner, H., Pavenstädt, H., Büssemaker, E., Hausberg, M. (2009) Nebivolol decreases endothelial cell stiffness via the estrogen receptor beta: a nano-imaging study.J Hypertens. Vol 27, No 3, pp 517-26.

Hiratsuka, M., Katayama, T., Uematsu, K., Kiyomura, M., Ito, M. (2009) In vivo visualization of nitric oxide and interactions among platelets, leukocytes, and endothelium following hemorrhagic shock and reperfusion. Inflamm Res. Vol 58, No 8, pp 463-471.

Hong, H., Sun, J., Cai, W. (2009) Multimodality imaging of nitric oxide and nitric oxide synthases. Free Radic Biol Med. Vol 47, No 6, pp 684-698.

Horman, S., Browne, G., Krause, U., Patel, J., Vertommen, D., Bertrand, L., Lavoinne, A., Hue, L., Proud, C. & Rider, M. (2002) Activation of AMP-activated protein kinase leads to the phosphorylation of elongation factor 2 and an inhibition of protein synthesis. Curr. Biol. 12, pp 1419–1423.

Hortelano, S., Zeini, M., Traves, P.G., Bosca, L. (2005) Nitric Oxide and Cell Signaling: In Vivo Evaluation of NO-Dependent Apoptosis by MRI and Not NMR Techniques. Method in Enzymol Vol 396, pp 579-584.

Hsiao, J.K., Chu, H.H., Wang, Y.H., Lai, C.W., Chou, P.T., Hsieh, S.T., Wang, J.L., Liu, H.M. (2008) Macrophage physiological function after superparamagnetic iron oxide labeling. NMR Biomed. Vol 21, No 8, pp 820-9

Huang, K. J., Zhang, M., Xie, W. Z., Zhang, H. S., Feng, Y. Q., Wang, H. (2007) Sensitive determination of nitric oxide in some rat tissues using polymer monolith microextraction coupled to high-performance liquid chromatography with fluorescence detection. Anal. Bioanal. Chem. Vol 388, pp 939–946.

Huang, KJ.,Wang, H., Ma, M., Zhang, X., Zhang, H.S. (2007b) Real-time imaging of nitric oxide production in living cells with 1,3,5,7-tetramethyl-2,6-dicarbethoxy-8-(3',4'-diaminophenyl)-difluoroborad iaza-s-indacence by invert fluorescence microscope. Nitric Oxide, Vol 16, No 1, pp 36-43.

Itoh, K., Watanabe, M., Yoshikawa, K., Kanaho, Y., Berliner, L.J., Fujii, H. (2004) Magnetic resonance and biochemical studies during pentylenetetrazole-kindling development: the relationship between nitric oxide, neuronal nitric oxide synthase and seizures. Neuroscience. Vol 129, No 3, pp 757-66.

Itoh, Y., Ma, F.H., Hoshi, H., Oka, M., Noda, K., Ukai, Y., Kojima, H., Nagano, T., Toda, N. (2000) Determination and bioimaging method for nitric oxide in biological specimens by diaminofluorescein fluorometry. Anal Biochem, Vol 287, No 2, pp 203-9.

Janssen-Heininger, Y.M., Mossman, B.T., Heintz, N.H., Forman, H.J., Kalyanaraman, B., Finkel, T., Stamler, J.S., Rhee, S.G., van der Vliet, A.(2008) Redox-based regulation of signal transduction: principles, pitfalls, and promises. Free Radic Biol Med. Vol. 45, No 1, pp 1-17.

Jares-Erijman, E. A.; Jovin, T. M. FRET imaging. Nat. Biotechnol. 21:1387–1395.

Jordan, B.F., Misson, P.D., Demeure, R., Baudelet, C., Beghein, N., Gallez, B. (2000) Changes in tumor oxygenation/perfusion induced by the no donor, isosorbide dinitrate, in comparison with carbogen: monitoring by EPR and MRI. Int J Radiat Oncol Biol Phys, Vol 48, No 2, pp 565-70.

Jordan, B.F., Sonveaux, P., Feron, O., Gregoire, V., Beghein, N., Dessy, C., Gallez, B., (2004) Nitric oxide as radiosensitizer:Evidence for an intrinsic role in addition to its effect on oxygen delivery and consumption. Int J Cancer.Vol 109, No 5, pp 768-773.

Kakiailatu, F.A.(2000) The role of nitric oxide in the mechanism of penile erectionClinical Hemorheology and Microcirculation. Vol 23, No 2-4, pp 283-286.

Kashiwagi, S., Izumi, Y., Gohongi, T., Demou, Z. N., Xu, L., Huang, P. L., Buerk, D. G., Munn, L. L., Jain, R. K., Fukumura, D. (2005) NO mediates mural cell recruitment and vessel morphogenesis in murine melanomas and tissue-engineered blood vessels. J. Clin. Invest. Vol 115, pp 1816–1827.

Keelan, J., Vergun, O., Duchen, M.R. (1999) Excitotoxic mitochondrial depolarisation requires both calcium and nitric oxide in rat hippocampal neurons. J Physiol, Vol 520 Pt 3, pp 797-813.

Khoo, J. P., Alp, N. J., Bendall, J. K., Kawashima, S., Yokoyama, M., Zhang, Y. H., Casadei, B., Channon, K. M. (2004) EPR quantification of vascular nitric oxide production in genetically modified mouse models. Nitric Oxide. Vol.10, pp 156–161.

Kimura, C., Koyama, T., Oike, M., Ito, Y. (2000) Hypotonic stress-induced NO production in endothelium depends on endogenous ATP. Biochem Biophys Res Commun, Vol 274, No 3, pp 736-40.

Kimura, C., Oike, M., Koyama, T., Ito, Y. (2001) Impairment of endothelial nitric oxide production by acute glucose overload. Am J Physiol Endocrinol Metab,Vol 280, No 1, pp 171-8.

Kleschyov, A. L., Mollnau, H., Oelze, M., Meinertz, T., Huang, Y., Harrison, D. G., Munzel, T. (2000) Spin trapping of vascular nitric oxide using colloid Fe(II)-diethyldithiocarbamate. Biochem. Biophys. Res. Commun. Vol. 275, pp 672–677.

Kleschyov, A. L., Muller, B., Keravis, T., Stoeckel, M. E., Stoclet, J. C. (2000) Adventitia derived nitric oxide in rat aortas exposed to endotoxin: cell origin and functional consequences. Am. J. Physiol, Heart Circ. Physiol. Vol 279, pp H2743–H2751.

Kleschyov, A. L.,Wenzel, P., Munzel, T. (2007) Electron paramagnetic resonance (EPR) spin trapping of biological nitric oxide. J. Chromatogr., B Analyt. Technol. Biomed. Life Sci. Vol 851, pp 12–20.

Kojima, H., Hirata, M., Kudo, Y., Kikuchi, K., Nagano, T. (2001a) Visualization of oxygen-concentration-dependent production of nitric oxide in rat hippocampal slices during aglycemia. J Neurochem, Vol 76, No 5, pp 1404-10.

Kojima, H., Hirotani, M., Nakatsubo, N., Kikuchi, K., Urano, Y., Higuchi, T., Hirata, Y., Nagano, T. (2001b) Bioimaging of nitric oxide with fluorescent indicators based on the rhodamine chromophore.Anal Chem. Vol 73, No 9, pp 1967-73.

Kolodziejski, P. J., Musial, A., Koo, J. S. & Eissa, N. T. (2002) Ubiquitination of inducible nitric oxide synthase is required for its degradation.Proc. Natl. Acad. Sci. U. S. A. 99, 12315–12320

Komarov, A.M. Lai, C.S.(1995) Detection of nitric oxide production in mice by spin-trapping electron paramagnetic resonance spectroscopy. Biochim Biophys Acta, Vol 1272, No 1, pp 29-36.

Komeima, K., Hayashi, Y., Naito, Y., Watanabe, Y. (2000) Inhibition of neuronal nitric-oxide synthase by calcium/calmodulin dependent protein kinase II through Ser847 phosphorylation in NG108-15 neuronal cells.J. Biol. Chem. Vol.275, No.36, pp 28139-28143

Konorev, E. A.,Joseph, J.,Kalyanaraman, B. (1996) S-Nitrosoglutathione induces formation of nitrosylmyoglobin in isolated hearts during cardioplegic ischemia—an electron spin resonance study. FEBS Lett. Vol 378:111–114.

Kosaka, H., Sawai, Y., Sakaguchi, H., Kumura, E., Harada, N.,Watanabe, M., Shiga, T. (1994) ESR spectral transition by arteriovenous cycle in nitric oxide hemoglobin of cytokine-treated rats. Am. J. Physiol. Vol 266, pp C1400–C1405.

Kozlov, A. V., Bini, A., Iannone, A., Zini, I., Tomasi, A. (1996) Electron paramagnetic resonance characterization of rat neuronal nitric oxide production ex vivo. Methods Enzymol. Vol 268, pp 229–236.

Kubrina, L.N., Caldwell, W.S., Mordvintcev, P.I., Malenkova, I.V., Vanin, A.F. (1992) EPR evidence for nitric oxide production from guanidino nitrogens of L-arginine in animal tissues in vivo. Biochim Biophys Acta, Vol 1099, No 3, pp 233-7.

Kuppusamy, P., Chzhan, M., Vij, K., Shteynbuk, M., lefer, D.J., Giannella, E., Zweier, J.L. (1994) Three-dimensional spectral-spatial EPR imaging of free radicals in the heart: a technique for imaging tissue metabolism and oxygenation. Proc Natl Acad Sci U S A, Vol 91, No 8, pp 3388-92.

Kuppusamy, P., Shankar, R.A., Roubaud, V.M., Zweier, J.L. (2001) Whole body detection and imaging of nitric oxide generation in mice following cardiopulmonary arrest: detection of intrinsic nitrosoheme complexes.Magn Reson Med. Vol 45, No 4, pp 700-7.

Kuppusamy, P.,Wang, P., Samouilov, A., Zweier, J. L. (1996) Spatial mapping of nitric oxide generation in the ischemic heart using electron paramagnetic resonance imaging. Magn. Reson. Med.Vol 36, pp 212–218.

Labet, V., Grand, A., Morell, C., Cadet, J., Eriksson, L.A. (2009) Mechanism of nitric oxide induced deamination of cytosine. Phys. Chem. Chem. Phys., Vol. 11, pp 2379 - 2386.

Lai, C.S., Komarov, A.M.(1994) Spin trapping of nitric oxide produced in vivo in septic-shock mice. FEBS Lett, Vol 345, No 2-3, pp 120-4.

Lajoix, A. D., Badiou, S., Peraldi-Roux, S., Chardes, T., Dietz, S., Aknin, C.,Tribillac, F., Petit, P., Gross, R. (2006) Protein inhibitor of neuronal nitric oxide synthase (PIN) is a new regulator of glucose-induced insulin secretion. Diabetes Vol 55, pp 3279–3288;

Laszlo, F. and Whittle, B. J. (1997) Actions of isoform-selective and non-selective nitric oxide synthase inhibitors on endotoxin-induced vascular leakage in rat colon. Eur. J. Pharmacol. *334*, pp 99-102.

Li, L., Storey, P., Kim, D., Li, W., Prasad, P. (2003) Kidneys in hypertensive rats show reduced response to nitric oxide synthase inhibition as evaluated by BOLD MRI. J Magn Reson Imaging. Vol 17, No 6, pp 671-5.

Li, L.P., Ji, L., Santos, E.A., Dunkle, E., Pierchala, L., Prasad, P.(2009) Effect of nitric oxide synthase inhibition on intrarenal oxygenation as evaluated by blood oxygenation level-dependent magnetic resonance imaging. Invest Radiol. 2009 Vol.44, No 2, pp 67-73.

Lim, M. H.(2007) Preparation of a copper-based fluorescent probe for nitric oxide and its use in mammalian cultured cells. Nat. Protoc. Vol. 2, pp 408–415.

Lin, S., Fagan, K.A., Li, K.X., Shaul, P.W., Cooper, D.M.F., Rodman, D.M. (2000) Sustained endothelial nitric-oxide synthase activation requires capacitative Ca2+ entry. J Biol Chem, Vol 275, No 24, pp 17979-85.

Liu, G., Li, Y., Pagel, M.D. (2007) Design and characterization of a new irreversible responsive PARACEST MRI contrast agent that detects nitric oxide. Magn Reson Med. Vol 58, No 6, pp 1249-56.

Lopez-Figueroa, M.O., Caamano, C., Morano, M.I., Ronn, L.C., Akil, H., Watson, S.J. (2000) Direct evidence of nitric oxide presence within mitochondria. Biochem Biophys Res Commun, Vol 272, No 1, pp 129-33.Hanke, C.J., O'Brien, T., Pritchard, K.A., Campbell, W.B. (2000) Inhibition of adrenal cell aldosterone synthesis by endogenous nitric oxide release. Hypertension, Vol 35, No 1 Pt 2, pp 324-8.

López-Figueroa, M.O., Caamaño, C.A., Inés Morano, M.A., Stanley, H., Watson, J. (2002) Fluorescent imaging of mitochondrial nitric oxide in living cells. Methods in Enzymology, Vol 352, pp 296-303Maril, N., Margalit, R., Rosen, S., Heyman, H., Degani, H. (2006) Detection of evolving acute tubular necrosis with renal 23Na MRI: studies in rats. Kidney Int, Vol 69, No 4, pp 765-8.

Lopez-Figueroa, M.O., Day, H.W., Lee, S., Rivier, C., Akil, H., Watson, S.J. (2000) Temporal and anatomical distribution of nitric oxide synthase mRNA expression and nitric oxide production during central nervous system inflammation. Brain Res, Vol 852, No 1, pp 239-46.

Lundberg, J. O., Weitzberg, E. (2005) NO generation from nitrite and its role in vascular control. Arterioscler. Thromb. Vasc. Biol. Vol 25, pp 915–922.

Mader, K. (1998) Pharmaceutical applications of in vivo EPR. Phys. Med. Biol. Vol. 43, pp 1931–1935.

Madhunapantula, S. V., Desai, D., Sharma, A., Huh, S. J., Amin, S., Robertson, G. P.(2008) PBISe, a novel selenium-containing drug for the treatment of malignant melanoma. Mol. Cancer Ther. Vol 7, pp 1297–1308; 2008.

Massoud, T. F., Gambhir, S. S. (2003) Molecular imaging in living subjects: seeing fundamental biological processes in a new light. Genes Dev. Vol.17, pp 545–580.

Matsuo, T.(2000) Basal nitric oxide production is enhanced by hydraulic pressure in cultured human trabecular cells. Br J Ophthalmol, Vol 84, No 6, pp 631-5.

Matter, H., Kotsonis, P.(2004) Biology and chemistry of the inhibition of nitric oxide synthases by pteridine-derivatives as therapeutic agents. Med Res Rev Vol 24, pp 662-684.

McCarthy, T. J., Dence, C. S., Holmberg, S.W., Markham, J., Schuster, D. P.,Welch, M. J. (1996) Inhaled [13N]nitric oxide: a positron emission tomography (PET) study.Nucl. Med. Biol. Vol 23, pp 773–777.

Metukuri, M.R., Beer-Stolz, D., Namas, R.A., Dhupar, R., Torres, A., Loughran, P.A., et al.(2009) Expression and subcellular localization of BNIP3 in hypoxic hepatocytes and liver stress. Am J Physiol Gastrointest Liver Physiol. Vol 296, No 3, pp G499-509.

Moore, P.K., Handy, D.W. (1997) Selective inhibitors of neuronal nitric oxide synthase--is no NOS really good NOS for the nervous system? Trends Pharmacol Sci, Vol 18, No 6, pp 204-11.

Moriyama, Y., Moriyama, E. H., Blackmore, K., Akens,M. K., Lilge, L.(2005) In vivo study of the inflammatorymodulating effects of low-level laser therapy on iNOS expression using bioluminescence imaging. Photochem. Photobiol. Vol 81, pp 1351–1355.

Mulsch, A.,Lurie, D.J., Seimenis, I., Fichhtischerer, B., Foster, M.A. (1999) Detection of nitrosyl-iron complexes by proton-electron-double-resonance imaging. Free Radic Biol Med, Vol 27, No 5-6, pp 636-46.

Nagano, T., Yoshimura, T. (2002) Bioimaging of nitric oxide. Chem Rev. Vol 102, pp 1235-1269.

Nagata, S.(1997) Apoptosis by death factor. Cell, Vol 88, No 3, pp 355-65.

Narayanan, K., Spack, L., McMillan, K., Kilbourn, R.G. (1995)S.Alkyl-L-thiocitrullines. Potent stereoselective inhibitors of nitric oxide synthase with strong pressor activity in vivo. J. Biol. Chem. Vol.270, No.19, pp 11103-11110.

Nie F, Mai XL, Chen J, Gu N, Shi HJ, Cao AH, Ge YQ, Zhang Y, Teng GJ. [In vitro MR imaging of Fe2O3-arginine labeled heNOS gene modified endothelial progenitor cells]. Zhonghua Xin Xue Guan Bing Za Zhi. 2008 Aug;36(8):695-701.

Nishida, C., R., Ortiz de Montellano. (1999) Autoinhibition of endothelial nitric oxide synthase. J. Biol. Chem. Vol. 274, No. 21, pp 14692-14698]

Ny, L., Li, H., Mukherjee, S., Persson, K., Holmqvist, B., Zhao, D., Shtutin, V., Huang, H., Weiss, L.M., Machado, F.S., Factor, S.M., Chan, J., Tanowitz, H.B., Jelicks, L.A. (2008) A magnetic resonance imaging study of intestinal dilation in Trypanosoma cruzi-infected mice deficient in nitric oxide synthase. Am J Trop Med Hyg. Vol 79, No 5, pp 760-7.

Ockaili, R., Emani, V.R., Okubo, S., Brown, M., Krottapalli, K., Kukreja, R.C. (1999) Opening of mitochondrial K_{ATP} channel induces early and delayed cardioprotective effect: role of nitric oxide. Am J Physiol Heart Circ Physiol Vol 277, pp 2425-2434.

Ouyang, J., Hong, H., Shen, C., Zhao, Y., Ouyang, C., Dong, L., Zhu, J., Guo, Z., Zeng, K., Chen, J., Zhang, C., Zhang, J. (2008) A novel fluorescent probe for the detection of nitric oxide in vitro and in vivo. Free Radic. Biol. Med. Vol. 45, pp 1426–1436.

Palombo, F., Cremers, S.G., Weinberg, P.D., Kazarian, S.G. (2009) Application of Fourier transform infrared spectroscopic imaging to the study of effects of age and dietary L-arginine on aortic lesion composition in cholesterol-fed rabbits.J R Soc Interface. Vol 6, No 37, pp 669-680.

Panda, K., Chawla-Sarkar, M., Santos, C., Koeck, T., Erzurum, S. C., Parkinson, J. F., Pomper, M. G.,Musachio, J. L., Scheffel, U., Macdonald, J. E., McCarthy, D. J., Reif, D. W., Villemagne, V. L., Yokoi, F., Dannals, R. F., Wong, D. F. (2000) Radiolabeled neuronal nitric oxide synthase inhibitors: synthesis, in vivo evaluation, and primate PET studies. J. Nucl. Med. Vol 41, pp 1417–1425.

Pechkovsky, D.V., Zissel, G., Stamme, C., Goldmann, T., Ari Jaffe, H., Einhaus, M., Taube, C., Magnussen, H., Schlaak, M., MullerQuernheim, J. (2002) Human alveolar epithelial cells induce nitric oxide synthase-2 expression in alveolar macrophages. Wur Respir J. Vol 16, pp 672-683.

Pedroso, M.C., Magalhaes, J.R., Durzan, D. (2000a) A nitric oxide burst precedes apoptosis in angiosperm and gymnosperm callus cells and foliar tissues. J Exp Bot, Vol 51, No 347, pp 1027-36.

Pedroso, M.C., Magalhaes, J.R., Durzan, D. (2000b) Nitric oxide induces cell death in Taxus cells. Plant Sci, Vol 157, No 2, pp 173-180.

Perrier, E., Fournet-Bourguignon, M.P., Royere, E., Molez, S., Reure, H., Lesage, L., Gosgnach W., Frapart, Y., Boucher, J.L., Villeneuve, N., Vilaine, J.P. (2009) Effect of uncoupling endothelial nitric oxide synthase on calcium homeostasis in aged porcine endothelial cells. Cardiovasc Res. Vol 82, No 1, pp 133-42.

Piknova, B.; Gladwin, M. T.; Schechter, A. N.; Hogg, N.(2005) Electron paramagnetic resonance analysis of nitrosylhemoglobin in humans during NO inhalation.J. Biol. Chem. Vol 280, pp 40583–40588.

Pilon G, Dallaire P, Marette A. Inhibition of Inducible Nitric-oxide Synthase by Activators of AMP-activated Protein Kinase: A New mechanism of action of insulin of insulin-sensitizing drugs. J. Biol. Chem. 2004 279: 20767-20774.

Pittner, J., Liu, R., Brown, R., Wolgast, M., Persson, A.E.G. (2003) Visualization of nitric oxide production and intracellular calcium in juxtamedullary afferent arteriolar endothelial cells. Acta Physiol Scand, Vol 179, No 3, pp 309-17.

Quaresima, V., Takehara, H., Tsushima, K., Ferrari, M., Utsumi, H. (1996) In vivo detection of mouse liver nitric oxide generation by spin trapping electron paramagnetic resonance spectroscopy. Biochem. Biophys. Res. Commun. Vol. 221, pp 729–734.

Raman, C. S., Li, H., Martasek, P., Kral, V., Masters, B. S. and Poulos, T. L. (1998) Crystal structure of constitutive endothelial nitric oxide synthase : a paradigm for pterin function involving a novel metal center. Cell (Cambridge, Mass.) *95*, pp 939-950.

Rast, S., Borel, A., Helm, L., Belorizky, E., Fries, P.H. et al.(2001) EPR spectroscopy of MRI-related Gd(III) complexes: simultaneous analysis of multiple frequency and temperature spectra, including static and transient crystal field effects. J Am Chem Soc, Vol 123, No 11, pp 2637-44.

Ratovitski, E.A., Alam, M.R., Quick, R.A., McMillan, A., Bao, C.,Kozlovsky, C.,Hand,T.A., Johnson, R.C., Mains, R.E., Eippers, B.A.(1999) Kalirin inhibition of inducible nitric oxide synthase. J.Biol. Chem. Vol. 274, No. 2, Issue of January 8, pp. 993–999.

Rees, D. D., Monkhouse, J. E., Cambridge, D. and Moncada, S. (1998) Nitric oxide and the haemodynamic pro®le of endotoxin shock in the conscious mouse. Br. J. Pharmacol. *124*, pp 540-546.

Regehr, W.G., Connor, J.A., Tank, D.W. (1989) Optical imaging of calcium accumulation in hippocampal pyramidal cells during synaptic activation. Nature, Vol 341, No 6242, pp 533-6.

Reif, A., Jacob, C.P., Rujescu, D., Herterich, S., Lang, S., Gutknecht, L., Baehne, C.G., Strobel, A., Freitag, C.M., Giegling, I., Romanos, M., Hartmann, A., Rösler, M., Renner, T.J., Fallgatter, A.J., Retz, W., Ehlis, A.C., Lesch, K.P. (2009) Influence of functional variant of neuronal nitric oxide synthase on impulsive behaviors in humans. Arch Gen Psychiatry. Vol.66, No 1, pp 41-50.

Richards, D.A., T.V. Bliss, and C.D. Richards, Differential modulation of NMDA-induced calcium transients by arachidonic acid and nitric oxide in cultured hippocampal neurons. Eur J Neurosci, 2003; 17(11): 2323-8.

Robinson, J. K.; Bollinger, M. J.; Birks, J. W. (1999) Luminol/H_2O_2 chemiluminescence detector for the analysis of nitric oxide in exhaled breath. Anal. Chem. Vol 71, pp 5131–5136.

Salerno, L., Sorrenti, V., Giacomo, C., Romeo, G., Siracusa, M.A. (2002) Progress in the development of selective nitric oxide synthase (NOS) inhibitors. Curr Pharm Des, Vol 8, No 3, pp177-200.

Samouilov, A., Woldman, Y.Y., Zweier, J.L., Khramtsov, V.V. (2007) Magnetic resonance study of the transmembrane nitrite diffusion.Nitric Oxide. Vol 16, No 3, pp 362-70.

Sari-Sarraf, F., Pomposiello, S., Laurent, D. (2008) Acute impairment of rat renal function by L -NAME as measured using dynamic MRI. MAGMA. Vol 21, No 4, pp 291-7.

Schuppe, H., Cutte, M., Chad, J.E., Newland, P.L. (2002) 4,5-diaminofluoroscein imaging of nitric oxide synthesis in crayfish terminal ganglia. J Neurobiol, Vol 53, No 3, pp 361-9.

Sennequier, N., Wolan, D. and Stuehr, D. J. (1999) Antifungal imidazoles block assembly of inducible NO synthase into an active dimer. J. Biol. Chem. Vol *274*, pp 930-938.

Sergeant, G.P., Craven, M., Hollywood, M.A., McHale, N.G., Thornbury, K.D.(2009) Spontaneous Ca^{2+} waves in rabbit corpus cavernosum: modulation by nitric oxide and cGMP.J Sex Med. Vol 6, No 4, pp 958-66.

Sirmatel O, Sert C, Tümer C, Oztürk A, Bilgin M, Ziylan Z. Change of nitric oxide concentration in men exposed to a 1.5 T constant magnetic field. Bioelectromagnetics. 2007;28(2):152-4.

Southan, G.J., Szabo,C.(1996) Selective pharmacological inhibition of distinct nitric oxide synthase isoforms. Biochem Pharmacol, Vol 51, No 4, pp 383-94.

St Croix, C. M.; Bauer, E. M. Use of spectral fluorescence resonance energy transfer to detect nitric oxide-based signaling events in isolated perfused lung. Curr. Protoc. Cytom. Chap. 12 (Unit12):13; 2008.

Stuehr, D. J. (2005) Visualizing inducible nitric-oxide synthase in living cells with a hemebinding fluorescent inhibitor. Proc. Natl. Acad. Sci. USA 102:10117–10122; 2005.

Suzuki, N., Kojima, H., Urano, Y., Kikuchi, K., Hirata, Y., Nagano, T. (2002) Orthogonality of calcium concentration and ability of 4,5-diaminofluorescein to detect NO. J Biol Chem, Vol 277, No 1, pp 47-49.

Swartz, H. M., Khan, N., Khramtsov, V. V. (2007) Use of electron paramagnetic resonance spectroscopy to evaluate the redox state in vivo. Antioxid. Redox Signal. Vol. 9, pp 1757–1771.

Taha, Z. H. (2003) Nitric oxide measurements in biological samples. Talanta Vol. 61, pp 3–10.

Terashima M, Ehara S, Yang E, Kosuge H, Tsao PS, Quertermous T, Contag CH, McConnell MV. In Vivo bioluminescence imaging of inducible nitric oxide synthase gene expression in vascular inflammation. Mol Imaging Biol. 2010 Nov 6.

Thatte, H.S., He, X.D., Goyal, R.K. (2009) Imaging of nitric oxide in nitrergic neuromuscular neurotransmission in the gut.PLoS ONE. Vol 4, No 4, pp 4990.

Townsend, D. W., Beyer, T. (2002) A combined PET/CT scanner: the path to true image fusion. Br. J. Radiol. Vol 75 Spec. No., pp S24–S30.

Tsuchiya, K., Yoshizumi, M., Houchi, H., Mason, R. P. (2000) Nitric oxide-forming reaction between the iron-N-methyl-D-glucamine dithiocarbamate complex and nitrite. J. Biol. Chem. Vol. 275, pp 1551–1556.

Ueno, T., Suzuki, Y., Fujii, S., Vanin, A.F., Yoshimara, T. (2002) In vivo nitric oxide transfer of a physiological NO carrier, dinitrosyl dithiolato iron complex, to target complex. Biochemical Pharmacol.Vol 63, No 485-493.

Vandsburger, M.H., French, B.A., Helm, P.A., Roy, R.J., Kramer, C.M., Young, A.A., Epstein, F.H. (2007) Multi-parameter in vivo cardiac magnetic resonance imaging demonstrates normal perfusion reserve despite severely attenuated beta-adrenergic functional response in neuronal nitric oxide synthase knockout mice. Eur Heart J. Vol 28, No 22, pp 2792-8.

Vanin, A. F., Bevers, L. M., Mikoyan, V. D., Poltorakov, A. P., Kubrina, L. N., van Faassen, E. (2007) Reduction enhances yields of nitric oxide trapping by irondiethyldithiocarbamate complex in biological systems. Nitric Oxide Vol. 16, pp 71–81.

Vila Petroff M.G., Kim, S.H., Pepe, S., Dessy, C., Marban, E., Balligand, J.L., Sollott, S.J.(2001) Endogenous nitric oxide mechanisms mediate the stretch dependence of Ca++ release in cardiomyocytes. Nature Cell Biol. Vol 3, pp 867-873..

Vodovotz, Y., Lucia, M.S., Flanders, K.C., Chesler, L., Xie, Q.Q., Smith,W.W.,Weidner, J., Mumford, R.,Webber, R.,Nathan, C., Roberts, A.B., Lippa, C.F., Sporn, M.B. (1996) Inducible nitric oxide synthase in tangle-bearing neurons of patients with Alzheimer's disease. J Exp Med Vol 184, pp 1425–1433.

von Bohlen, U., Halbach, O.(2003) Nitric oxide imaging in living neuronal tissues using fluorescent probes. Nitric Oxide, Vol 9, No 4, pp 217-28.

von Bohlen, U., Halbach, O., Albrecht, D., Heinemann, U., Schuchmann, S. (2002) Spatial nitric oxide imaging using 1,2-diaminoanthraquinone to investigate the involvement of nitric oxide in long-term potentiation in rat brain slices. Neuroimage, Vol 15, No 3, pp 633-9.

Waller, C., Hiller, K.H., Rüdiger, T., Kraus, G., Konietzko, C., Hardt, N., Ertl, G., Bauer, W.R. (2005) Noninvasive imaging of angiogenesis inhibition following nitric oxide synthase blockade in the ischemic rat heart in vivo. Microcirculation. Vol 12, No 4, pp 339-47.

Wingrove, J.A., O'Farrell, P.H. (1999) Nitric oxide contributes to behavioral, cellular, and developmental responses to low oxygen in Drosophila. Cell, Vol 98, No 1, pp 105-14.

Xu, Y. C., Cao, Y. L., Guo, P., Tao, Y., Zhao, B. L. (2004) Detection of nitric oxide in plants by electron spin resonance. Phytopathology. Vol. 94, pp 402–407.

Yermolaieva, O., Brot, N., Weissbach, H., Heinemann, H. T. (2000) Reactive oxygen species and nitric oxide mediate plasticity of neuronal calcium signaling. Proc Natl Acad Sci U S A, Vol 97, No 1, pp 448-53.

Yokoyama, H., Fujii, S., Yoshimura, T., Ohya-Nishiguchi, H., Kamada, H. (1997) In vivo ESR-CT imaging of the liver in mice receiving subcutaneous injection of nitric oxide-bound iron complex. Magn. Reson. Imaging Vol 15, pp 249–253.

Yoshimura, T., Yokoyama, H., Fujii, S., Takayama, F. (1996) In vivo EPR detection and imaging of endogenous nitric oxide in lipopolysaccharide-treated mice. Nat Biotechnol, Vol 14, No 8, pp 992-4.

Young, R. J., Beams, R. M., Carter, K., Clark, H. A., Coe, D. M., Chambers, C. L., Davies, P. I., Dawson, J., Drysdale, M. J., Franzman, K. W. et al. (2000) Inhibition of inducible nitric oxide synthase by acetamidine derivatives of hetero-substituted lysine and homolysine. Bioorg. Med. Chem. Lett. Vol 10, pp 597-600.

Zhang, J., Xu, M., Dence, C. S., Sherman, E. L., McCarthy, T. J., Welch, M. J.(1997) Synthesis, in vivo evaluation and PET study of a carbon-11-labeled neuronal nitric oxide synthase (nNOS) inhibitor S-methyl-L-thiocitrulline. J. Nucl. Med. Vol 38, pp 1273–1278.

Zharikov, S.I., Sigova, A.A., Chen, S., Bubb, M.R., Block, E.R. (2001) Cytoskeletal regulation of the L-arginine/NO pathway in pulmonary artery endothelial cells. Am J Physiol Lung Cell Mol Physiol, Vol 280, No 3, pp 465-73.

Zhou, D., Lee, H., Rothfuss, J.M., Chen, D.L., Ponde, D.E., Welch, M.J., Mach, R.H. (2009) Design and synthesis of 2-amino-4-methylpyridine analogues as inhibitors for inducible nitric oxide synthase and in vivo evaluation of [^{18}F]6-(2-fluoropropyl)-4-methyl-pyridin-2-amine as a potential PET tracer for inducible nitric oxide synthase.J Med Chem. Vol 52, No 8, pp 2443-53.

Ziaja, M., Pyka, J., Machowska, A., Maslanka, A., Plonka, P. M. (2007) Nitric oxide spin-trapping and NADPH-diaphorase activity in mature rat brain after injury.J. Neurotrauma Vol. 24, pp 1845–1854.

6

Transcriptional Bursting in the Tryptophan Operon of *E. coli* and Its Effect on the System Stochastic Dynamics

Emanuel Salazar-Cavazos and Moisés Santillán
Centro de Investigación y de Estudios Avanzados del IPN, Unidad Monterrey, Parque de Investigación e Innovación Tecnológica, Apodaca NL México

1. Introduction

Transcriptional bursting, also known as transcriptional pulsing, is a fundamental property of genes from bacteria to humans (Chubb et al., 2006; Golding et al., 2005; Raj et al., 2006). Transcription of genes, the process which transforms the stable code written in DNA into the mobile RNA message can occur in "bursts" or "pulses". This phenomenon has recently come to light with the advent of new technologies, such as MS2 tagging, to detect RNA production in single cells, allowing precise measurements of RNA number, or RNA appearance at the gene. Other, more widespread techniques, such as Northern Blotting, Microarrays, RT-PCR and RNA-Seq, measure bulk RNA levels from homogeneous population extracts. These techniques lose dynamic information from individual cells, and give the impression transcription is a continuous smooth process. The reality is that transcription is irregular, with strong periods of activity, interspersed by long periods of inactivity. Averaged over millions of cells, this appears continuous. But at the individual cell level, there is considerable variability, and for most genes, very little activity at any one time.

The bursting phenomenon, as opposed to simple probabilistic models of transcription, can account for the high variability in gene expression occurring between cells in isogenic populations (Blake et al., 2003). This variability in turn can have tremendous consequences on cell behaviour, and must be mitigated or integrated. In certain contexts, such as the survival of microbes in rapidly changing stressful environments, or several types of scattered differentiation, the variability may be essential (Losick & Desplan, 2008). Variability also impacts upon the effectiveness of clinical treatment, with resistance of bacteria to antibiotics demonstrably caused by non-genetic differences (Lewis, 2010). Variability in gene expression may also contribute to resistance of sub-populations of cancer cells to chemotherapy (Sharma et al., 2010).

Bursting may result from the stochastic nature of biochemical events superimposed upon a 2 or more step fluctuation. In its most simple form, the gene can exist in 2 states, one where activity is negligible and one where there is a certain probability of activation (Raj & van Oudenaarden, 2008). Only in the second state does transcription readily occur. Whilst the nuclear and signalling landscapes of complex eukaryotic nuclei are likely to favour more than two simple states—for example, there are over twenty post-translational modifications of nucleosomes known, this simple two step model provides a reasonable framework for understanding the changing probabilities affecting transcription. What do the restrictive and permissive states represent? An attractive idea is that the repressed state is a closed chromatin conformation whilst the permissive state is an open one. Another hypothesis is that the fluctuations reflect transition between bound pre-initiation complexes (permissive) and dissociated ones (restrictive) (Blake et al., 2003; Ross et al., 1994). Bursts may also result from bursty signalling, cell cycle effects or movement of chromatin to and from transcription factories. Nonetheless, to the best of our knowledge, there is no generally accepted explanation for this phenomenon. Transcriptional bursting in prokaryotic cell is particularly puzzling given the simplicity of the transcription initiation process, as opposed to eukaryotic cells.

There is evidence that the cooperative interaction of distant operators through a single repressor molecule has a strong influence on the transcriptional bursting observed in the *lac* operon of *E. coli* (Choi et al., 2008). Since the repression regulatory mechanism in *E. coli*'s *trp* operon involves cooperativity between two repressor molecules bound to neighbouring operators, it is interesting to investigate whether such cooperative interaction has any effect upon the system transcriptional dynamics. The present chapter is advocated to tackling this question from a mathematical modelling perspective. We also investigate the effects of the enzymatic feedback-inhibition regulatory mechanism (also present on the *trp* operon regulatory pathway) on the system dynamic behaviour.

2. Theory

2.1 The *trp* operon

The amino acid tryptophan can be synthesized by bacteria like *E. coli* through a series of catalysed reactions. The catalysing enzymes in *E. coli* are made up of the polypeptides encoded by the tryptophan operon genes: *trpE*, *trpD*, *trpC*, *trpB*, and *trpA*, which are transcribed from *trpE* to *trpA*. Transcription is initiated at promoter *trpP*, which is located upstream from gene *trpE*. The *trp* operon is regulated by three different negative-feedback mechanisms: repression, transcription attenuation, and enzyme inhibition. Below these regulatory mechanisms are briefly reviewed based on (Brown et al., 1999; Grillo et al., 1999; Jeeves et al., 1999; Xie et al., 2003; Yanofsky, 2000; Yanofsky & Crawford, 1987). The reader should consult Figure 1 for a better understanding.

Repression in the *trp* operon is mediated by three operators (*O1*, *O2*, and *O3*) overlapping with the operon promoter, *trpP* (see Figure 1A). When an active repressor is bound to an operator it blocks the binding of a RNA polymerase to *trpP* and prevents transcription initiation. The *trp* repressor normally exists as a dimeric protein (called the *trp* aporepressor) and may or may not be complexed with tryptophan (Trp). Each portion of the *trp* aporepressor has a binding site for tryptophan.

The *trp* aporepressor cannot bind the operators tightly when not complexed with tryptophan. However, if two tryptophan molecules bind to their respective binding sites the *trp* aporepressor is converted into a functional repressor. The resulting functional repressor complex can tightly bind to the *trp* operators and so the synthesis of the tryptophan producing enzymes is prevented. This sequence constitutes the repression negative-feedback mechanism: an increase in the concentration of tryptophan induces an increase in the concentration of the functional repressor, thus preventing the synthesis of tryptophan.

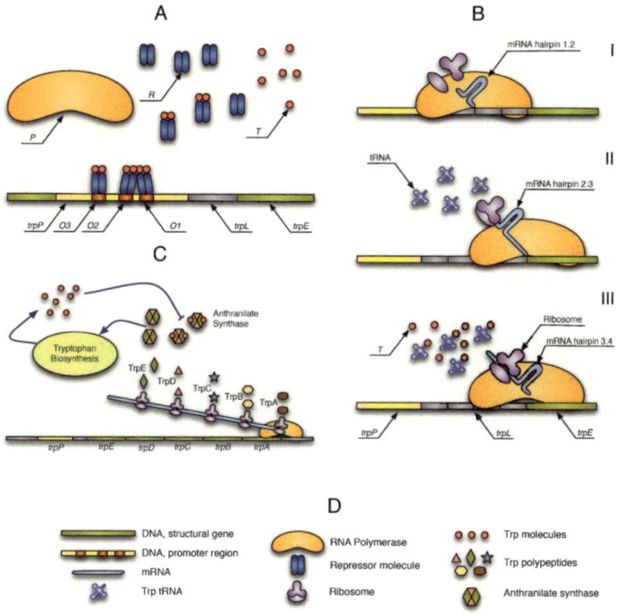

Fig. 1. Schematic representation of the three regulatory mechanisms found in the tryptophan operon: A) repression, B) transcription attenuation, and C) enzyme inhibition. A glossary with the meaning of all the geometric forms in this figure is shown in panel D. See the main text for further explanation.

Transcription attenuation works by promoting an early termination of mRNA transcription, see Figure 1B. The transcription starting site in the *trp* operon is separated from *trpE* by a leader region responsible for attenuation control. The transcript of this leader region consists of four segments (termed Segments 1, 2, 3, and 4) which can form three stable hairpin structures between consecutive segments. After the first two segments are transcribed they form a hairpin which stops transcription (c.f. Figure 1B-I). When a ribosome binds the nascent mRNA, it disrupts Hairpin 1:2 and transcription is re-initiated along with translation. Segment 1 contains two tryptophan codons in tandem. If there is scarcity of tryptophan, and thus of loaded $tRNA^{Trp}$, the ribosome stalls in the first segment. The development of Hairpin 2:3 (the antiterminator) is then facilitated, and transcription proceeds until the end (c.f. Figure 1B-II). However, if tryptophan is abundant, the ribosome rapidly finishes translation of Segments 1 and 2 and promotes the formation of a stable hairpin structure between Segments 3 and 4 (c.f. Figure 1B-III). RNA polymerase molecules recognize this hairpin structure as a termination signal and transcription is prematurely terminated.

Enzyme inhibition takes place through anthranilate synthase, the first enzyme to catalyse a reaction in the catalytic pathway that leads to the synthesis of tryptophan from chorismate. This enzyme is a hetero-tetramer consisting of two TrpE and two TrpD polypeptides. Anthranilate synthase is inhibited by tryptophan through negative feedback. This feedback inhibition is achieved when the TrpE subunits in anthranilate synthase are individually bound by a tryptophan molecule, see Figure 1C. Therefore, an excess of intracellular tryptophan inactivates most of the anthranilate synthase protein avoiding the production of more tryptophan.

3. Methods

3.1 Model development

There are three different repressor binding sites (operators) overlapping the *trp* promoter. Hence, the promoter can be in eight different states, with each operator being either free or bound by a repressor molecule. Furthermore, when two repressor molecules are bound to the first and second operators, they do it cooperatively (Grillo et al., 1999). As discussed in Appendix A, this cooperativity allows the grouping of the promoter states into two different sets that we term the permissive and the restrictive global states. The transitions within each global state and those from the permissive global state to the restrictive global state being much faster than those from the restrictive global to the permissive global states. This fact justifies the assumption that the system "instantaneously" reaches a stationary probability distribution for the states within every global state. This supposition in turn permits the derivation of the following expressions for the transition rates from the permissive into the restrictive global states (k^+), and vice versa (k^-):

$$k^+ = \frac{k_i^+ R_A/K_j + k_j^+ R_A/K_i}{1 + R_A/K_i + R_A/K_j}, \tag{1}$$

$$k^- = \frac{k_i^- + k_j^-}{k_c}, \tag{2}$$

where k_i^+ and k_j^+ are the rates for the reactions where a repressor molecule binds to the first and second operators, respectively; k_i^- and k_j^- are the rates for the reactions in which a repressor molecule detaches from the first and second operator; R_A is the number of active repressors; $K_i = k_i^-/k_i^+$; and $K_j = k_j^-/k_j^+$.

On the other hand, when the promoter is in the permissive global state, the probability that it is not bound by any repressor and so it is free to be bound by a polymerase to start transcription is

$$P_R = \frac{1}{(1 + R_A/K_k)(1 + R_A/K_i + R_A/K_j)}. \tag{3}$$

Repressor molecules are activated when they are bound by a couple of tryptophan molecules. The kinetics of repressor activation were analysed in (Santillán & Zeron, 2004), where the number of active repressors is demonstrated to be given by

$$R_A = R_T \left(\frac{T}{T + K_T} \right)^2, \tag{4}$$

where R_T stands for the total number of repressor molecules, while T is the tryptophan molecule count. Substitution of Eqn. (4) into Eqns. (1) and (3) permits the calculation of the promoter inactivation rate (k^+) and the probability P_R in terms of the tryptophan level (T).

Due to transcriptional attenuation, only a fraction of the polymerase molecules that initiate transcription reach the end of the *trp* genes and produce functional mRNA molecules, which in turn are translated to produce the proteins coded by the *trp* genes. Santillán and Zeron

(2004) found that the probability that transcription is not prematurely terminated due to transcriptional attenuation is:

$$P_A = \frac{1 + 2\alpha T}{(1 + \alpha T)^2}, \tag{5}$$

with α a parameter to be estimated.

It follows from the considerations in the previous paragraphs that a promoter in the restrictive global state is completely incapable of being expressed, but its activity level when it is in the permissive global state is a function of the tryptophan level T and is given by the product of P_R and P_A. Therefore, if k_E denotes the rate of enzyme synthesis by a fully active promoter, the enzyme synthesis rate when the promoter is in the permissive state at a given tryptophan level turns out to be:

$$k_E P_R(T) P_A(T). \tag{6}$$

Tryptophan is synthesized by proteins which are assembled from the polypeptides coded by the *trp* genes. Conversely, tryptophan is mainly consumed in the synthesis of all kinds of proteins in *E. coli*. Thus, the equation governing the tryptophan-level dynamics is:

$$\frac{dT}{dt} = k_T E P_I(T) - \rho(T) - \mu T,$$

where μ is the bacterial growth rate, k_T is the tryptophan rate of synthesis per enzyme molecule

$$P_I(T) = \frac{K_I^n}{T^n + K_I^n}$$

is the probability that an enzyme molecule is not feedback inhibited by tryptophan, and

$$\rho(T) = \rho_{max} \frac{T}{T + K_\rho}$$

is the rate of tryptophan consumption associated to protein synthesis. If we assume that these processes are much faster than those associated to gene expression and protein degradation, then we can make the following quasi-steady state approximation: $dT/dt = 0$, and the tryptophan molecule count can be uniquely calculated in terms of the enzyme molecule count as the root of the following algebraic equation:

$$k_T E P_I(T) - \rho(T) - \mu T = 0. \tag{7}$$

Following previous modelling studies we assume that the enzyme degradation rate is negligible as compared with the bacterial growth rate, μ. On the other hand, instead of considering a cell that grows exponentially, we assume that we have a constant-volume cell and that the effective enzyme degradation rate is μ.

The facts previously discussed in the present section provide enough information to develop a model for the *trp* operon regulatory pathway. This model consists of four chemical reactions: promoter activation, promoter inactivation, enzyme synthesis, and enzyme degradation, whose rates are k^-, $k^+(T(E))R_A(T(E))$, $k_E P_R(T(E)) P_A(T(E))$, and μE, respectively. Figure 2 provides a schematic representation of such a model. It is worth emphasizing that the repression regulatory feedback loop is implicitly accounted for by functions $k^+(T(E))$ and $P_R(T(E))$, that function $P_A(T(E))$ corresponds to the attenuation feedback regulatory mechanism, and the feedback enzyme inhibition is implicit in the function $T(E)$, obtained after solving Eqn. (7).

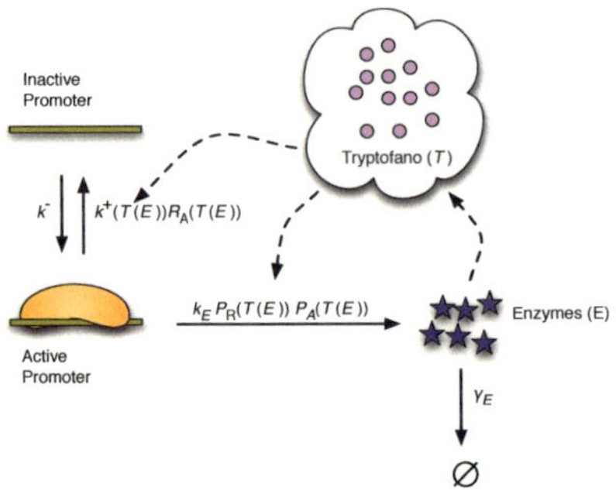

Fig. 2. Schematic representation of the mathematical model here developed for tryptophan operon gene regulatory circuit.

3.2 Parameter estimation

We paid special attention to the estimation of all the model parameters from reported experimental data. The parameter values we employ in the present work and the detailed procedure to estimate them are presented in Appendix B.

3.3 Numerical methods

The time evolution of the reaction network that models the tryptophan operon regulatory pathway was simulated by means of the Gillespie algorithm, which we implemented in `Python`.

4. Results

We carried out stochastic simulations with the model described in the previous section. As formerly stated, we made use of Gillespie's algorithm to mimic the system dynamic evolution for 200,000 min. In the first simulation we employed the parameter values estimated in Appendix B, which correspond to a wild-type bacterial strain. The results are summarized in Figure 3. In Figure 3A the cumulative sum of the promoter activity is plotted vs. time. We can appreciate there the existence of alternated activity and inactivity periods, just like it has been observed in transcriptional bursting. To further investigate this phenomenon we calculated the histograms of the permissive and restrictive period lengths. The results are shown in Figures 3B and 3C, respectively. Observe that both histograms are well fitted by exponential distributions, in agreement with the reported experimental data on transcriptional bursting. Finally, we present in Figures 3D and 3E the histograms for the enzyme and tryptophan molecule counts, respectively.

One feature worth noticing is that the histogram for the enzyme abundance is well fitted by a gamma distribution with parameters $k = 32.5$ and $\theta = 63$. This last fact is in agreement with the existence on transcriptional bursting in the *trp* operon of *E. coli*. It has been proved that in such a case, the protein count obeys a gamma distribution, with parameters k and θ respectively interpreted as the average number of transcriptional bursts occurring during an average protein lifetime and the mean number of proteins produced per burst (Shahrezaei & Swain, 2008). It is also interesting to point out that the coefficient of variation in the tryptophan molecule count is similar to that of the enzyme molecule count.

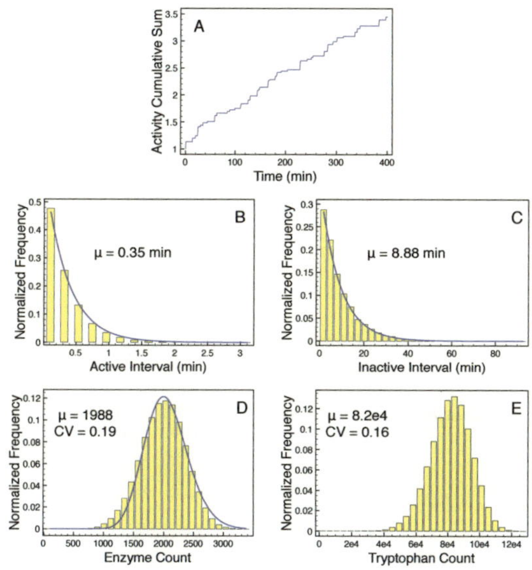

Fig. 3. Statistical analysis of the simulation corresponding to the wild-type strain. A) Plot of the cumulative promoter activity vs. time. B) Histogram of the promoter permissive time intervals and best fit to an exponential distribution. C) Histogram of the promoter restrictive time intervals and best fit to an exponential distribution. D) Histogram of the enzyme molecule count and best fit to a gamma distribution. E) Histogram of the tryptophan molecule count.

In order to understand the influence of the slow promoter gating between the global permissive and restrictive states on the operon dynamics, we increased the value of parameters k_x^+ and k_x^- $(x = I, j, k)$ by a factor of 100. In this way, the promoter switching rate among all its available states gets faster, without altering each state's stationary probability. We repeated the simulation described in the previous paragraph with this new parameter set and the results are condensed in Figure 4.

We observe by comparing Figures 3A and 4A that there are many more alternated activity and inactivity periods in the fast-switching model than in that corresponding to the wild type strain. Concomitantly, the periods are shorter in the former case. Interestingly, the accumulated promoter activity is quite similar for both models. This last result comes from our increasing parameters k_x^+ and k_x^- by the same factor, and is in agreement with the fact that the mean enzyme and tryptophan counts are quite similar in both models (see below).

In Figures 4B and 4C we present the histograms for the activity and inactivity periods, and the corresponding fits to exponential distributions. By comparing with the wild-type period distributions we can see that the mean values of both the activity and inactivity periods decrease. However the decrease of the activity-period average is about twice as large as that of the average of the inactivity period.

Finally, the histograms for the enzyme and tryptophan molecule counts are plotted in Figures 4D and 4E. Notice that the histogram for the enzyme count is well approximated by a gamma distributions with parameters $k = 2079$ and $\theta = 1$. Therefore, by making the promoter switching rate faster we increased the frequency of bursting, but decreased in the same

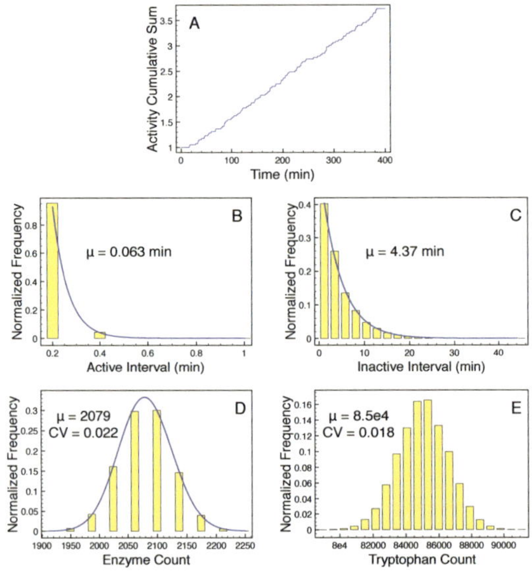

Fig. 4. Statistical analysis of the simulation corresponding to the strain with fast promoter-transition rates. A) Plot of the cumulative promoter activity vs. time. B) Histogram of the promoter permissive time intervals and best fit to an exponential distribution. C) Histogram of the promoter restrictive time intervals and best fit to an exponential distribution. D) Histogram of the enzyme molecule count and best fit to a gamma distribution. E) Histogram of the tryptophan molecule count.

amount the number of proteins synthesized per burst. In that way, the average enzyme count remains the same (compare with Figure 3D). However, the variation coefficient is much smaller in the fast promoter switching model than in the wild-type one. Recall that, in the gamma distribution, the mean and the standard deviation are: $\mu = \theta k$ and $\sigma = \theta\sqrt{k}$, while the variation coefficient is given by $CV = \sigma/\mu = 1/\sqrt{k}$.

We further simulated a bacterial mutant strain lacking the feedback inhibition regulatory mechanism. To mimic this mutation we increased the value of parameter K_I by two orders of magnitude, up to $K_I = 500,000$ molecules. We analysed this last simulation in a similar way than the previous ones and present the results in Figure 5.

A comparison of Figures 3A and 5A reveals that the promoter level of activation is generally smaller in the inhibition-less mutant strain than in the wild-type strain, because the accumulated activity is about three times smaller in the former case. Nonetheless the length of the activity and inactivity periods seem to be similar. This last assertion is corroborated by the plots in Figures 5B and 5C, where we can see the the activity and inactivity period histograms are well fitted by exponential distributions, and that the corresponding mean values are similar to the corresponding ones in the wild-type strain.

In agreement with the fact that the promoter level of activity is smaller in the inhibition-less than in the wild-type strain, the mean protein count is smaller in the first case (compare Figures 3D and 5D). On the other hand, the coefficient of variation (CV) is similar in both cases. To understand why this happens when one would expect a larger CV in the inhibition-less

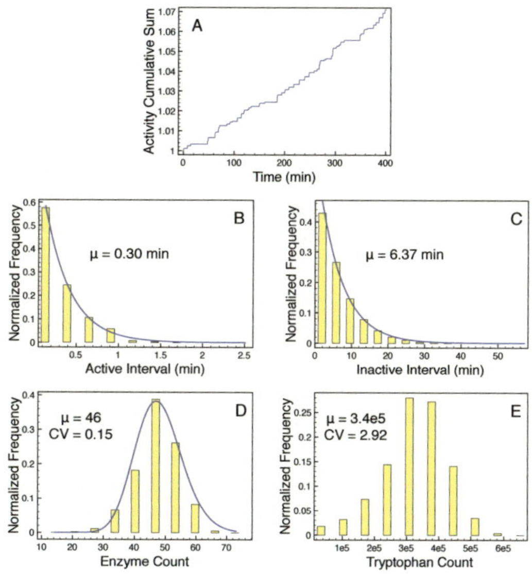

Fig. 5. Statistical analysis of the simulation corresponding to the inhibition-less strain. A) Plot of the cumulative promoter activity vs. time. B) Histogram of the promoter permissive time intervals and best fit to an exponential distribution. C) Histogram of the promoter restrictive time intervals and best fit to an exponential distribution. D) Histogram of the enzyme molecule count and best fit to a gamma distribution. E) Histogram of the tryptophan molecule count.

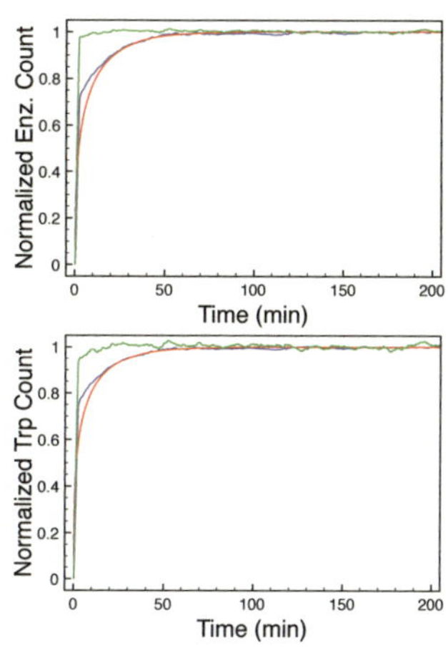

Fig. 6. Plots of the normalized enzyme count and the normalized tryptophan count, averaged over 1,000 independent simulations, vs. time for the wild-type (blue line), the fast promoter-transition (red line), and the inhibition-less (green line) *E. coli* strains.

mutant because of the reduced enzyme count, we fitted the histogram in Figure 5D and found that the best fit is obtained with parameters $k = 40$ and $\theta = 1.2$. Since the values of k for the inhibition-less and the wild-type strains are similar, we conclude that the burst frequency is comparable in both cases. However, the number of proteins produced per burst is notably smaller in the inhibition-less strain due to the reduced promoter activity, as well as to the increased level of transcriptional attenuation.

We note that, in contrast with the two previous simulations, the variation coefficient of the tryptophan molecular count in the inhibition-less strain is much larger than the coefficient of variation for the enzyme count. To our understanding, this happens because, being enzyme inhibition absent, there is a highly non-linear relation between the enzyme and tryptophan molecule counts.

In order to investigate the effect of the inhibition-less mutation and of the increased promoter switching rate on the *trp* operon response time, we carried out 1,000 simulations with each bacterial strain (including the wild type), starting with the promoter in the restrictive state and zero enzyme and tryptophan molecules as initial conditions. Then, we averaged the enzyme and the tryptophan counts over all the simulations, and plotted the results in Figure 6 to compare how fast each strain approaches the steady state. We can see there that the inhibition-less strain is the one with the shortest response time, followed by the wild type and the fast promoter switching strains, respectively.

5. Concluding remarks

In this work we have introduced a stochastic mathematical model for the tryptophan operon. Our objectives were twofold: 1) to investigate whether the reported reaction rates of the interaction between repressor and the operators can give rise to transcriptional bursting; and 2) to study the dynamic effects of transcriptional bursting, if it exists, and of the feedback enzyme-inhibition regulatory mechanism.

Regarding the first objective our results indicate that, indeed, the reported reaction rates make the promoter switching between its available states slow enough so as to give rise to transcriptional bursting. As previously discussed, this assertion is supported by the agreement between our model results and a number of reported experimental facts. Interestingly, experiments on the *lac* operon also suggest that transcriptional bursting has its origin in the kinetics of the repressor-operator interaction. As a matter of fact, Choi et al. (2008) demonstrated that the cooperativity present when a single repressor molecule is bound to two distant operators is responsible of generating two different types of bursts in the expression of *lac* operon.

In our model we have assumed that the trp promoter activation rate k^{-} is independent of the tryptophan concentration. This assumption is supported by direct and indirect experimental measurements of the promoter activation and deactivation rates on the *lac* and several other promoters (Choi et al., 2008; So et al., 2011). Those reports demonstrate that modulation of gene expression is mainly achieved by changing the promoter deactivation rate.

Regarding our second objective, our results allow us to put forward the following conclusions:

- Transcriptional bursting increases the noise level and decreases the system response time after a nutritional shift. To the best of our understanding, the noise level is increased because the promoter-transition events become less frequent as the promoter-repressor interactions slow down, thus enhancing the concomitant stochastic effects. On the other hand, the faster system response can be explained by a single burst of intense transcriptional activity, occurring during the first couple of minutes after the nutritional shift. This burst allows most of the cells to reach tryptophan levels superior to two thirds of the steady-state level.

- Enzyme inhibition also has important dynamic effects. It increases the noise level in the enzyme count, but decreases the noise level in the number of tryptophan molecules. Furthermore, this regulatory mechanism also increases the system response time. Knowing that the presence of a strong negative feedback is capable of reducing the noise in a biological system (Austin et al., 2006; Becskei & Serrano, 2000; Dublanche et al., 2006), we can explain the above observations as follows: the fact that the wild-type *E. coli* strain has lower tryptophan levels than the inhibition-free strain means that the transcriptional-attenuation and the repression feedback loops are weaker in the first strain; this weakening of both negative feedback loops is responsible for the increment of the noise level in the enzyme count. For the same reasons, the presence of the enzyme-inhibition feedback loop reduces the noise level in the tryptophan count, but makes it necessary to produce much more enzymes to fulfil the required tryptophan production. This increased enzyme synthesis requirement lengthens the system response time.

Finally, if we assume that having a tryptophan operon with short response times and low noise levels in the tryptophan molecular count are beneficial traits for *E. coli* then we can speculate from the previously discussed facts that evolution has bestowed this system with an optimal trade-off between short response times and low tryptophan noise.

6. Acknowledgements

This research was partially supported by Consejo Nacional de Ciencia y Tecnología (CONACyT, MEXICO) under Grant: 55228.

7. Appendix

A. Promoter dynamics modelling

Three different operators are overlap with the *trp* promoter and each of them can be bound by a repressor molecule. Therefore, the eight promoter states can be denoted as $\{i, j, k\}$, where $i, j, k = 0, 1$ represent the binding state of the first, second, and third repressors, respectively. A zero (one) value means that the corresponding operator is free from (bound by) a repressor molecule.

Let k_i^+, k_j^+, and k_k^+ respectively denote the rates of binding of a repressor molecule to the first, second, and third operators, when the other two are free. Similarly, let k_i^-, k_j^-, and k_k^- respectively represent the dissociation rate of a repressor molecule solely bound to the first, second, and third operators.

It is known that the first and second operators are bound by repressor molecules cooperatively and the cooperativity constant $k_c > 1$ has been measured. Here we assume that this cooperativity means that the rate of dissociation for a repressor molecule bound to either the first or the second operator is respectively given by k_i^- / k_c or k_j^- / k_c, when both operators are bound by repressor molecules. Under this assumption, the eight different promoter states can be grouped into two sets that we call the permissive and the restrictive global states. The permissive global state consists of states $(0, 0, 0)$, $(1, 0, 0)$, $(0, 1, 0)$, $(0, 0, 1)$, $(1, 0, 1)$, and $(0, 1, 1)$, while the restrictive global state consists of $(1, 1, 0)$, and $(1, 1, 1)$. From the way the were constructed, the transitions within each global state and the transitions from the permissive to the restrictive global state are much faster than the transitions from the restrictive to the permissive global state. From the above considerations we make an adiabatic approximations for all the transitions within the permissive state, and thus:

$$
\begin{aligned}
P(0,0,0)\tfrac{R_A}{K_i} &= P(1,0,0),\\
P(0,0,0)\tfrac{R_A}{K_j} &= P(0,1,0),\\
P(0,0,0)\tfrac{R_A}{K_k} &= P(0,0,1),\\
P(0,0,1)\tfrac{R_A}{K_i} &= P(1,0,1),\\
P(0,0,1)\tfrac{R_A}{K_j} &= P(0,1,1),
\end{aligned}
$$

where $P(i,j,k)$ stands for the probability of state i,j,k within the restrictive global state, R_A denotes the number of active repressors, and $K_x = k_x^-/k_x^+$ $(x = i,j,k)$. It follow from the equations above and the constraint $P(0,0,0) + P(1,0,0) + P(0,1,0) + P(0,0,1) + P(1,0,1) + P(0,1,1) = 1$ that

$$
\begin{aligned}
P(0,0,0) &= \tfrac{1}{(1+R_A/K_k)(1+R_A/K_i+R_A/K_j)},\\
P(1,0,0) &= \tfrac{R_A/K_i}{(1+R_A/K_k)(1+R_A/K_i+R_A/K_j)},\\
P(0,1,0) &= \tfrac{R_A/K_j}{(1+R_A/K_k)(1+R_A/K_i+R_A/K_j)},\\
P(0,0,1) &= \tfrac{R_A/K_k}{(1+R_A/K_k)(1+R_A/K_i+R_A/K_j)},\\
P(1,0,1) &= \tfrac{R_A/K_i \times R_A/K_k}{(1+R_A/K_k)(1+R_A/K_i+R_A/K_j)},\\
P(0,1,1) &= \tfrac{R_A/K_j \times R_A/K_k}{(1+R_A/K_k)(1+R_A/K_i+R_A/K_j)}.
\end{aligned}
$$

Let k^+ (k^-) be the transition rate from each of the states in the permissive (restrictive) global state to each if the states in the restrictive (permissive) global state. From Zeron and Santillán (2010), these rates are given by

$$
k^+ = k_i^+(P(0,1,0) + P(0,1,1)) + k_j^+(P(1,0,0) + P(1,0,1)) = \frac{k_i^+ R_A/K_j + k_j^+ R_A/K_i}{1 + R_A/K_i + R_A/K_j},
$$

and

$$
k^- = \frac{k_i^- + k_j^-}{k_c}(P(1,1,0) + P(1,1,1)) = \frac{k_i^- + k_j^-}{k_c},
$$

since $P(1,1,0) + P(1,1,1) = 1$.

Finally, when the promoter is in the permissive global state, the probability that it is not bound by any repressor and so it is free to be bound by a polymerase to start transcription is

$$
P_R = P(0,0,0) = \frac{1}{(1 + R_A/K_k)(1 + R_A/K_i + R_A/K_j)}.
$$

Thus P_R can be interpreted as the operator activity level.

B. Estimation of the model parameters

In this work we consider a bacterial doubling time of 40 min, and thus

$$\mu \simeq 0.017 \text{ min}^{-1}.$$

From the website `E. coli Statistics`, the total number of proteins in *E. coli* is about 3.6 millions, while the average protein size is 360 residues. On the other hand, we have from the website `B1ONUMB3RS` (`http://bionumbers.hms.harvard.edu/`) that the abundance of tryptophan in the *E. coli* proteins is around 1.1%. The data above imply that there are of the order of 14.256 million tryptophan molecules assembles in the *E. coli* proteins at any given time. If we further consider that all the proteins in a bacterium have to be doubled before it duplicates (every 40 min), then the average tryptophan consumption rate is

$$\rho_{max} \simeq 360,000 \text{ molecules/min}.$$

Since this consumption rate cannot be maintained when the tryptophan level is too low, we assumed that the consumption rate for this amino acid is given by

$$\rho(T) = \rho_{max} \frac{T}{T + K_\rho},$$

with

$$K_\rho = 1,000 \text{ molecules}.$$

This choice for K_ρ guarantees that the tryptophan consumption rate is most of the time very close to ρ_{max}, except when there are of the order of a few thousand molecules of the amino acid.

According to the website `E. coli Statistics` (`http://redpoll.pharmacy.ualberta.ca/CCDB/cgi-bin/STAT_NEW.cgi`), the average tryptophan molecule count in this bacterium is

$$T^* \simeq 80,000 \text{ molecules}.$$

From Morse et al. (1968), the average number of anthranilate synthase enzymes in *E. coli* is

$$E^* \simeq 2,000 \text{ molecules}.$$

Caligiuri & Bauerle (1991) found from their experimental data that the probability that an anthranilate synthase enzyme is not feedback inhibited by tryptophan can be approximated by the following function:

$$P_I(T) = \frac{K_I^n}{K_I^n + T^n},$$

with

$$K_I \simeq 2,500 \text{ molecules}, \quad \text{and} \quad n \simeq 1.2.$$

Let k_T be the tryptophan synthesis rate per anthranilate synthase molecule. Given, that the average tryptophan synthesis rate must equal the consumption rate for this amino acid, we can solve for k_T from the following equation $k_T E^* P_I(T*) = \rho_{max}$. After doing the math we obtain

$$k_T \simeq 12,000 \text{ molecules/min}.$$

Consider the promoter model discussed in Section 3. Let P_{act} be the probability that the promoter is in the global permissive state. We have thus that, in the stationary state, $k^+ R_A P_{act} = k^-(1 - P_{act})$. Then, by solving for P_{act} in the previous equation and taking into consideration that, when the promoter is in the global permissive state, the probability that it is ready to be bound by a polymerase is P_R. The stationary promoter activity level as a function of the tryptophan molecule count is

$$\frac{1}{1 + k^+(T) R_A(T)/k^-} P_R(T).$$

It is straightforward to test that the promoter activity level equals one when $T = 0$ molecules. On the other hand, Yanofsky & Horn (1994) measured that the operon expression level is maximal under conditions of tryptophan starvation, and that the repression regulatory mechanism decreases the promoter activity level by 60 times when the tryptophan level reaches its normal value. Thus, we must have that

$$\frac{1}{1 + k^+(T^*) R_A(T^*)/k^-} P_R(T^*) = 1/60.$$

This last result can then be used to estimate parameter K_T—see Eqn. (4). Thus, from Eqn. (4) and given that (Gunsalus et al., 1986)

$$R_T \simeq 400 \text{ molecules},$$

we obtain after some algebra that

$$K_T \simeq 1.7 \times 10^6 \text{ molecules}.$$

Grillo et al. (1999) estimated the following values for the promoter-repressor interaction rates:

$$\begin{aligned}
k_i^+ &\simeq 8.1 \text{ molecules}^{-1}\text{min}^{-1}, \\
k_i^- &\simeq 6.0 \text{ min}^{-1}, \\
k_j^+ &\simeq 0.312 \text{ molecules}^{-1}\text{min}^{-1}, \\
k_j^- &\simeq 0.198 \text{ min}^{-1}, \\
k_k^+ &\simeq 0.3 \text{ molecules}^{-1}\text{min}^{-1}, \\
k_k^- &\simeq 36.0 \text{ min}^{-1}, \\
k_c &\simeq 40.0
\end{aligned}$$

The probability that transcription is not prematurely terminated due to transcriptional attenuation is given by Eqn. (5). On the other hand, Yanofsky & Horn (1994) measured that one of every ten polymerases that have initiated transcription finish transcribing the operon genes when the tryptophan level is at its normal value. This means that $P_A(T) = 0.1$. We obtain from this that

$$\alpha \simeq 2.3 \times 10^{-4} \text{ molecules}^{-1}.$$

Finally, the value of parameter k_E is chosen so that, when $T = T^*$, the average enzyme molecule count is E^*. We found by inspection that

$$k_E \simeq 30,000 \text{ molecules/min}.$$

complies with this requirement.

8. References

Austin, D., Allen, M., McCollum, J., Dar, R., Wilgus, J., Sayler, G., Samatova, N., Cox, C. & Simpson, M. (2006). Gene network shaping of inherent noise spectra, *Nature* 439(7076): 608–611.

Becskei, A. & Serrano, L. (2000). Engineering stability in gene networks by autoregulation, *Nature* 405(6786): 590–593.

Blake, W., Kaern, M., Cantor, C. & Collins, J. (2003). Noise in eukaryotic gene expression, *Nature* 422(6932): 633–637.

Brown, M. P., Grillo, A. O., Boyer, M. & Royer, C. A. (1999). Probing the role of water in the tryptophan repressor-operator complex, *Protein Sci* 8(6): 1276–85.

Caligiuri, M. G. & Bauerle, R. (1991). Identification of amino acid residues involved in feedback regulation of the anthranilate synthase complex from *Salmonella typhimurium*, *J. Biol. Chem.* 266: 8328–8335.

Choi, P. J., Cai, L., Frieda, K. & Xie, S. (2008). A stochastic single-molecule event triggers phenotype switching of a bacterial cell, *Science* 322(5900): 442–446.

Chubb, J., Trcek, T., Shenoy, S. & Singer, R. (2006). Transcriptional pulsing of a developmental gene, *Current Biology* 16(10): 1018–1025.

Dublanche, Y., Michalodimitrakis, K., Kuemmerer, N., Foglierini, M. & Serrano, L. (2006). Noise in transcription negative feedback loops: simulation and experimental analysis, *Molecular Systems Biology* 2: 41.

Golding, I., Paulsson, J., Zawilski, S. & Cox, E. (2005). Real-time kinetics of gene activity in individual bacteria, *Cell* 123(6): 1025–1036.

Grillo, A. O., Brown, M. P. & Royer, C. A. (1999). Probing the physical basis for *trp* repressor-operator recognition, *J. Mol. Biol.* 287: 539–554.

Gunsalus, R. P., Miguel, A. G. & Gunsalus, G. L. (1986). Intracellular *trp* repressor levels in *Escherichia coli*, *J Bacteriol* 167(1): 272–8.

Jeeves, M., Evans, P. D., Parslow, R. A., Jaseja, M. & Hyde, E. I. (1999). Studies of the *Escherichia coli trp* repressor binding to its five operators and to variant operator sequences, *Eur J Biochem* 265(3): 919–28.

Lewis, K. (2010). Persister cells, *Annual Review of Microbiology, Vol 64, 2010* 64: 357–372.

Losick, R. & Desplan, C. (2008). Stochasticity and cell fate, *Science* 320(5872): 65–68.

Morse, D. E., Baker, R. F. & Yanofsky, C. (1968). Translation of the tryptophan messenger RNA of *Escherichia coli*, *Proc. Natl. Acad. Sci.* 60(4): 1428–1435.

Raj, A., Peskin, C. S., Tranchina, D., Vargas, D. Y. & Tyagi, S. (2006). Stochastic mRNA synthesis in mammalian cells, *Plos Biology* 4(10): 1707–1719.

Raj, A. & van Oudenaarden, A. (2008). Nature, nurture, or chance: Stochastic gene expression and its consequences, *Cell* 135(2): 216–226.

Ross, I. L., Browne, C. M. & Hume, D. A. (1994). Transcription of individual genes in eukaryotic cells occurs randomly and infrequently, *Immunol Cell Biol* 72(2): 177–85.

Santillán, M. & Zeron, E. S. (2004). Dynamic influence of feedback enzyme inhibition and transcription attenuation on the tryptophan operon response to nutritional shifts, *J. Theor. Biol.* 231: 287–298.

Shahrezaei, V. & Swain, P. S. (2008). Analytical distributions for stochastic gene expression, *Proceedings of the National Academy of Sciences of the United States of America* 105(45): 17256–17261.

Sharma, S. V., Lee, D. Y., Li, B., Quinlan, M. P., Takahashi, F., Maheswaran, S., McDermott, U., Azizian, N., Zou, L., Fischbach, M. A., Wong, K.-K., Brandstetter, K., Wittner, B.,

Ramaswamy, S., Classon, M. & Settleman, J. (2010). A chromatin-mediated reversible drug-tolerant state in cancer cell subpopulations, *Cell* 141(1): 69–80.

So, L.-H., Ghosh, A., Zong, C., Sepulveda, L. A., Segev, R. & Golding, I. (2011). General properties of transcriptional time series in escherichia coli, *Nature Genetics* 43(6): 554–U84.

Xie, G., Keyhani, N. O., Bonner, C. A. & Jensen, R. A. (2003). Ancient origin of the tryptophan operon and the dynamics of evolutionary change, *Microbiol. Mol. Biol. Rev.* 67: 303–342.

Yanofsky, C. (2000). Transcription attenuation, once viewed as a novel regulatory strategy, *J. Bacteriol.* 182: 1–8.

Yanofsky, C. & Crawford, I. P. (1987). The tryptophan operon, *in* F. C. Neidhart, J. L. Ingraham, K. B. Low, B. Magasanik & H. E. Umbarger (eds), *Escherichia coli and Salmonella thyphymurium: Cellular and Molecular Biology, Vol. 2,* Am. Soc. Microbiol., Washington, DC, pp. 1453–1472.

Yanofsky, C. & Horn, V. (1994). Role of regulatory features of the *trp* operon of *Escherichia coli* in mediating a response to a nutritional shift, *J. Bacteriol.* 176: 6245–6254.

7

Mechanisms of Hepatocellular Dysfunction and Regeneration: Enzyme Inhibition by Nitroimidazole and Human Liver Regeneration

Rakesh Sharma[1,2]
[1]*Center of Nanomagnetics and Biotechnology, Florida State University, Tallahassee, Florida*
[2]*Amity Institute of Nanotechnology, Amity University, NOIDA UP,*
[1]*USA*
[2]*India*

1. Introduction

Enzymes so called (Enz'-a-ai-am) are biologically active protein molecules responsible of biochemical reactions in the bacteria, cells and organs. Enzymes regulate the rates of biochemical reactions to maintain the metabolism to keep active physiological actions in the body. High enzyme activities or high rates of reactions cause higher product formation or deposits to initiate disease. Drugs are used as enzyme competitors to normalize the reactions to correct disease. Most of diseases are cured by 'enzyme inhibition'. Enzyme inhibition can be three types: competitive, non-competitive, and uncompetitive. Enzyme inhibition also provides a kind of defense in cells by regulation of metabolism to inhibit or stimulate the biochemical processes. Most of the drugs undergo detoxification and biotransformation in liver to regenerate or formation of new hepatocytes and Kupffer cells. Major biochemical events in liver regeneration are regulated by enzymes in energy metabolism, growth factors and cytokine molecules.

Present chapter describes an example of liver cell enzyme battery to regulate the carbohydrate, lipid and protein metabolism in liver cells, mainly hepatocytes and Kupffer cells in the light of liver damage due to amoebic liver abscess and role of enzyme inhibition in liver regeneration by 2'-nitroimidazole. Liver damage by amoeba is manifested by elevated enzymes in cells. As a result, two major clinical manifestations of *Entamoeba histolytica* infection are amoebic colitis and amoebic liver abscesses. To cure amoebic liver abscess, liver regeneration and amoebic killing by 2'-nitroimidazole therapy is routine in clinical practice. 2'-nitroimidazole acts in liver to perform enzyme inhibition at the level of carbohydrate, lipid, and protein metabolic regulatory steps. Earlier, nitroimidazole derivatives were considered drug of choice in treatment of hepatic hypoxia (low oxygen) conditions in parasitic infections, cancer and recently nitroimidazole derivatives are emerging as hypoxia markers and radiosensitizers in tumor treatment [Sharma 2001, Sharma 2011a, 2011b].

Since 1960, hepatocytes were investigated rich in major enzymes regulating the energy metabolism located in cytoplasm and mitochondria for glycolysis, TCA cycle while drug metabolizing and redox enzymes are located in lysosomes and microsomes. Other enzymes also participate in defense processes such as respiratory burst, HMP shunt, oxygenase pathways and inflammatory cytokines. Present chapter reviews the ongoing developments with clear and complete information on action of new nitroimidazole derivatives as 'enzyme inhibitors' in liver selective cytotoxicity, oxygen depletion, hepatocellular DNA and enzyme normalization before enzyme can be used in hypoxia monitoring and therapy [Sharma, 2011b]. The paucity of information on nitroimidazole-liver tissue interaction is poor and available data of initial sequential biochemical changes in liver cells is scanty which further leads to detectable hypoxia [Sharma 2011a, 2011b]. It is believed that initially enzyme regulated glucose and calcium hemostasis are the primary targets of hepatic hypoxia followed by enzyme regulated induced metabolic integrity loss leads to apoptosis, regulatory failure in glycolytic, TCA cycle, gluconeogenesis and Ca^{++} mediated cAMP related biodegradation of molecules [Sharma et al.2011c].

The present chapter focuses on hepatic enzyme inhibition by nitroimidazole at different levels of energy metabolism, respiratory burst and drug metabolizing enzymes. In following section, a molecular basis of enzyme inhibition and hepatocellular hypoxia and dysfunction criteria is described in detail. In following section, regulatory behavior of rate limiting enzymes of glycolysis, TCA cycle, phagocytosis is described with examples: glucokinase, phosphofructokinase, pyruvate kinase, lactate dehydrogenase, NADPH cytochrome P450 reductase, phosphodieaterase, lysosomal enzymes in serum, liver biopsy samples. In next section, role of enzymes in liver abscess is described.

1.1 Enzymes in liver abscess

Entamoeba histolytica is a human-specific pathogen and reproducible animal models of intestinal amoebiasis have proved elusive to describe interaction between *E. histolytica* and liver cells. The most common extraintestinal manifestation of disease is amoebic liver-abscess. Amoebic liver-abscesses arise from haematogenous spread (probably via the portal circulation) of amoebic trophozoites that have breached the colonic mucosa.[Thompson et al. 1985, Abuabara et al. 1982, Shandera et al. 1998, Barnes et al. 1987, Adams 1977, Lancet 2003]. Effective nitroimidazole treatment and rapid diagnosis showed the mortality rates to 1–3%. *Entamoeba histolytica* trophozoites can lyse neutrophils *in vitro*, causing them to release toxic substance killer enzymes such as superoxide dismutases, collagenases, elastases and cathepsins [Jarumilinta et al. 1964, Guerrant et al. 1981]. On these lines, author proposed a 'Hepatocellular dysfunction criteria' to make systematic observations in abscess development and step by step method of medical/surgical intervention [Sharma et al.2011]. In previous reports, patient symptoms, leucocytosis without eosinophilia, mild anaemia, raised concentrations of alkaline phosphatase, and a high rate of erythrocyte sedimentation were the most common laboratory findings.[Thompson et al. 1985, Abuabara et al. 1982, Shandera et al. 1998, Barnes et al. 1987, Adams 1977, Lancet 2003].

Amoeba cause amoebic liver abscesses, which are circumscribed regions of dead hepatocytes, liquefied cells and cellular debris surrounded by a rim of connective tissue, some inflammatory cells and few amoebic trophozoites [Sharma et al. 2011]. The adjacent liver parenchyma is usually completely normal. At same sites in liver so called 'acini' fixed sinusoidal phagocytic cells (Kupffer cells) provide defense against any bacteria, virus or foreign drug. Kupffer cells have two intracellular structures filled with enzymes: lysosomes (Lai-zo-somes) and microsomes (Mai-kro-zomes). Lysososmal bags filled with a large number of lysosomal enzymes acting as destroyers actually lyse and digest the bacteria, virus and toxicants. Microsomal enzymes are active in detoxification of drugs or drug biotransformaton to interact (stimulation or inhibition of enzymes) with metabolic regulation in liver cells. Let us introduce a bit of enzymes in liver cells performing different metabolic regulatory functions in following description. Abscesses occupying large areas of the liver can be cured without drainage, and even by one single dose of nitroimidazole.[Powell et al. 1969, Lasserre et al. 1983, Akgun et al. 1999]. Today, role of ultrasound or percutaneous therapeutic aspiration guided by CT in the treatment of uncomplicated amoebic liver-abscess by surgical drainage is controversial.

Enzyme	Amoeba	Hepatocyte	Kupffer cell
Glucose 6 Pase	++++	++	+
Glucose 6 PDH	++++	++	+
Phoshogluconate DH	+++	+	+
Phosphofructokinase	+++	++	+
Aldolase	++	++	+
Pyruvate kinase	++	+	+
Pyruvate DH	++	++	+
LDH	+	++	+++
Citrate Synthase	++++	+	++
Isocitrate DH	+++	++	++
Succinate DH	++	+	+
Malate DH	+++	++	++
Cytochrome C Oxidase	+	++	++
NADPH Cyt C Reducatase	+	+++	++
NADH Oxidase	+	++	++
Tyrosine Aminotransferase	++	+	++++
Aniline Hydroxylase	++	+	++++
Aminopyrine demethylase	++	+	++++
Acid DNAase	++	+	++++
Glutathione Reductase	++	+	++++
Peroxidase	+++	+	++++
Catalase	++	+	++++
Superoxide dismutase	++	+	++++
Guanase	++	++	+++
Adinosine Deaminase	+	+	+++
Leucine aminopeptidase	++	+	++++
Ca++ ATPase	++	+	++++
5′ Nucleotidase	++	+	++++
Acid Lipase	++	+	++++
Acid Phosphatase	++	+	+++++
Alkaline Phosphatase	+++	+	+++++
Beta glucuronidase	++	++	+++

Source: Sharma R. Effect of Nitroimidazole on isolated liver cells in development of amoebic liver abscess. Ph.D dissertation submitted to Indian Institute of Technology, Delhi, 1995.

Table 1. Carbohydrate metabolizing enzymes in amoeba, hepatocytes and Kupffer cells. Abundance of enzyme activities in cells are shown. Comparative enzyme database of amoeba and isolated liver cells is shown and sketched in Figures 3,5, and 6. Strengths of different enzyme activities in amoeba, liver cells are shown as weak(+), moderate(++), high(+++), extreme(++++ or more).

Characterization of enzymes in amoeba has been a long time quest to explore the possibility of amebic enzyme inhibition by new drugs in drug discovery. Several antiamoebic drugs are in market as potent enzyme inhibitors since last four decades. Author proposed a comparative enzyme database of amoeba and hepatocellular cells in culture (see Table 1 and Figure 1). Of specific mention, cysteine proteinase enzymes are secreted in large quantities by the parasite during immune response and can cleave extracellular matrix proteins, which might facilitate amoebic invasion[Scholze et al. 1988, Keene 1986]. Mainly carbohydrate metabolizing, energy metabolizing enzymes and lyssomal enzymes are abundant and participate in amoebic -host liver cell interaction as shown in Figures 2,3a,3b and 3c. These proteins act as virulence factors in animal models of amoebic liver abscess [Stanley et al.1995, Ankri et al.1999]. A family of six genes encodes *E. histolytica* cysteine proteinases (*ehcp1–6*) but ~90% of the proteinase activity is related to three proteinase enzymes, EhCP1, EhCP2 and EhCP5 [Bruchhaus, et al. 2003]. Although targeted disruption of selected *E. histolytica* genes has not yet been achieved, stable episomal expression of foreign DNA is possible in amoebae by maintaining continuous selective pressure [Hamann et al. 1995, Vines et al. 1995]. This has enabled the investigators to target specific molecules in *E. histolytica* by the episomal expression of antisense mRNA or of genes encoding dominant negative mutants [Ankri et al. 1998]. The antisense approach has been applied to *E. histolytica* cysteine proteinases and episomal expression of an antisense RNA to *ehcp5* could reduce total amoebic proteinase activity by 80–90% [Ankri et al. 1998]. To assess the role of *E. histolytica* cysteine proteinases in amoebiasis, in earlier study, human xenografts in SCID-HU-INT mice were infected with amoebic trophozoites expressing the *ehcp5* antisense RNA(proteinase-deficient amoebae) or amoebic trophozoites containing the same plasmid without the antisense insert (the control group)[Zhang et al. 2000]. Major findings were: 1. no obvious defect was apparent in the ability of cysteine-proteinase deficient amoebae to inhabit and survive within the colonic lumen; 2. post-24 h after infection, cysteine proteinase-deficient amoeba had, in contrast to the control group, failed to induce significant amounts of human IL-1α or IL-8 from infected intestine; 3. gut inflammation was also reduced in human intestine infected with cysteine-proteinase-deficient *E. histolytica* trophozoites; 4. control *E. histolytica* trophozoites damaged the intestinal permeability barrier at 24 h but there was only a minimal increase in intestinal permeability in human intestinal xenografts infected with cysteine-protease-deficient amoeba; 5. histological studies at 24 h, human xenografts infected with control amoebae showed damage to the colonic mucosa, invasion of amoebic trophozoites into submucosal tissues and neutrophil-predominant inflammation; 6. by contrast, xenografts infected with cysteine-proteinase-deficient amoebae showed less mucosal damage, almost no evidence for amoebic invasion into submucosal tissue and little inflammation. Authors speculated that cysteine-proteinase-deficient amoeba might have a defect in their ability to induce gut inflammation and to invade into the submucosal tissues [Zhang et al. 2000]. How amoebic cysteine proteinases contribute to gut inflammation and tissue damage in amoebiasis? Possibly, *Entamoeba histolytica* trophozoites expressing the *ehcp5* antisense RNA might have reduced the phagocytic capabilities or reduced virulence in amoebiasis [Ankri et al.1998]. However, protease-deficient amoebae maintain their ability to lyse target cells, so a defect in cell killing does not underlie the reduced virulence of cysteine proteinase-deficient amoebae [Seydel et al.1997]. In addition, it is unlikely to be a direct effect of cysteine proteases on intestinal epithelial cells, because amoebic lysates rich in cysteine protease activity fail to induce high levels of cytokine production or inflammation when placed in human intestinal xenografts [Ankri et al.1998]. It suggests that amoebic cysteine proteinases must exert their effects on gut inflammation and tissue damage after an initial step that requires live active trophozoites.

1.2 Entamoeba histolytica and programmed cell death

Other important role of caspase enzymes was explored when *E. histolytica* trophozoites were incubated with common types of mammalian cells. Normally trophozoites kill mammalian cells in a contact dependent manner by two processes: by lytic necrosis [Berninghausen et al.1997] or undergo apoptosis[Ragland et al.1994]. Apoptosis, or programmed cell death, is an ordered system of cell death with an initiator or signaling phase followed by an effector stage that causes cell death by degrading various cellular components. The effector stage involves the activation of caspase enzymes, cysteine proteinases with specificity for aspartate residues that form a cascade, converging on caspase 3. A key biochemical marker of apoptosis is endonuclease cleavage of chromatin between histone bodies, which can be seen as a DNA ladder on separating gels.

Detection of DNA fragmentation *in situ* by the endlabeling of single- or double-stranded DNA breaks with terminal deoxyribonucleotidyltransferase (TUNEL assay) has also been used as a biomarker of apoptosis [Schulte-Hermann et al.1994]. However, *E. histolytica* can induce apoptosis *in vitro* raises doubt if apoptosis is a component of the cell death commonly seen in amoebic liver abscess. To test this fact, amoebae were inoculated into the livers of SCID mice and evidence for apoptosis was sought by gel separation of DNA from infected livers and TUNEL staining of amoebic liver abscesses [Seydel et al.1998]. DNA ladder formation was detected in samples obtained from SCID mice by 1 h after amoebic inoculation. TUNEL staining revealed significant areas of apoptosis within amoebic liver abscesses common in inflammatory cells and hepatocytes close to amoebic trophozoites [Stenley, 2001]. However, some distant liver cells also were TUNEL positive. Thus, in the murine model of amoebic liver abscess, *E. histolytica* induced cellular apoptosis appears to be a significant component of cell death as shown in Fig. 3. Interestingly, *E. histolytica*-induced apoptosis was not blocked by Bcl-2 [Ragland et al.1994] and did not appear to involve either of the two of the major pathways for the induction of apoptosis: ligation of the Fas receptor and ligation of TNFα1 receptor 1 (TNFR1). In both C57Bl/6 MRL-lpr/lpr mice with no Fas receptor, and C57Bl/6.C^{3}H-gld/gld mice with no Fas ligand, apoptosis was detected in amoebic liver abscesses by DNA ladder formation and TUNEL staining [Seydel et al. 1998]. Apoptosis was also present in hepatocytes in amoebic liver abscesses from both TNFR1 knockout mice and the heterozygote controls[Seydel et al. 1998].]. These data indicate that *E. histolytica* causes apoptosis in mouse liver by a mechanism that is independent of Fas–Fas-ligand interactions and TNFR1.

Genetic approaches and new models of amoebic liver abscesses have provided good prospects on the complex interactions between *E. histolytica* and the host liver cells [Hamann et al.1995]. Well-known *E. histolytica* enzymes lyse cells and digest extracellular matrix proteins. Enzymes have capability to induce apoptosis in hepatocytes and inflammatory cells. Enzymes can stimulate, and perhaps enhance, an NF-κB-mediated human liver inflammatory response that contributes to tissue damage. These new enzyme regulated pathways might offer an explanation for some of the clinical and pathological differences between amoebiasis and amoebic liver abscess. The hepatocyte inhabited apoptosis would predominate in amoebic liver abscesses. If these mechanisms prove to be important in human disease, they will provide new targets in both the host and the parasite for interventions designed to ameliorate or inhibit amoebiasis and amoebic liver abscesses. Recently, author in team established that apoptosis results with hypoxia in liver cells and evaluated nitroimidazole cytotoxicity [Kwon et al. 2009]. In following section, a criterion of hepatocellular hypoxia is described based on liver cell enzymes and cytomorphology.

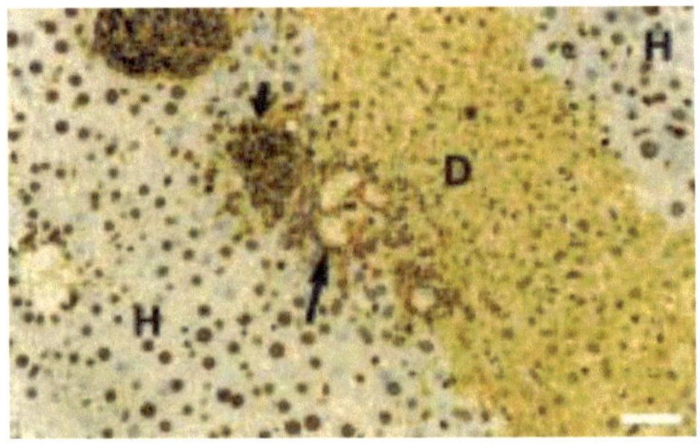

Fig. 1. Entamoeba histolytica trophozoites induce apoptosis in inflammatory cells and hepatocytes in amoebic liver abscesses. Section of liver stained for apoptotic cells with the TUNEL method from an amoebic liver abscess in a C57Bl/6 MRL-lpr/lpr mouse. Amoebic trophozoites (long arrow) are adjacent to a cluster of inflammatory cells (short arrow), which show marked TUNEL staining. TUNEL staining in nuclei and cytoplasm is also evident in a band of dead hepatocytes and inflammatory cells (D). TUNEL-positive nuclei (brown staining) are also visible in hepatocytes in regions (H) flanking the dead cells. Reproduced, with permission, from Ref. Seydel et al. 1998.

2. Liver enzymes and enzyme inhibition in human liver: Hepatocellular criteria

The liver is made of parenchymal hepatocytes and nonparenchymal Kupffer cells as sole targets that exhibit their intracellular biochemical changes as hepatocellular enzyme biomarker profile. Liver cells are rich in enzymes. Enzyme inhibition in liver is characterized by two ways: 1. intact hepatocellular enzymes in hypoxia serum and liver homogenates; 2. Enzyme regulatory behavior by characterizing enzymes in presence of additives added to cultures of isolated liver cells. Additives and drugs are mainly biotransformed and detoxified in liver by a battery of drug metabolizing enzymes: lysosomal and microsomal enzymes. Drugs inhibit enzymes. In following description, readers are introduced with the importance of enzyme inhibition as research tool for two purposes: 1. drug testing and hypoxia disease monitoring (Hepatocellular Hypoxia Criteria) to understand the hypoxia and liver cell interactions with amoeba and nitroimidazole; 2. Enzyme regulatory behavior in presence of additives in isolated liver cells in culture (hepatocytes and Kupffer cells).

2.1 Hepatocellular hypoxia criteria

Author proposed a "Hepatocellular Hypoxia Criteria". 'Hepatocellular hypoxia criteria' assumes that initially liver cells loose metabolic integrity (ATP and NADPH insufficiency from glucose to cause oxygen insufficiency in mitochondria) and undergoes apoptosis followed by detectable necrosis in liver. Hypothesis was that liver cells undergo a series of changes - hepatocytes undergo metabolic energy loss and oxygen depletion (hypoxia) while Kupffer cells undergo phagocytosis, respiratory burst and proteolysis. Nitroimidazole induces enzyme inhibition in liver cells to normalize the elevated enzymes during metabolic energy loss and oxygen depletion (hypoxia) in liver cell (in mitochondria) and lysosomal enzyme inhibition to combat the phagocytosis. The approach is applicable possibly in design of drugs to treat oxygen starved tumor cells or infected liver lesions and abscesses.

Previous reports on first initial stage of nitroimidazole induced liver cell cytotoxicity indicated the apoptosis, glutathione, DNA interaction, oxygen depletion, lactate inhibition, enhanced NADPH cytochrome reductase and superoxide dismutase enzymes [Ersoz et al.2001, Brezden et al.1997, Noss et al.1988, Sugiyama et al.1993, Berube et al. 1992, Sapora et al.1992, Moselen et al. 1995, Stratford et al.1981, Adams et al.1980, Noss et al. 1989, Mulcahy et al.1989, Widel et al. 1982, Rauth et al. 1981, Chao et al. 1982, Whitmore et al.1986]. Still nitroimidazole structure-function relationship between hypoxia, radiosensitization and cytotoxicity or liver regeneration remain less understood perhaps due to nitroimidazole low sensitivity to clinical investigations and significant enzyme alterations only at the very late necrotic stage [Cowan et al.1994, Melo et al. 2000, Moller et al. 2002, Ballinger et al. 1996, Edwards et al. 1981, Edwards et al.1982, Ballinger et al.2001, Riche et al.2001, Melo et al.2000]. However, nitroimidazole cytotoxicity was reported continuously in last three decades indicating many altered enzymes involved in nitric oxide production, cytokines, oxygen depletion (hypoxia) and immunoactive substance release from both hepatic infections and cancer tumor tissues [Su et al.1999, Rumsey et al.1994, Koch et al.2003, Hodgkiss et al.1997, Sharma 2009a, Sharma et al.2009b]. In following section, author puts an evidence of hepatic metabolic integrity loss (in mitochondria and cytoplasm) and lysosomal stimulation associated with hypoxia in liver. Nitroimidazole is still considered clearly hepatotoxic, renal and neurotoxic drug with several reports of diffused damage of parenchymal and nonparenchymal liver cells as initial stage of hepatic damage observed by electron microscopy and biochemical markers in serum with emphasis of pathophysiology [Mahy et al.2004]. Our previous observation on nitroimidazole induced effects to act against hepatic amoebiasis and amoebic abscess development indicated the possibility of nitroimidazole concentration-dependent effects upon oxygen depletion, nitric oxide production, cytokine synergy and liver cell enzyme alterations as inter-related consequences of liver regeneration or bringing back elevated enzyme levels [Chu et al.2004]. The major players were the energy metabolizing and drug metabolizing lysosomal enzymes in initial liver damage with assumption that liver cell enzyme profile defines the hypoxia sensitivity of nitroimidazole by 'Hepatocellular Hypoxia Criteria'. The novelty of hepatocellular hypoxia criteria was detail cytomorphic-biochemical information extracted from altered enzyme activities during initial cytomorphic changes in hypoxic hepatic cells simultaneously with microstructure changes by electron microscopy to confirm the role of cell organelle in hypoxia development. The criterion can be used to evaluate common liver hypoxia conditions such as hepatic tumors, hepatic infections etc. Several new nitroimidazole derivatives are emerging as potential cancer chemosensitizers, hypoxia markers and hypoxia imaging contrast agents acting as enzyme inhibitors [Gronroos et al.2004, Ziemer et al.2003, Jankovic et al. 2006, Papadopoulou et al.2003, Eschmann et al.2005, Heimbrook et al.1988, Berube et al.1992, Sharma et al. 2009].

Components of Hepatocellular Hypoxia Criteria: The criterion is described a step by step sequence of oxygen depletion in hepatocellular damage: 1. initial loss of metabolic integrity; 2. programmed regulatory failure of cell oxygen and energy metabolism; 3. liver tissue inflammation and immunity loss; 4. necrosis and active death. The purpose of metabolic integrity in hepatocytes is keeping cells intact by maintaining balance between glucose formation and glucose breakdown to maintain energy flow of NADPH and ATP molecules. Sequence: The nitroimidazole induced liver cell cytotocity leads to enhanced glucose breakdown and more demand of ATP and NADPH. With course of time, high energy demand at the cost of cell metabolic resources leads to metabolic integrity loss or energy loss and oxygen deprivation followed by possibility of step by step programmed cell death (only in tumor cells). In normal liver cells, the nitroimidazole cytotoxicity effect cause less damage and cells sustain the effect while struggling oxygen starved infected or tumor cells further loose capability to stay alive (such cells die after nitroimidazole induced infected or tumor cell killing). In normal cells, liver regeneration recovers the damage.

Hepatic hypoxia criterion has 3 major enzyme components: 1. major event of high glycolytic rate is immediate result of high glucose turnover such as glycolysis followed by secondary metabolic cycles viz. tricarboxylic acid, glycogenolysis, gluconeogenesis, pentose phosphate pathways; 2. changes in peripheral biomarkers such as cytokine immunity, altered glutathione reductase, nitric oxide production, super oxide dismutase enzymes; 3. after severe energy loss, changes occur in liver cellular morphology and tissue shrinkage [Sharma et al.2011, Kwon et al. 2009]. The sensitivity of different enzymes as cytomorphic manifestation of hypoxia in liver abscess and nitroimidazole induced liver regeneration is a simple scheme of "Hepatocellular Hypoxia Criteria" in steps is shown in Table 2. In following section, different enzymes are described.

Morphological changes	Clinical changes	Liver biochemical changes
1. Physical examination:		
--	abdominal pain	--
intestinal damage	fever	--
hepatic infiltration	hepatomegaly	liver function tests(elevated)
	⇩	
	Loss of cellular metabolic integrity	
2.Electron microscopy:	hepatomegaly with	altered hepatocyte enzymes of:
Mitochondria(M)	diffused injury	low ATP/ADP; NADPH/NADP
endoplasmic reticulum(ER)		gluconeogenesis
peroxisome (P)	⇩	glycogenolytic
lysosome (L)		lysosomal enzymes
nuclear changes (N)		oxygen flux related
	Hepatocellular enlargement (Apoptosis)	

Table 2. Continued

Morphological changes	Clinical changes	Liver biochemical changes
3. Cellular organelle damage:	poor drug response	slow metabolic disorder:
mitochondria(M)		oxidative phosphorylation
microsomes (MI)	⇩	drug metabolizing enzymes
lysosome (L)		initial phagocytosis
nuclear (N)		DNA fragmentation(beads)
cytosol (C)		glucose/protein/respiratory burst
	Hepatocellular Oxygen insufficiency (inflammation)	
4. Liver pathology changes:	raised diaphragm	
mitochondrial damage	amebic liver scan +ve	stimulation of Kupffer cells
exfoliative ER		hyperplasia of Kupffer cells
anisonucleosis		loss of metabolic control
autophagy & lysosomal irritation	⇩	increased water accumulation
cytosolic granulation		increased molecule imbalance
fatty liver appearance with membrane damage		increased lipid synthesis
	Hepatocellular degeneration and necrosis	
5. Hepatocytology:		
Cell proliferation	advancing tumor vascularization	surgical aspirates (altered
Cell debris	tissue growth on ultrasound	proteins, lipids, enzymes)
	⇩	
	Nitroimidazole single dose therapy schedule	
6. Liver cell recovery	negative liver scan/ultrasound	normal liver function test
Tissue shrinkage	Hypoxia monitoring positive	-ve Hypoxia biomarkers
OR	if unchanged or poor recovery	abnormal ELISA,enzymes
	⇩	
	Surgical intervention	

Source: Sharma R. Effect of Nitroimidazole on isolated liver cells in development of amoebic liver abscess. Ph.D dissertation submitted to Indian Institute of Technology, Delhi, 1995.

Table 2. A step by step scheme of "hepatocellular hypoxia criteria" to evaluate liver hypoxia damage in infected hepatitis or hepatic tumors. Different clinical methods suggest composite picture of hepatic hypoxia and associated biochemical and cytomorphic changes.

2.2 Origin of hypoxia and enzyme regulation

Oxygen insufficiency or hypoxia begins with low NADPH and ATP supply from glucose. Glucokinase, aldolase, pyruvate kinase and lactate dehydrogenase possibly serve as initial glycolytic regulatory enzymes for energy flow to generate tricarboxylic acid cycle (TCA) precursors. TCA cycle being as main source of energy molecule synthesis and gluconeogenesis, it serves as sole source of central ATP pool in glucose homeostasis during liver cytototoxicity or regenerated liver cell [Sharma et al.2007, Sharma et al.2010]. The citrate synthase, malate dehydrogenase, isocitrate dehydrogenase and succinate dehydrogenase enzymes serve as TCA cycle regulatory enzymes. These enzymes in hepatocytes characterize the tumor hypoxia and hepatocyte response after nitroimidazole therapy [Horlen et al.2000, Tabak et al.2003, Sapora et al.1992, Moselen et al.1995, Stratford et al.1981]. At the cost of ATP from TCA cycle and oxygen from the cell, oxidative phosphorylation maintains the metabolic integrity and continuous flow of energy. In the state of low NADPH and low oxygen cell undergoes state of metabolic integrity loss and "hypoxia". NADPH dependent cytochrome redox enzymes are biomarker of oxidative phosphorylation status and oxygen insufficiency (hypoxia) in liver. Other glutathione reductase and superoxide dismutase enzymes are biomarkers of hypoxic state of liver cells [Larrey et al.2000, Michaopoulos et al.1997]. Any energy imbalance and oxygen insufficiency (hypoxia) in liver cell are indicated by these enzymes [Michaopoulos et al.1997, Ramirez-Emiliano et al.2007]. The behavior of these enzymes in isolated liver cells is described in following section on isolated liver cell enzymes.

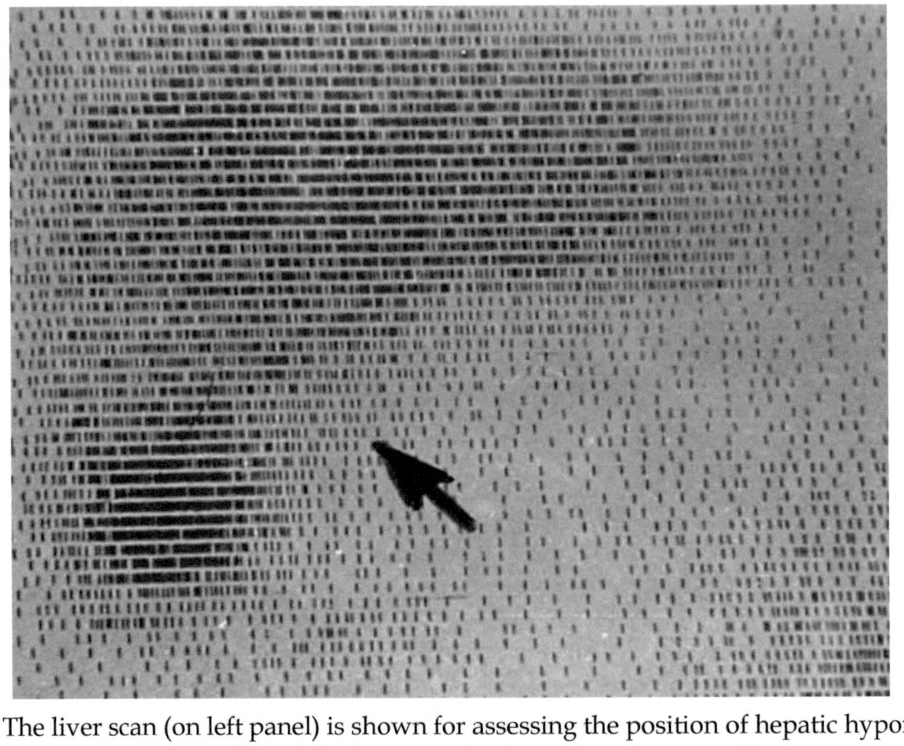

Fig. 2. The liver scan (on left panel) is shown for assessing the position of hepatic hypoxia and associated hepatomegaly and for biopsy collection site as shown with arrow.

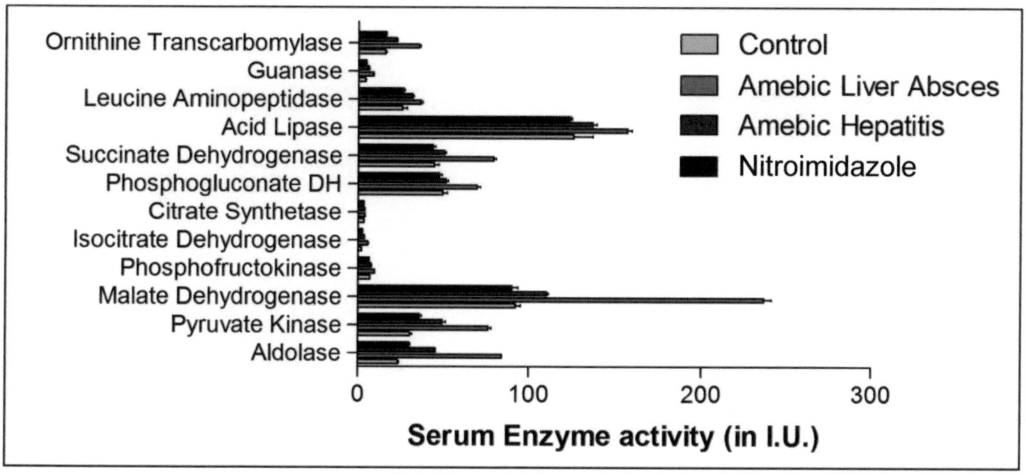

Fig. 3a. The histogram bars show the effect of nitroimidazole on biomarker enzymes and comparison of different enzymes in hepatocytes in control vs nitroimidazole treated subjects

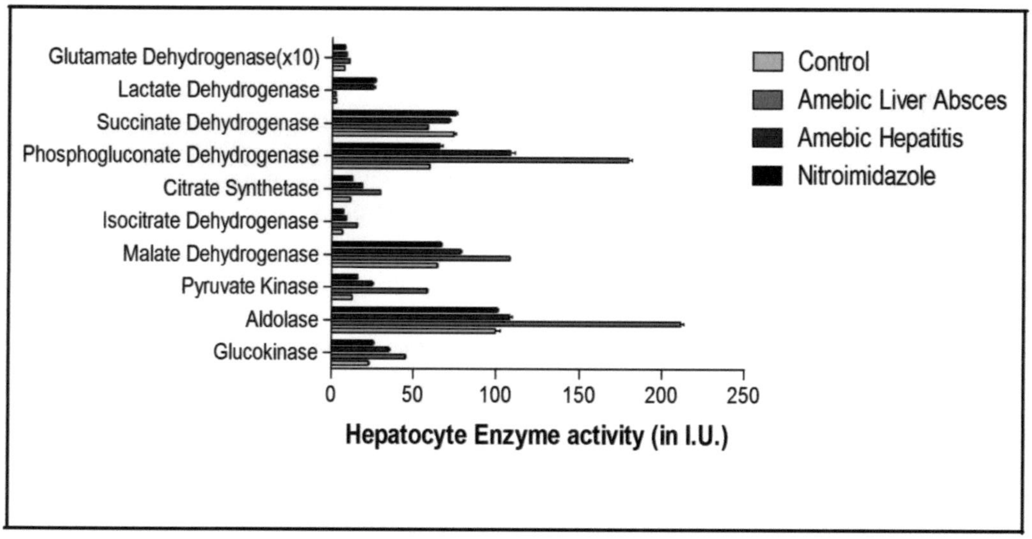

Fig. 3b. The histogram bars show the effect of nitroimidazole on biomarker enzymes and comparison of different enzymes in Kupffer cells in control vs nitroimidazole treated subjects in control vs nitroimidazole treated subjects

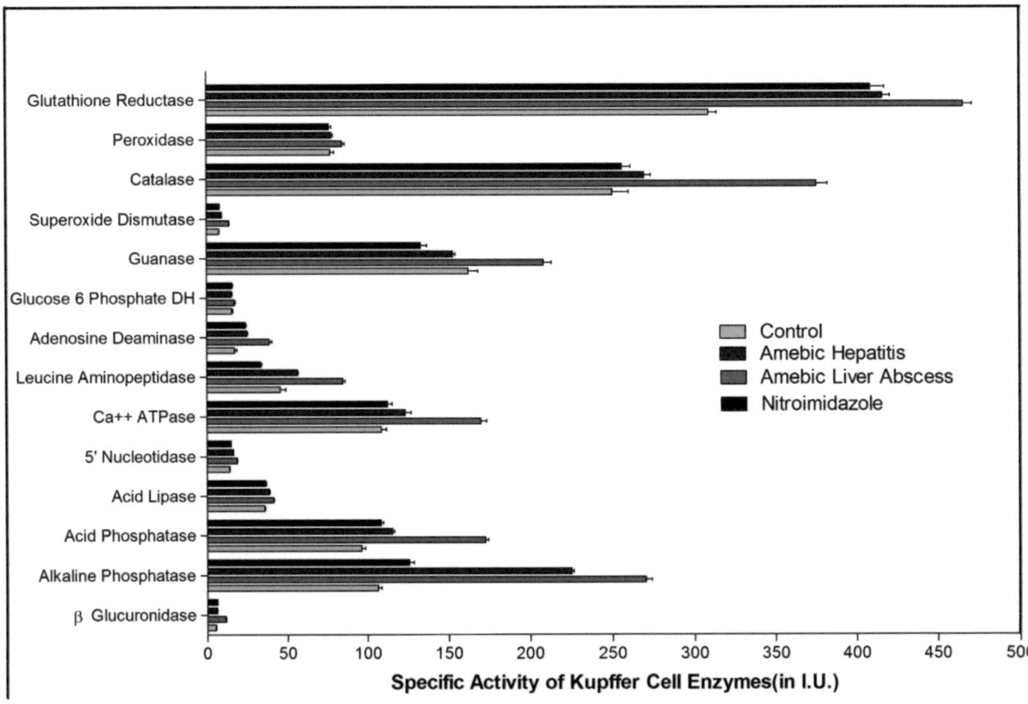

Fig. 3c. The histogram bars show the effect of nitroimidazole on biomarker enzymes and comparison of different enzymes in serum biomarker enzymes.

3. Enzyme inhibition in initial loss of metabolic integrity, glycolysis and ATP in hepatocytes

Enzymes in serum and liver cells from positive control amoebic liver abscess patients exhibited more or less specific characteristic changes in enzymes [Sharma et al.2008]. First goal was to establish elevated enzyme levels in patients and to prove subsequent enzyme inhibition after nitroimidazole therapy. Author reported ten control subjects with 2′-nitroimidazole therapy follow up for their carbohydrate metabolizing enzymes in serum and hepatocellular enzymes in liver biopsy tissues. Proven ten hepatic hypoxia control subjects were studied for hypoxia by enzyme assays [Sharma et al.2008]. The 2′-nitroimidazole treated paired ten subjects were studied for hypoxia related enzyme inhibition using enzyme assays and hepatocellular cytomorphology by electron microscopy. Out of ten positive control subjects, nine patients showed elevated carbohydrate metabolizing and lysosomal enzyme levels in serum: Glucokinase (in 80% samples), aldolase (in 80% samples), phosphofructokinase (in 80% samples), malate dehydrogenase (in 75% samples), isocitrate dehydrogenase (ICDH) (in 60% patients) were elevated while succinate dehydrogenase and lactate dehydrogenase (LDH) levels remained unaltered. Lysosomal β-glucuronidase, alkaline phosphatase, acid phosphatase enzymes showed enhanced levels in the all patient serum samples. In ten control liver biopsies, the isolated hepatocytes and Kupffer cell preparations showed altered liver cell enzyme levels. Hepatocytes showed reduced glucokinase (in 80%), LDH (in 80%), and higher content of aldolase (in 80%), pyruvate kinase (in 100%), malate dehydrogenase (in 80%), ICDH (in 80%), citrate dehydrogenase (in 70%), phosphogluconate dehydrogenase (in 80%). Kupffer cells showed higher enzyme levels of β-glucuroronidase (in 80%), leucine aminopeptidase (in 70%), acid phosphatase (in 80%) and aryl sulphatase (in 88%). In these 10 biopsy samples from subjects, the electron microscopy cytomorphology observations showed swollen bizarre mitochondria, proliferative endoplasmic reticulum, and anisonucleosis. In 2′-nitronidazole clinical trial patients, after 2′-Nitroimidazole effect in liver cell damage was manifested as enzyme inhibition in liver cells as shown in Figure 3[Sharma 2008]. Nitroimidazole treated patients shows altered enzymes: glucokinase levels were more or less normal in both serum and liver biopsy (100%); phosphofructokinase levels were nonspecific in nitroimidazole treated subjects or normal (80%); elevated lactate dehydrogenase levels were reversed to normal. In nitroimidazole treated subjects the elevated enzyme levels due to amoebic abscess brought back to normal after nitroimidazole treatment due to enzyme inhibitory effect of nitroimidazole: isocitrate dehydrogenase enzyme normal levels (80 %); the citrate synthase normal levels (80%); phosphogluconate dehydrogenase normal levels (80 %). In nitroimidazole treated subjects, the succinate dehydrogenase levels were normal (60 %) in both. The enzymes are shown in Figure 3 and Tables 2-4.

3.1 Enzyme inhibition in Kupffer cell lysosomal enzymes and Hypoxia

In nitroimidazole treated patients showed elevated enzymes brought back to normal: β-glucuronidase normal levels (50 %) in Kupffer cells and normal in serum (60 %); acid phosphatase normal levels in both; leucine aminopeptidase levels were normal in serum (80 %) and remained high in Kupffer cells (60%); Guanase levels were normal in both Kupffer cells and serum (80%). The enzymes are shown in Figure 2 and Table 3.

Kupffer cell hyperplasia was observed with swollen lysosomal contents as shown in Figure 4. Altered glycolytic enzymes in cytosol, TCA cycle enzymes in mitochondria, lysosomes and increased synthesis of enzymes by endoplasmic reticulum showed correlation with clear liver cell degeneration of microbodies. After nitroimidazole treatment, observations of both electron microscopy and biochemical parameters suggested the reversed hepatocellular changes towards normal recovery or liver regeneration. In following section, I describe the outcome of nitroimidazole effect in liver regeneration and enzyme inhibition as defense.

The enzyme biomarkers could be analyzed in serum and liver cells (hepatocytes and Kupffer cells) as clinically significant indicator of hepatic damage in disease or cytotoxicity of drug. We assume that initially metabolic integrity loss leads to over-secretion of liver cell enzymes including lysosomal enzymes. Soon after, the ultrastructural changes in liver cells become evident by electron microscopy suggestive of acute organelle degeneration. Ultrastructural hepatocyte cytotoxicity was associated with nitroimidazole overdosage (normal dosage is 2 × 3 gm one time in amoebic hepatitis and thrice in amoebic abscess). The regenerative changes were consistently observed after nitroimidazole therapy to reverse liver damage. Few pathology reports are available to show nitroimidazole cytotoxicity and no electron microscopy study is available to support asymptomatic inflammatory or unequivocal diffuse parenchymal injury exhibiting diffused sinusoidal and portal infiltration events [Tabak et al.2003, Larrey et al.2000]. Biomarker liver cell enzymes with electron microscopy put evidence of amoebic cytotoxicity in liver biopsy samples and liver regeneration after nitroimidazole treatment by 'hepatic hypoxia criteria'. Degenerative changes of hepatocytes suggested possible necrosis.

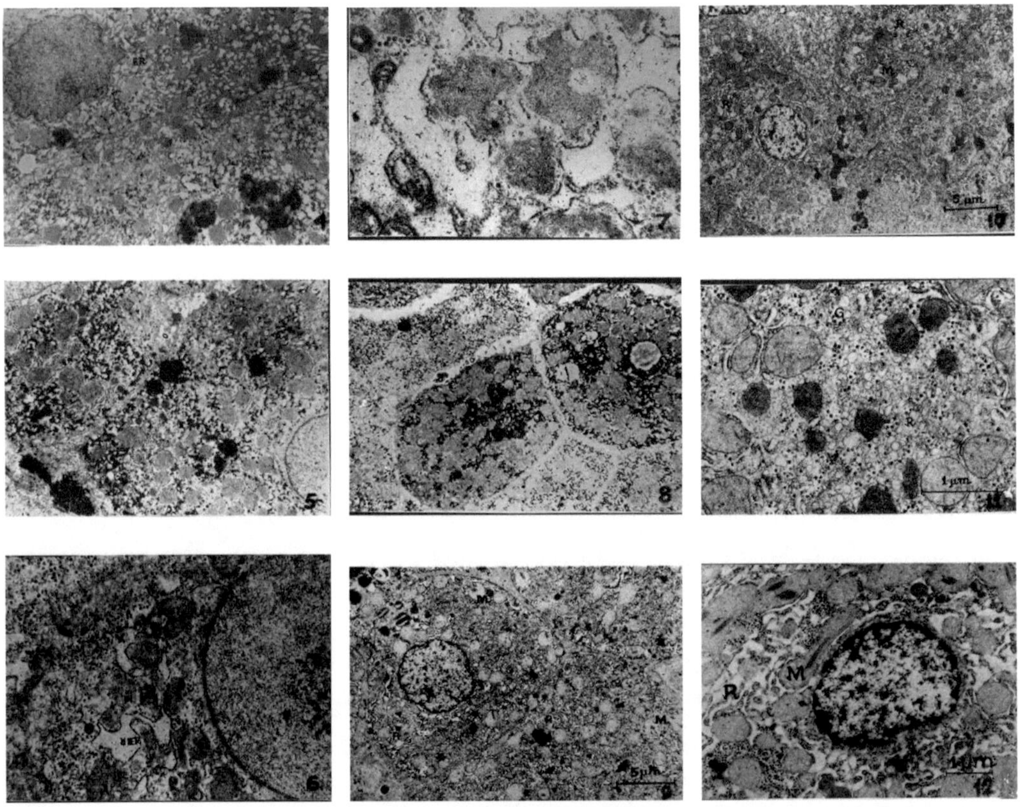

Fig. 4. Ultrastructural changes are shown in hepatocyte organelles during hypoxia: exfoliation of endoplasmic reticulum (top on left); anisonucleosis (mid and bottom on left); intercellular junction gaps (top and mid panels in center); nuclear inclusions (bottom on center); swollen and bizarre mitochondria(top on right); inclusions in peroxisome (mid on right); mitochondrial atrophy with lipid vesicles (bottom on right). Hepatocyte mitochondria became swollen, bizarre with dense matrix and showed destorted cristae, endoplasmic reticulum dilated vesicles, giant nuclei with diffuse proliferation of endoplasmic reticulum and clear anisonucleosis features. The ultrastructure of liver cells showed characteristic organelle changes in nitroimidazole treated liver.

Evidence of hepatic regeneration and nitroimidazole cytotoxicity (hypoxia induced changes) was characterized previously in terms of cytokine synergy, unusual degree of anisonucleosis, nitric oxide production and the presence of giant nucleus in hepatocytes after liver regenerative therapy [Michalopoulous et al.1997]. Endoplasmic reticulum showed a diffuse and intense proliferative activity in these liver cells with normal appearance of mitochondria as earlier reported [Ramirez-Emiliano et al.2007, Das et al.1999]. However, intramitochondrial inclusion bodies were absent while they were very prominent features

[Das et al.1999]. The cause of the ultrastructural changes after nitroimidazole cytotoxicity were supported by initial enzyme alterations reflecting loss of metabolic integrity probably induced by free radical formation from nitroimidazole [Das et al.1999]. Since ultrastructural changes in liver were completely reversible after nitroimidazole therapy within 7-9 days, it is quite reasonable that pathogenesis of diffused hepatocyte damage was due to nitroimidazole breakdown products.

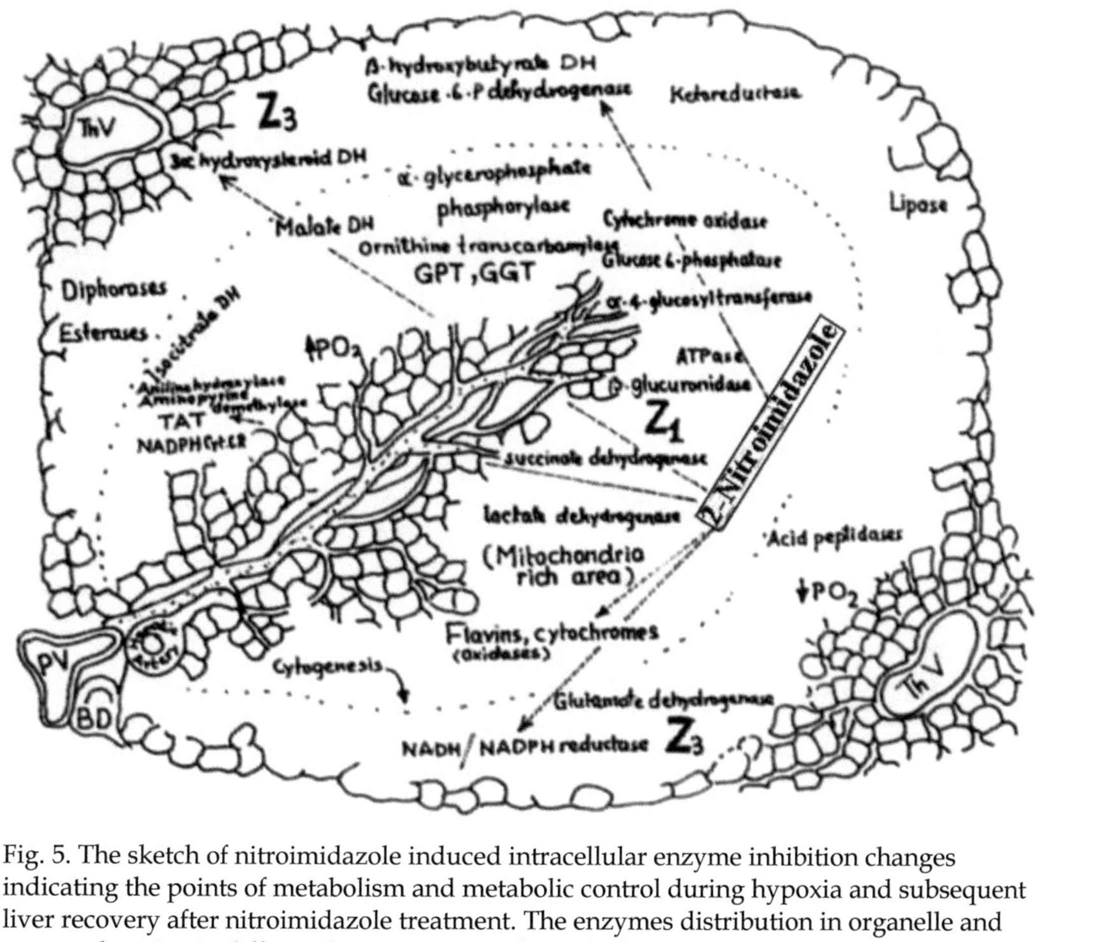

Fig. 5. The sketch of nitroimidazole induced intracellular enzyme inhibition changes indicating the points of metabolism and metabolic control during hypoxia and subsequent liver recovery after nitroimidazole treatment. The enzymes distribution in organelle and enzyme location in different hepatic sites explains the liver damage and enzymatic basis of hepatocellular hypoxia criteria [reproduced from Sharma R. 1990. Ph.D dissertation].

As indicated in previous section, 'Hepatocellular hypoxia criteria' is a sequence of events in hepatic cell injury at molecular level. The initial loss of hepatocellular metabolic integrity leads to hepatic injury. Glucose-energy metabolic integrity loss and nitric oxide formation with Ca^{++} homeostasis are main initial determinants of hypoxia [Ramirez-Emiliano et al.2007]. In following section, present chapter extends the enzyme regulatory behavior during glucose metabolizing pathways viz. glycolysis, TCA cycle and gluconeogenesis to establish the value of enzyme inhibition in liver regeneration. In isolated liver cells, metabolic alterations of energy metabolism pathways explain the events in hepatitis and hepatic tumors as best correlated respective enzyme dysfunctions and ultrastructural changes observed in cells in biopsy. The following description is broad explanation of different enzymes secreted from hepatocytes as a result of nitroimidazole induced cytotoxicity and oxygen depletion. From biochemical stand point, different regulatory enzymes are discussed as biochemical events of hypoxia development as shown in Figures 3a,3b,3c and 5.

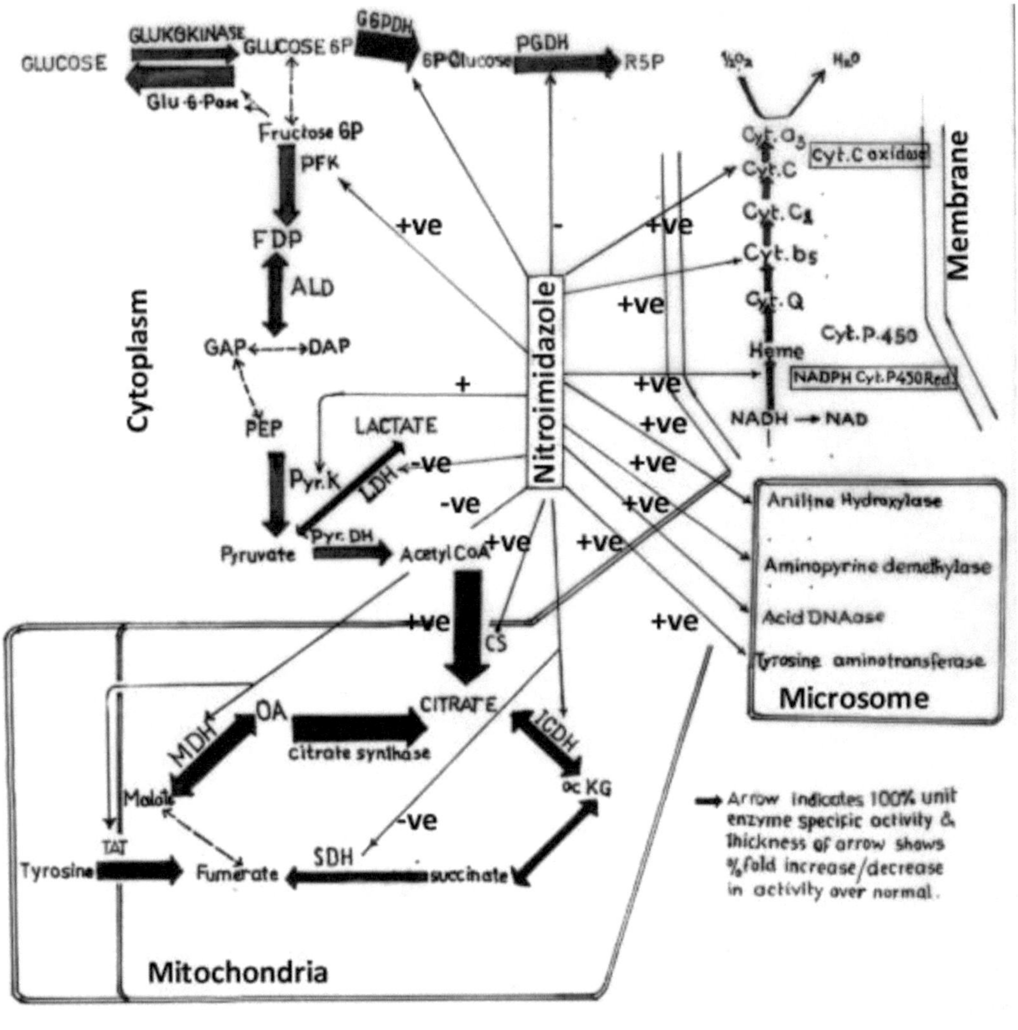

Fig. 6. The sketch of relative biomarker enzyme changes (thickness of arrow) in liver hypoxia. The +ve sign denotes the relative increase in enzyme activity and –ve sign denotes the decrease in enzyme activity in liver cells in different organelles due to ameba induced hypoxia. The figure sketch also represents a relative enzyme activities in sequence of metabolic steps as 'hepatocellular hypoxia criteria'. Notice the liver regenerative or enzyme inhibitory action of nitroimidazole at different enzyme reactions is shown [Reproduced from Sharma R. 1995. Ph.D dissertation].

3.2 Role of enzymes in energy metabolic integrity in hypoxia

Hexokinase, a rate limiting enzyme in cytosol for glucose turnover was estimated due to high glucose conversion into glucose 6P showed high enzyme secretion from hepatocytes. Aldolase, rate controlling enzyme which splits fructose-1, 6 Diphosphate into two 3-phosphoglyceric acid and dehydroxyacetone phosphate, showed high enzyme secretion from heptocytes. Phosphofructokinase, rate controlling enzyme which phosphorylates fructose 6P into fructose 1, 6 DiP, enhanced as its more demand in hepatocellular glucose turnover which probably induced higher secretion of phosphofructokinase. Pyruvate kinase, transferring phosphoryl group from phosphoenolpyruvate to ADP with pyruvate formation, showed high enzyme secretion from hepatocytes. It may be attributed due to high concentration of glycolytic intermediates as terminal step was blocked or due to 2, 3 Di-Phosphoglycerate regulated abnormal oxygen dissociation. Malate dehydrogenase and isocitrate dehydrogenase enzymes, catalyzing malate oxidation into oxaloacetate and oxidative decarboxylation of isocitrate into α-ketoglutarate respectively both require NAD^+. Both enzymes were elevated in damaged cells due to reducing equivalent low ratio of $NADH/NAD^+$ pushing in forward direction. Succinate dehydrogenase, oxidizing succinate to fumerate using FAD, showed decreased enzyme levels due to low iron sulphur proteins resulting with lowered electron transport system in inner mitochondrial membrane i.e. supply of electrons to molecular oxygen by electron transfer from $FADH_2$ to Fe^{+++} (SDH) in amoebic recruited liver cells. Citrate synthase, synthesizing citrate from oxaloacetate and acetyl CoA by aldol condensation followed by hydrolysis, showed elevated levels due to rapid turnover of oxaloacetate and acetyl CoA molecules during cytolysis. Moreover, high TCA cycle activity in hepatocyte during liver damage conditions was described earlier [Sharma 2010]. In serum, the enzymes exhibit their significance but aldolase, pyruvate kinase and LDH, MDH, ICDH observed as distinguishing the diffused injury or abscess formation [Virk et al.1989]. Phosphogluconate dehydrogenase elevated levels may be attributed due to ribulose 5P formation and transaldolase and transketolase control, for phosphogluconate pathway. Nitroimidazole induced cytotoxicity perhaps have insignificant impact on ATP supply. In following section, different enzyme regulatory behaviors of nitroimidazole induced enzyme inhibition are described during liver cell regeneration. A strategy was developed to isolate hepatocytes and Kupffer cells in monoaxenic cultures from excised liver samples. Mixing different inhibitors or additives in liver cell cultures stimulate the enzyme regulatory behavior in presence of specific growth factors, metabolite analogues.

3.3 Role of lysosomal enzymes in hepatocellular damage

In previous section, role of secretory lysosomal enzymes was described in defense of liver cells in context of intestinal amoebiasis and amoebic liver abscess. Conceptually, lysosomal hydrolases cause foreign body damage by several reactions including redox reaction, phosphorylation reaction, glucuronidation reaction, hydroxylation reaction, dehydroxylation reaction, nucleation reaction, terminal breakdown reaction, trans-carbamylation reaction. Other studies also reported molecular mechanisms of enzymes in pathogenesis [Das et al.1999], liver cell damage [Virk et al.1989, Virk et al.1988], phagocytosis [Sharma 2009], cyclooxygenase-2 expression [Sanchez-Ramirez et al.2000].

4. Enzyme inhibition properties and enzyme regulatory behavior in isolated liver cells

Isolated liver cells with characteristic enzyme properties are research tools. In our previous study, the hepatocyte yield from livers was 2.5× 10^8 per mg liver. The cytosol fraction showed the removal of all cell organelle from supernatant and resultant supernatant as cytosol was free of any remnant by microscopic observation by trypan blue exclusion. Isolation of human Kupffer cells was developed by a modified pronase-collagenase enzymatic isolation and purity of Kupffer cells by using biomarker enzyme assay. Earlier reports used similar practice of enzymatic digestion and enzymatic assays as indicators of cell viability and yield [Hendriks et al.1990, Neaud et al.1995, McCuskey et al.1990]. A new rapid method was used for the isolation and fractionation of both rat and human Kupffer cells without the need of liver perfusion techniques. The study used rat livers or small human liver wedge biopsies obtained peroperatively and incubated with pronase under continuous pH registration. Kupffer cells were subsequently separated from other nonparenchymal cells by Nycodenz gradient centrifugation and purified by counterflow centrifugal elutriation [Neaud et al.1995]. Identification of Kupffer cells was achieved on the basis of ultrastructural analyses and immunophenotyping [Neaud et al.1995]. The fractionation of Kupffer cells gave good yield comparable with other studies [Hendriks et al.1990, Neaud et al.1995, McCuskey et al.1990]. For details of protocol of liver cell isolation, see the Appendix in the end of this chapter. Human liver cell enzymes in isolated liver cells (hepatocytes and Kupffer cells in cultures) showed characteristic regulatory behavior in presence of additives.

Liver cell hypoxia is represented as a state of oxygen depletion in cell. Initially liver cell gets its ATP and NADPH supply from glycolysis and TCA cycle. Consequently, electron transfer chain (oxidative phosphorylation) through series of cytochrome redox reactions converts available cell oxygen to water to produce high energy metabolites in the cell (metabolic integrity). Infected liver cells show loss of metabolic integrity and less available oxygen (oxygen starved or hypoxia). Nitroimidazole was established two decades ago as potential oxygen quenching drug in most of the infected and tumor cells. "Nitroimidazole oxygen quenching action" makes oxygen depleted or starved cells that further undergo worse to die and leaving normal cells fully functional that can be observed as potential tumor hypoxia therapy and antiparasitic treatment by nitroimidazole.

In following section, enzyme inhibition characteristics in energy metabolism regulation including glycolysis, TCA cycle regulatory enzyme behavior and drug metabolizing proteolytic lysosomal enzymes induced by additives in liver cell cultures are described.

4.1 Glucokinase

The actinomycin D showed non-significant inhibitory change in glucokinase enzyme activity as shown in Figure 7 at the top histogram bars. The nitroimidazole showed enhanced glucokinase activity and reversed the inhibitory effect of actinomycin D. The glucokinase activity in hepatic abscess hepatocytes did not show any significant effect of nitroimidazole. Overall actinomycin D alone did not show any change in glucokinase enzyme activity in hepatic abscess hepatocytes or nitroimidazole treated hepatocytes.

The addition of triamcinolone with insulin in combination showed significant increase in glucokinase activity in all hepatocyte groups. It pointed out to the enhancement in hormone dependent glucokinase enzyme synthesis in hepatocytes and enzyme activity showed highest enzyme synthesis rate in hepatic abscess recruited hepatocytes while nitroimidazole treated hepatocytes showed lesser increase in hormone dependent glucokinase in comparison with hepatocytes as shown in Figure 7. The increase in enzyme activity or enhanced glucokinase enzyme synthesis in hepatocytes showed that glucokinase enzyme behavior was hormone dependent. Different concentration of insulin and progesterone hormones further indicated the regulatory behavior of glucokinase enzyme as hormone dependent.

The addition of progesterone hormone 0.1 μgm/ml in hepatocyte cultures showed maximum enhanced glucokinase enzyme activity and nitroimidazole treated hepatocytes showed less enhanced glucokinase activity but more than control hepatocyte enzyme activity as shown in histogram bars (see Figure 7).

In control group, glucokinase enzyme activity in untreated hepatocytes did not show any chnage in their glucokinase enzyme activity in presence of nitroimidazole additive. The yield of hepatocytes and glucokinase enzyme activities from hepatic abscess liver biopsy were similar as yield of cells and glucokinase enzyme activities in untreated and nitroimidazole treated hepatocytes. So, it was clear that addition of actinomycin-D and insulin with triamicilone both enhanced the glucokinase enzyme synthesis in hepatocytes as shown in Figure 7 histograms bars.

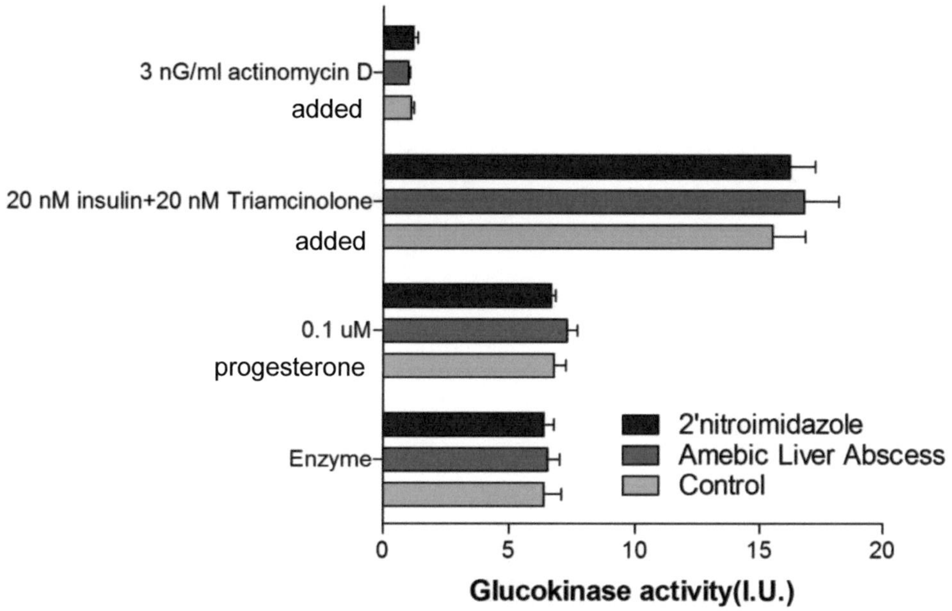

Fig. 7. The nitroimidazole effect on glucokinase enzyme activity in guinea pig normal and infected hepatocytes is shown in presence of different enzyme regulatory additives actinomycin D, insulin-triamcilone. Notice the enhanced effect of insulin-triamicilone on glucokinase enzyme (P value <0.0001; r^2 0.9872) in nitroimidazole treated vs liver abscess biopsy heaptocytes.

4.2 Pyruvate kinase

The altered behavior of pyruvate kinase regulation in isolated hepatocytes from nitroimidazole and progesterone induced cultures showed characteristic sigmoid behavior in normal group but towards linear relationship at increased pH of hepatocyte culture media. In other words, a pH increase alters the enzyme velocity in same proportion with ATP concentration changes in linear manner in liver cell medium. At high ATP concentration, enzyme synthesis in nitroimidazole stimulated cells remained significantly elevated while at pH 7.0-7.4 cells did not show sudden enzyme elevation and showed a fall in enzyme activity as observed in normal hepatocytes (see Figure 8). The higher ADP/ATP ratio kept the enzyme synthesis high at pH 7.4 in progesterone while nitroimidazole stimulated the hepatocyte cells in cultures (see Figure 8).

The intricacy was reported as ATP interacts with H^+ and Mg^{++} to give complex distribution of many ionized and metal complex species. However, Mg^{++} free ADP itself has three subspecies viz. ADP^{--}, $ADPK^{--}$, ADPH− whose relative proportions vary with change in pH. Out of which ADP− happens to be true substrate of pyruvate kinase which presumably was utilized by hepatocytes stimulated by nitroimidazole. It further suggested the binding of ADP to unbound and fully bound enzyme conformations involving different groups within active site. The binding of ADP and ADP free conformation of enzyme did not involve groups that ionized within applied pH range in nitroimidazole and progesterone stimulated hepatocytes but binding of ADP to the fully bound enzyme conformation (cooperativity) involved the ionizable groups with high pK values (at pH 6.6-7.4). Moreover, phosphoenol pyruvate bound enzyme ionization has been previously reported to promote cooperative binding of ADP in diseased cell. In present study, ADP binding becomes less cooperative by stimulated hepatocytes. It is possible that enzyme may have two active conformations. So, allosteric interactions actually might have altered the mode of ADP binding in normal cells while they were non-effective in pyruvate kinase enzyme of stimulated hepatocytes. Pyruvate kinase active site conformation perhaps seems to change the ADP binding.

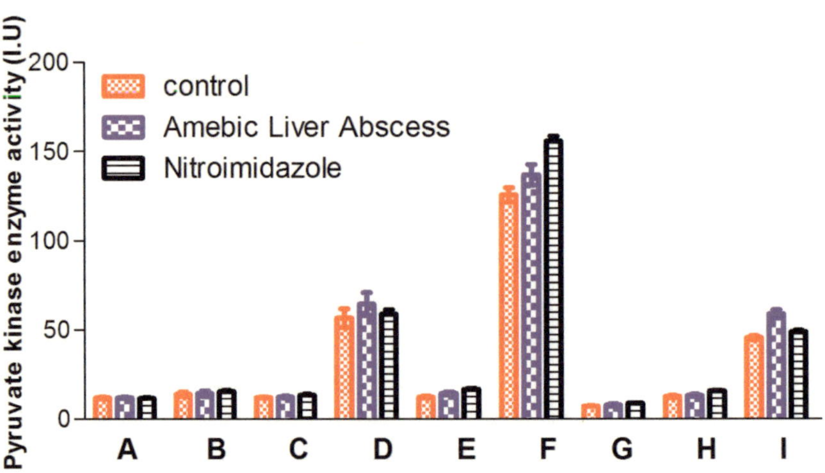

Fig. 8a. The pyruvate kinase enzyme activity in control, amoebic liver abscess and nitroimidazole treated hepatocytes is shown in presence of different enzyme regulatory additives (A-I).

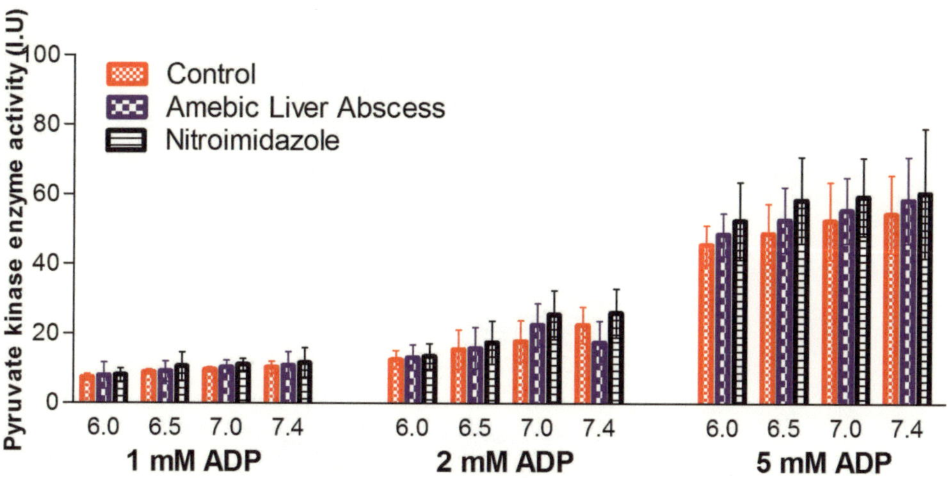

Fig. 8b. The pyruvate kinase enzyme activity in control, amoebic liver abscess and nitroimidazole treated hepatocytes is shown in presence of different enzyme regulatory ADP additives (2-5 mM ADP).

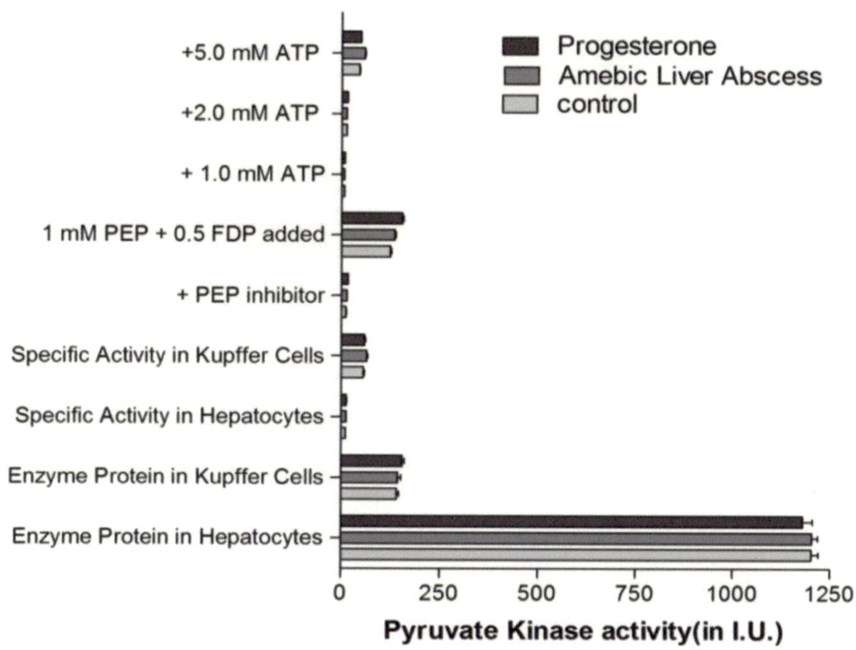

Fig. 8c. The pyruvate kinase enzyme activity in control, amoebic liver abscess and progesterone treated liver cells is shown in presence of different enzyme regulatory additives (ATP, PEP+FDP,PEP inhibitors).

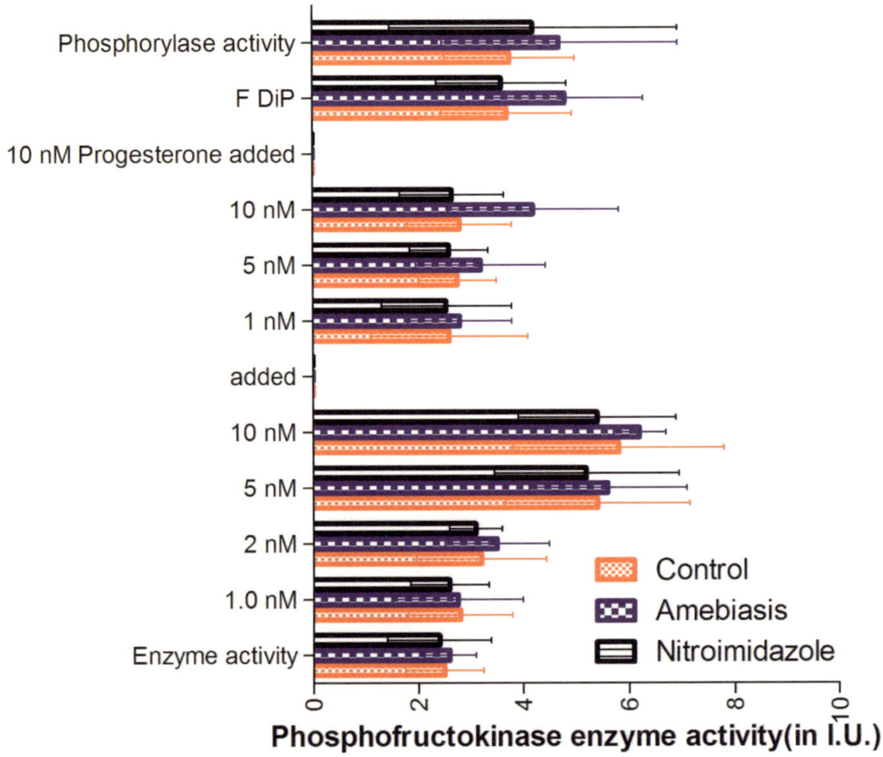

Fig. 9. The effect of nitroimidazole on phosphofructokinase enzyme activity of isolated hepatocytes from control, amebiasis and nitroimidazole groups. Hepatocytes from each group were sedimented and homogenized in Kreb Hanseleit buffer pH 8.0 containing 2.5 mM dithiothretiol, 0.1 mM EDTA and 50 mM NaF for 90 seconds followed by centrifugation at 27000 g for 10 minutes. One unit enzyme activity catalyzed 1 micromole fructose 6 phosphate to fructose 1,6 DiP per minute per mg enzyme protein.

4.3 Phosphofructokinase

Phosphofructokinase enzyme regulation in isolated hepatocytes cultures added with amoebic trophozoites, nitroimidazole and progesterone exhibited specific regulatory behavior. Addition of fructose 1,6 Diphosphate additive exhibited a stimulation of phosphofructokinase activity (see Figure 9). The effect of progesterone was seen as best response at 10 nM in hepatocytes of all groups while fructose 1,6 diphosphate showed maximum effect at concentration of 10 nM on the hepatocytes in presence of trophozoites added in hepatocyte cultures. However, hepatocyte cultures in other groups did not show any alteration (see Figure 9). Similarly, progesterone addition to hepatocytes in presence of trophozoites exhibited maximum fructose 1,6 diphosphate in the medium. The addition of progesterone concentration was within the physiological range (see Figure 9). Progesterone was presumed to effect enzyme activity by influencing intracellular level of various allosteric effectors of enzyme. Fructose 1,6 diphosphate acts as potent activator of enzyme activity. The phosphofructkinase enzyme activity was reported as cyclic AMP dependent

[Pilkis et al.1979, Sanchez-Martinez et al.2000]. However, nitroimidazole added hepatocytes did not respond or responded slow after progesterone addition.

4.4 Phosphodiesterase

Cyclic AMP dependent phosphodiesterase regulation over prostaglandin E_2 and $F_{2\alpha}$ synthesis was stimulated in Kupffer cells added with trophozoits and nitroimidazole. Phosphodiesterase activity in Kupffer cells was dependent upon calcium ion concentration which was regulatory factor for cyclic AMP pool and prostaglandin production. In amoebic trophozoite added cultures, cells exhibited insignificant elevated enzyme activity while nitroimidazole addition did not alter the enzyme activity as shown in Figure 10. The enzyme behavior was similar in nature for its dependence on cyclic AMP secretion, prostaglandin synthesis and calcium ion. However, tropozoites stimulated more cyclic AMP secretion and prostaglandins. It provides the probable mechanism of trophozoite stimulated macrophagal action. Earlier calmodulin dependent adenylate cyclase and phosphodiesterase regulation for cyclic AMP pool have been described [Hidi et al.2000]. Moreover, the stimulated prostaglandin synthesis was described as calmodulin dependent cyclic AMP regulated mechanism [Hidi et al.2000]. Thus it seems that amoebic trophozoite addition with hepatocytes regulates intracellular calcium ion concentration which controls prostaglandin synthesis via phospholipid and arachidonic acid precursors in hepatocytes. Thus synthesized prostaglandins regulate the cyclic AMP intracellular levels. The similar mechanism has been described as "self-limiting" mechanism [Martina et al.2006].

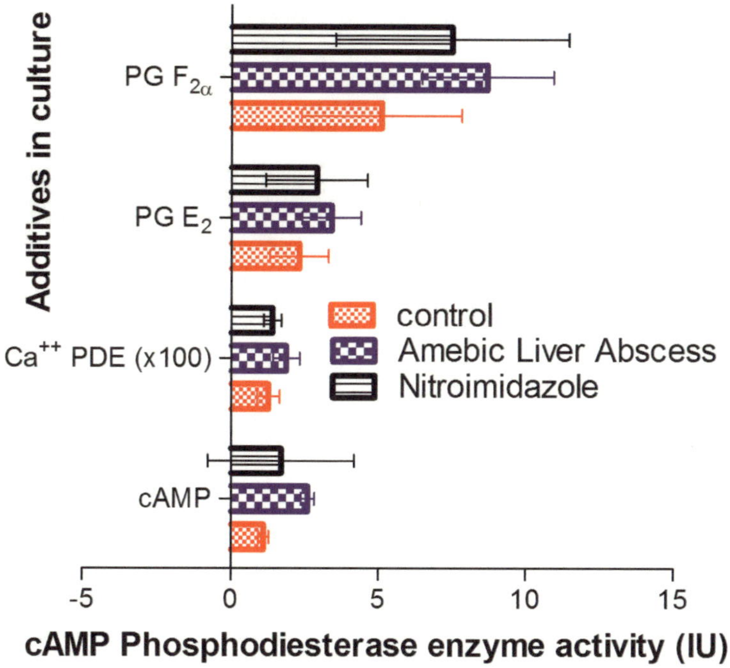

Fig. 10. Cyclic AMP dependent phosphodisesterase activity and secretion of prostaglandin E_2 and $F_{2\alpha}$ in isolated kupffer cells from different groups. The phosphodiesterase enzyme activity in amebic liver abscess group showed highest enzyme activity after additives were added.

4.5 Enzyme inhibition in respiratory burst of liver cells

An initial evidence of liver cell survival at different oxygen pressures is shown in Figure 11 to highlight respiratory burst of enzymes. The inhibited enzymes related with cell respiration are presumably considered responsible of induced respiratory burst (hypoxia) in liver cells. Respiratory burst process including enzyme alterations (superoxide dismutase, glutathione reductase, NADPH oxidase enzyme activity) was suggested as initial Kupffer cell stimulation or respiratory burst by additives and followed by energy and oxygen depletion as possible cause of hypoxia and cell viability loss as important findings in following section.

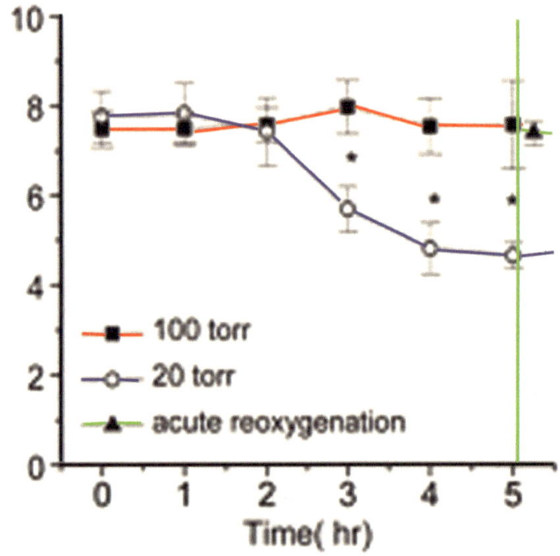

Fig. 11. Effect of oxygen supply (shown in different pressures) to liver cells. Notice the low oxygen supply depletes the oxygen content in cells while high oxygen supply keeps cells in in good survival. For long survival, reoxygenation of cells is needed(see green vertical line).

4.6 Superoxide dismutase enzyme activity in respiratory burst of Kupffer cells

The superoxide dismutase enzyme activity in isolated cells was comparable and showed minimal difference in both control and additive triggered Kupffer cells. Addition of sodium cyanide concentration showed small effect on elevated superoxide dismutase activities in both control and nitroimidazole triggered Kupffer cells. However, Kupffer cells showed elevated respiratory burst activity significantly high after adding 10 nM sodium cyanide in the cell culture medium as shown in Figure 12. However, the response of Kupffer cells with sodium cyanide addition was more in trophozoites added cell cultures at 10 mM sodium cyanide concentration. Zymosan stimulation on superoxide production was insignificantly high in the Kupffer cells in both groups. Earlier stimulation of zymosan induced superoxide anion production in Kupffer cells by sodium cyanide was reported to show the effect of cyanides on cytosolic superoxide dismutase enzyme activity [D'Alessandro et al.2011, Shahhosseini et al.2006]. Moreover, zymosan stimulated superoxide production was reported in Kupffer cells as reaction of cytochrome C reduction [Duan et al.2004].

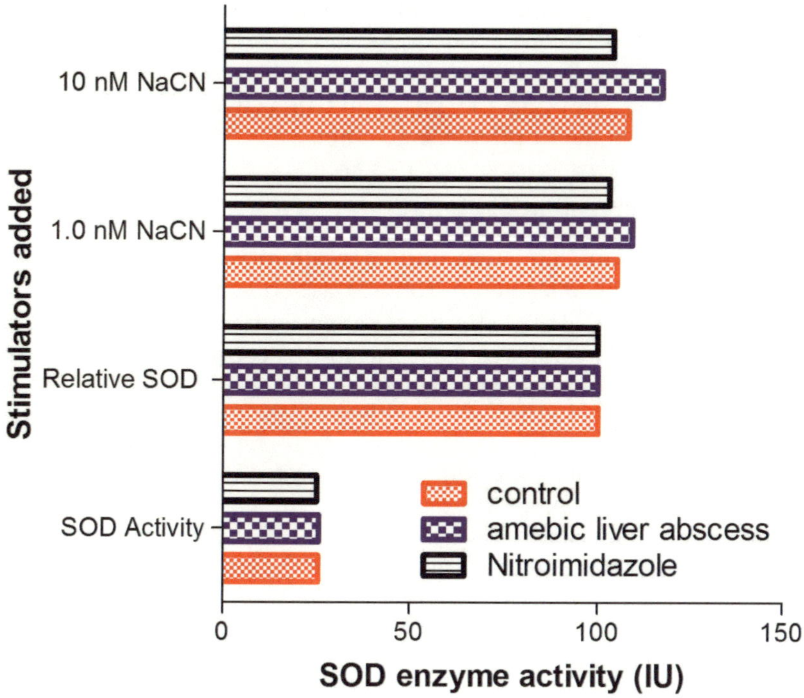

Fig. 12. The enzyme activity was estimated in Kupffer cells cultures added with sheep erythrocytes, trophozoites, nitroimidazole. The Kupffer cells were maintained for 48 hours or more to determine superoxide anion production in presence of 1.0 mM and 1.0 mM sodium cyanide additive added in culture medium. In brief, 3 x 10^5 Kupffer cells were suspended in Kreb's Henseleit buffer containing 5 mM glucose, 3 mM calcium chloride pH 7.4 in presence of 50 μM ferricytochrome in total 1.0 ml reaction mixture and cells were incubated for 15 minutes in presence of different sodium cyanide additives. The reaction was started by addition of zymosan 500 μg per ml to the medium after 10 minutes at 37 °C supernatant was taken, centrifuged at 8000 x g in cold for two minutes and assayed for cytochrome C reductase enzyme activity. Ref. Sharma, R. et al. 2010.

The O_2 superoxide anion radical production as result of respiratory burst activity showed proportionate elevated enzyme activity in isolated Kupffer cells from cultures added with trophozoites and nitroimidazole additives. The active role of reactive oxygen species and superoxide dismutase activity in cultured Kupffer cell damage during hypoxia is shown in Figure 12. Previous study supported these observations on cellular damage and cytotoxicity from reactive oxygen species from superoxide anion acting as a scavenger [Sharma et al.2010]. Another possibility of Kupffer cell toxicity by additives may be associated with cytokines, specifically TNF-α which was associated with the cytotoxicity [Sharma et al.2007]. Previous study also reported a decreased hepatocyte cell viability associated with altered cAMP phosphodiesterase and HMP shunt activity as a result of hypoxia [Carre et al.2010]. Authors suggested that the oxygen insufficiency is possibly a consequence of energy depletion and increased oxygen reactive species [Keeling et al.1982]. Other reports support this view of vasodilatory sensitive ATP depletion in Kupffer cells [Ana et al.2011]. The reactive oxygen species production also seems as another cause of hypoxia as suggested by the nature of superoxide dismutase as additive cyanide sensitive enzyme. We observed a good reason of association between superoxide anion scavenger, glutathione reductase and NADPH oxidase as cause of cell viability loss leading further to Kupffer cell damage. In this direction, ample evidence supports the possibility of activation of HMP shunt due to depleted ATP and NADH in the Kupffer cells [Grahm, 2000; Spolarics et al. 1991].

4.7 HMP shunt pathway in respiratory burst

Hexose monophosphate shunt activation during respiratory burst in isolated Kupffer cells exposed with amoebic trophozoites and nitroimidazole showed enhanced enzyme activities as shown in Figures 13A-13C. HMP shunt activation is measured in three ways: 1. HMP shunt activity in presence of additives such as latex, sheep erythrocytes, trophozoites; 2. HMP shunt activity as radiolabeled glucose conversion to carbon dioxide; 3. HMP shunt activity as enzyme inhibition.

a. The HMP shunt activity was maximum and it elevated in Kupffer cells after incubation with sheep erythrocytes and trophozoites while Kupffer cells exposed with inert latex particles did not show any activation or change in HMP shunt activity. Sheep erythrocytes and trophozoites both stimulated glucose metabolism via hexose monophosphate shunt activity. Perhaps latex particles being biologically inert nature could not change the cellular activity at all. So, latex particles were phagocytosed without altered HMP shunt activity. It suggested that HMP shunt activity may be a triggering event in Kupffer cell activation dependent not solely upon intracellular fate of ingested material or particles. However, other additives such as zymosan, erythrocytes and trophozoites also showed same response. Previous study reported similar response of zymosan stimulated and iodoacetate inhibited Kupffer cell glucose-6-phosphate dehydrogenase and phosphogluconate dehydrogenase enzymes.[Sharma et al.2010]
b. The HMP shunt activity as [^{14}C] glucose conversion to [^{14}C]-CO_2 liberated product indicated the active inflammation elicited by macrophage response of Kupffer cells [Schorlemmer et al. 1999]. The HMP shunt activity in control and triggered Kupffer cells was measured by [^{14}C]CO_2 release in dpm from labeled [^{14}C] glucose per minute per mg protein in presence of different effectors known as quenching the hexose monophosphates in glycolysis. The HMP shunt activity was similar with no difference (P value > 0.001) in control compared with triggered Kuffer cells with effector and inert latex particles did not alter the HMP shunt activity. The HMP shunt activity was significantly enhanced in triggered Kupffer cells in presence of trophozoites (P value < 0.001) but the HMP shunt enhancement was lesser in presence of erythrocytes. It indicated the dependance of HMP shunt activity on nature of effector either inert or surface active. The latex was inert, erythrocytes were nonvirulent and trophozoits were virulent as a result HMP shunt activity was enhanced in the order of latex < erythrocytes < trophozoits added in the medium of Kupffer cells.
c. The trophozoites further activated the Kupffer cell glucose-6-phosphate and phosphogluconate dehydrogenase enzymes in association of HMP shunt activity. In following section, enzyme inhibition in liver cells is reported in control, hypoxic cells in different groups. Hexose monophosphate shunt activation was enhanced during respiratory burst in isolated Kupffer cells in amoebic trophozoits and nitroimidazole added cultures. In the cultures added with trophozoits and nitroimidazole both, Kupffer cells exhibited significantly elevated HMP shunt activation. The HMP shunt

activity was highly elevated in all groups cultures incubated with sheep erythrocytes and amoebic trophozoits while cells with latex particles did not show any activation. Perhaps latex particles being biologically inert could not change the cellular sinusoidal function activity. Sheep erythrocytes and trophozoits perhaps stimulated the glucose metabolism via hexose monophosphate pathway but latex particles were phagocytosed without altered shunt activity. It suggested that HMP shunt may be a triggering event in Kupffer cell activation dependent not upon intracelllular fate of ingested material or particles [Knook et al. 1980]. The trophozoits added Kupffer cells showed direct interaction with increased glucose-6-phosphate dehydrogenase and 6-phosphogluconate dehydrogenase enzyme activities as described as a result of enhanced hexose monophosphate shunt activity. Earlier zymosan stimulated and iodoacetate additives inhibited Kupffer cell glucose 6 phosphate dehydrogenase and 6-phosphogluconate dehydrogenase activities have been reported [Heuff et al. 1994].

- **Inhibition of Glutathione Reductase enzyme activity in respiratory burst:** The glutathione reductase GSH specific enzyme activity was measured in hypoxic hepatocytes in presence of different GSH enzyme inhibitors such as DEM, antimycin, BSO as additive effectors. To demonstrate the dependence of enzyme activity on exposure time of inhibitors, the hypoxic hepatocytes were incubated in presence of effectors and enzyme activity was measured in medium. At 0 hour, the enzyme specific activity was same in hepatocytes in all groups. After 24 hours, the activity in hepatocytes was inhibited maximum by actinomycin while DEM and BSO inhibited the enzyme at similar extent and hypoxic hepatocytes exhibited no difference in enzyme activities (P value > 0.001) (see Figure 14). After 48 hours, the GSH enzyme activity was inhibited maximum by addition of BSO and minimum by antimycin in cell cultures. However, the behavior pattern of enzyme inhibition by DEM and BSO was similar. After 72 hours, the inhibition pattern of GSH enzyme activitity by inhibitors was same as after 48 hours. After 96 hours, all inhibitors showed maximal inhibition and hypoxic hepatocytes showed same specific activities in all groups. Antimycin inhibitor exhibited lesser inhibition activity of hypoxic GSH enzyme initially during 24-48 hours.
- **Inhibition of NADH oxidase in respiratory burst:** The respiratory burst activity was described as increased activity in liver macrophages during hepatic cell damage showing increased oxygen consumption, stimulation of HMP shunt and elevated superoxide production with activated membrane NADPH oxidase enzyme as one of the mode of phagocytosis [Dusting et al. 2005]. Moreover, our experiments suggested the specific roles of additives on respiratory burst biomarkers: 1. The increased sodium cyanide concentration showed stimulation of superoxide dismutase activity; 2. The live erythrocytes and trophozoites showed HMP shunt stimulation while inert latex particles showed no effect on HMP shunt activity suggesting the association of additive nature as determinant of the HMP shunt activity or trio-enzyme inhibition; 3. The glutathione plays a critical role in surviving Kupffer cells during hypoxia and indicates the stimulated reactive oxygen species in respiratory burst activity [Bhatnagar et al.1981]. All these respiratory burst biomarker activities showed dependence on the nature of additives in corroboration with earlier reports [Kumar et al.2002; Forman et al. 2002; Ding et al. 1988]. In present chapter, both HMP shunt activity and respiratory burst activity (superoxide dismutase, glutathione reductase, NADPH oxidase enzyme activity) are outlines for suggested initial Kupffer cell stimulation by additives and later followed by energy and oxygen depletion as possible cause of hypoxia and cell viability

loss as important finding. However, association of Kupffer cell stimulation and hypoxia is not conclusive and needs further investigation.

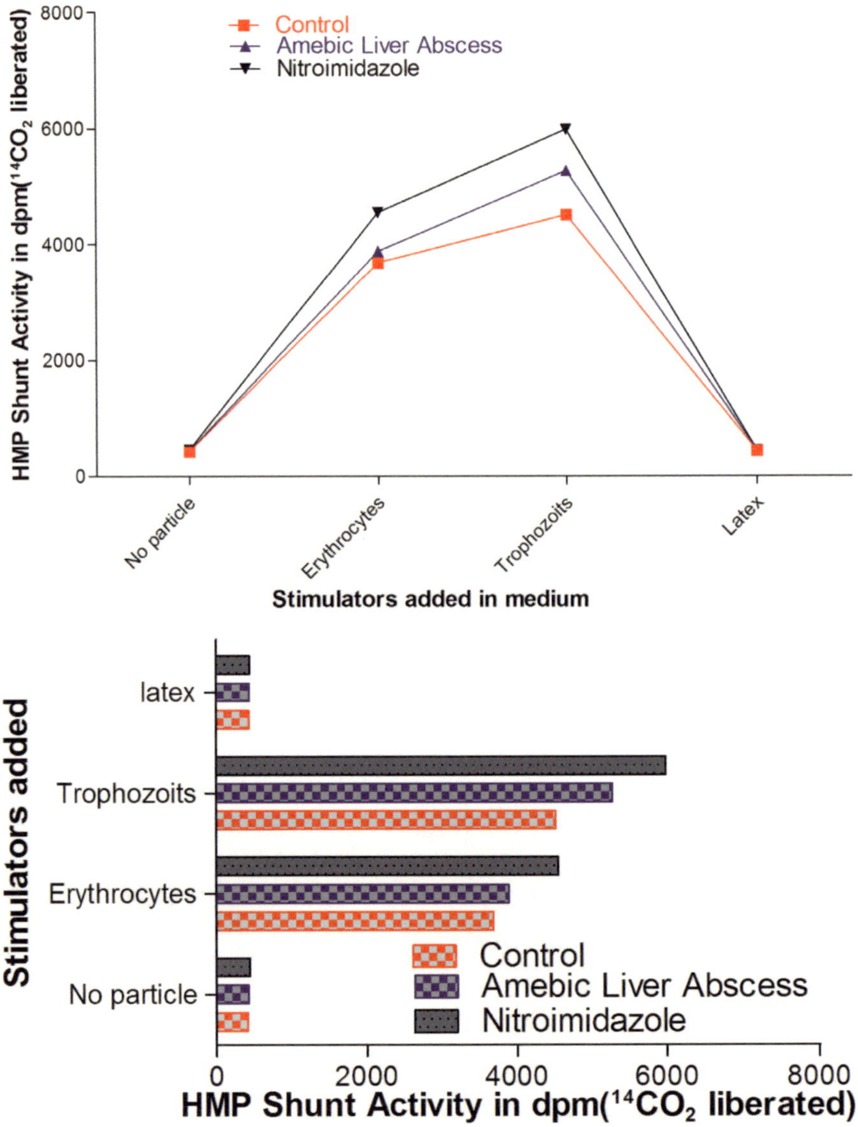

Fig. 13. Hexose monophosphate shunt activity is shown in isolated Kupffer cells from ameba infected and nitroimidazole treated cells in presence of erythrocytes, trophozoits and latex additives (see panel on top). In 3.0 ml medium, 1 micro Curie ^{14}C glucose was added and simultaneously particles added with Kupffer cells in ratio 5:1(particle:cell) for erythrocytes, 2:1 for amebic trophozoits and 40:1 for latex. After one hour, 2 ml of cell free medium soaked by filter paper wetted with 100 μL of 10% KOH in counter well which further was added with 200 μL 1.0 N H_2SO_4 and after 15 minutes filter paper strip elute in tritosol were measured for scientillation counts. The trophozoits showed maximum activity while inert latex beads did not show any change in activity(see panel at bottom).Ref. Sharma, R. et al. 2010.

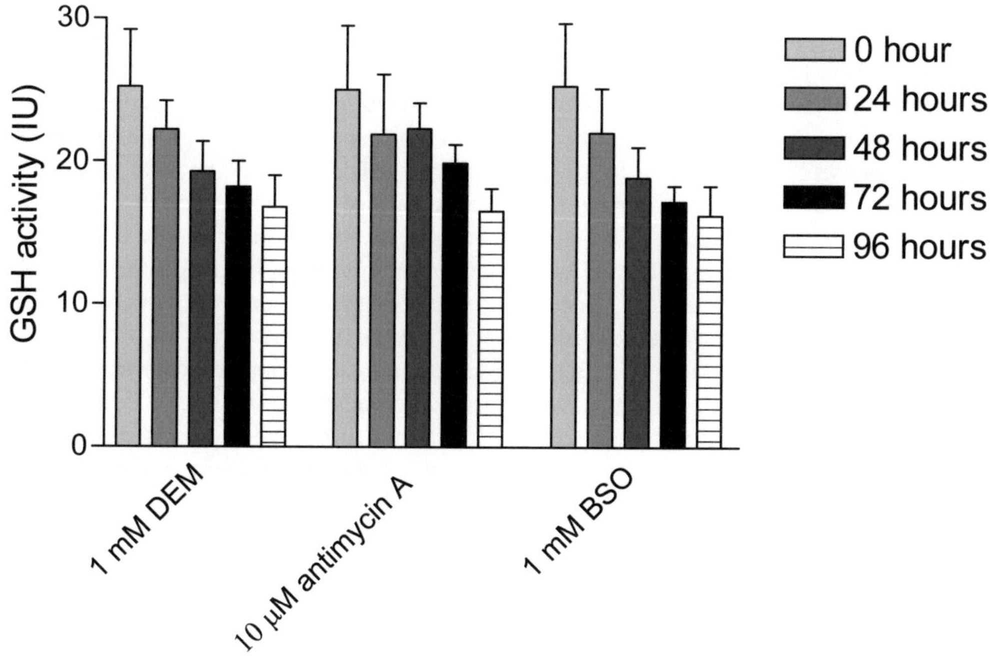

Fig. 14. Effect of added stimulators on Glutathione reductase enzyme activity in hypoxic hepatocytes in culture. Ref. Sharma, R. et al. 2010.

4.8 Cytochrome oxidase

Cytochrome C oxidase was determined based upon oxidation of ferrocytochrome C by decrease in absorbance at 550 nm on spectrophotometer. In 3 ml reaction mixture containing 66 mM potassium phosphate buffer pH 7.4. 20μM reduced cytochrome C and 20μl mitochondrial protein reaction was followed at 30°C for two minutes. Reading was recorded after two minutes for absorbance of completely oxidised cytochrome C by adding 0.05 M potassium ferricyanide. The activity of cytochrome C oxidase was expressed in n moles of cytochrome C oxidase/mg protein/min which is equal to concentration of cytochrome C x K divided by concentration of protein. For calculation of cytochrome C concentration a millimolar extinction coefficient of 19.1 was employed for reduced vs oxidized cytochrome C at 550 nm as described by Emmelot et al.1964.

4.9 Inhibition of NADPH cytochrome P450 reductase

Cytochrome P 450: Cytochrome P 450 was measured by spectrophotometer in absence or presence of carbon monoxide and determined by the method [Guengerich et al.2009]. In brief, 0.1 ml microsomal fraction in two cuvettes test and standard which mere added 1,5 mg sodium dithionite and carbon monoxide passed in only test sample cuvette for 25 seconds and same repeated again and ΔA recorded at 450 nm for cytochrome P 450 content in n moles per mg protein calculated by extinction coefficient 91 mM/cm path length.

4.10 Inhibition of adenylate cyclase

Adenylate cyclase regulatory properties: Adenylate cyclase enzyme in isolated hepatocytes has been reported as tool of regulatory studies of hormones, enzymes in the presence of Gs bound receptors [Small et al. 2000]. Membrane adenylate cyclase activity is defined due to its multi-pass transmembrane protein made of two cytosolic segments as shown in Figure 15. These segments play active roles in association of $G_{s\alpha}$-GTP. The GTP induced $G_{s\alpha}$ binding with adenylate cyclase can be a key mechanism of adenylate cyclase stimulation or inhibition during progesterone bound α-adrenergic receptor [Sunahara et al. 1997]. The possible mechanism of adenylate cyclase activation or inhibition by progesterone is shown in Figure 15. The membrane adenylate cyclase activity can be an indicator of membrane viability. However, the enzyme activity depends on the type and nature of effecter. The behavior of liver cells in culture in the presence of effectors may indicate the regulatory effect of stimulators *in vivo*. However, in cultures, hepatocytes and Kupffer cells may likely experience different physiological environment different from *in vivo*. In controlled liver cell culture conditions of medium, the liver cells showed adenylate cyclase stimulation in a specific manner by effectors such as progesterone, nitroimidazole, GTP, GITP, and prostaglandins [Aoshiba et al. 1997]. The receptors for prostaglandins interact with $G_{s\alpha}$ subunits. $G_{s\alpha}$-GTP complex binds with α-adrenalgic receptor made of seven subunits as shown in Figure 5. Binding of receptor with $G_{s\alpha}$-GTP protein inhibits adenylate cyclase activity and cAMP formation. The present study highlights the behavior of liver cells and the characteristic features of their cyclic AMP dependent adenylate cyclase regulatory system in culture.

Progesterone modulation of adenylate cyclase activity: Progesterone modulates the activity of adenylate cyclase and adenylate cyclase response is hormone sensitive [Gilman et al. 1984]. The progesterone stimulates the receptor activating the adenylate cyclase by stimulatory $G_{s\alpha}$-GTP complex resulting with formation of cAMP. Hepatocytes with progesterone added in medium exhibited the enhanced enzyme activity. The addition of GTP stimulator did not change the adenylate cyclase activity. In contrast, GITP stimulated cells exhibited an apparent adenylate cyclase activation. Unlike the adenylate cyclase enzyme activity in the presence of different effectors in stimulated hepatocytes, the rates of cyclic AMP production was delayed in control hepatocytes. However, progesterone addition to hepatocytes did not exhibit significant delay in GITP stimulated adenylate cyclase activity. Thus reason can be speculated that progesterone binds with stimulatory receptors on hepatocytes which lead to increase in intracellular cyclic AMP accumulation through adenylate cyclase activation via guinine nucleotide regulated mechanism [Cohen-Tannoudji et al. 1991]. The Kupffer cells also showed similar guanine regulatory enzyme properties.

The progesterone response of adenylate cyclase activity is a two phase process [Ko, et al.1999; Martin et al. 1987]. First, progesterone receptor and catalytic subunit (C complex) of adenylate cyclase occupy the place on membrane followed by $G_{s\alpha}$ protein interaction with enzyme to make G protein C complex [Ko et al. 1999]. However, the addition of GITP stimulated the adenylate cyclase to a greater extent in hepatocytes more than pre-exposed hepatocytes to progesterone. Recently, G protein-enzyme complex formation by GTP/GITP stimulation has been reviewed [Niu et al. 2003].

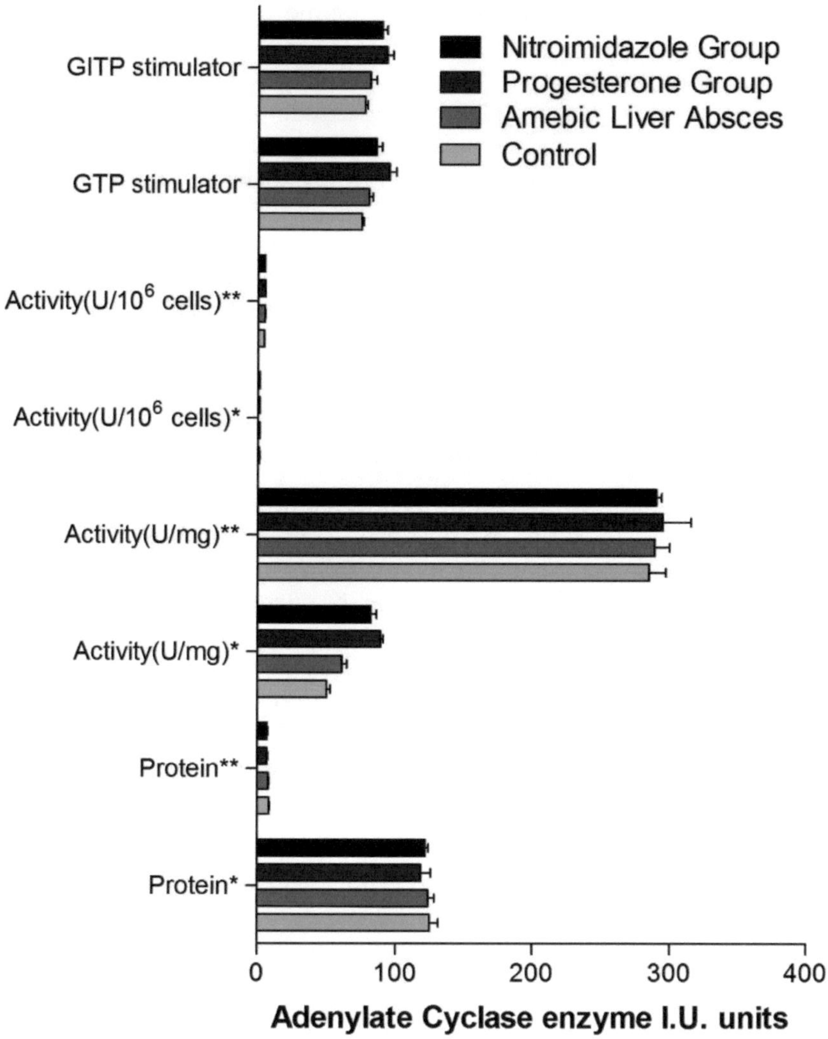

Fig. 15. Adenylate cyclase enzyme activity is shown in control, nitroimidazole, progesterone and amoebic liver abscess recruited liver cells in presence of GITP/GTP stimulators, protein* additives. Ref. Sharma, R. et al. 2010.

Nitroimidazole stimulation of adenylate cyclase activity: Nitroimidazole is recently emerged as anti-tumor and radiosensitizer compound. It is used in tumor imaging, therapy and anti-amoebic applications. Nitroimidazole has phenolic rings and phenolic side chains as shown in Figure 6. So, it is obvious that the side chains interfere with the dephosphorylation of ATP by phosphodiesterase enzyme. It appears to result with inhibition of hepatocyte adenylate cyclase catalytic subunit function in the presence of nitroimidazole. The cytotoxicity of nitroimidazole is poorly studied and reported. However, its anti-amoebic effect has been proven to decrease hepatic and intestinal infection rate. The nitroimidazole is widely reported as single dose in treatment of amoebic liver abscess. Moreover, nitroimidazole induced cytotoxic effect on human hepatocytes are still not known. DNA strand breakage, energy regulatory enzymes, phagocytosis, and immune response have been reported as nitroimidazole induced cytotoxic effects in animals [Zheng et al. 1997; Whitaker et al. 1992; Singh et al. 1991]. The role of nitroimidzole is understood to restore energy regulatory process involving adenylate cyclase, phosphodiesterase, cAMP formation in hepatocytes and reported as hormone and GTP specific in action. The nitroimidazole derivatives inhibited anterior pituitary cell function apparently by their direct effect on the catalytic subunit of the adenylate cyclase holoenzyme [Stalla et al. 1989]. The nonparenchymal Kuffer cells play a significant role of phagocytosis sensitive to prostaglandins. Recently, nitroimidazole derivatives have been identified as anticancer potential agents. Still the role of imidazoles as anticancer agent is not established [Anderson et al. 2006]. The indicated study the association of imidazole ring structure with the cyclic AMP dependent adenylate cyclase catalytic subunit regulatory mechanism in liver cells.

5. Inhibition of drug metabolizing enzymes in isolated liver cells

Drug metabolizing enzymes are abundant in microsomes and lysosomes. Most of these enzymes are elevated or secreted at the sites of inflammation, disease conditions and in presence of foreign bodies. An elevated enzyme in cells and secretion from lysosomes or microscomes action is in defense by 'suicide action' of microbody against disease or foreign bodies. Drugs inhibit these enzymes or bring back any disease or inflammation induced effect on enzyme activity in liver cell. Characteristic inhibition of enzyme activities in different groups by nitroimidazole are shown in Figure 15. Recently, we reported the enzyme inhibition by nitroimidazole in detail [Sharma 2011a, Sharma 2011b]. In general, lysosomal enzymes are peptidases, act at acidic pH, and participate as terminal scavengers of harmful anions, radicals and peptides. For details, read section 2.

Nitroimidazole inhibits the liver enzymes by making adduct for glyoxyl-glutamine metabolism in liver cells as author described in detail [Sharma, et al. 2011].

6. Phagocytosis

Although the clearance and distribution of ligand molecules in circulation represent the function of hepatic sinusoidal cells, these mechanisms reveal a network that is more intricate than would at first seem, since several receptors are common to not only one type of cell, but also to two or three types of cells in the liver. In the case of latex particles in which their uptake by a particular cell type seems to be determined by their size, sinusoidal endothelial cells are able to internalize particles up to 0.23 microns under physiologic conditions, in vivo, and larger particles are taken up by Kupffer cells. However, when the phagocytic function of Kupffer cells were impaired by frog virus or alcohol, the endothelial cells were reported to take up particles larger than 1 micron in diameter after the injection of an excess amount of latex particles. Endothelial cells would thus constitute a second line of defense in the liver in that they remove foreign materials from the blood when Kupffer cell phagocytic function is totally disturbed [Schoremmer et al. 1990]. This potential role may not, however, be fully expressed under physiologic conditions when Kupffer cells are active in clearing foreign substance from the circulation. The functions of liver sinusoidal cells are varied and complex and these cells can be regarded as "a sinusoidal cell unit." This cellular interaction must be taken into account for any quantitative analysis [Kampas et al. 1999].

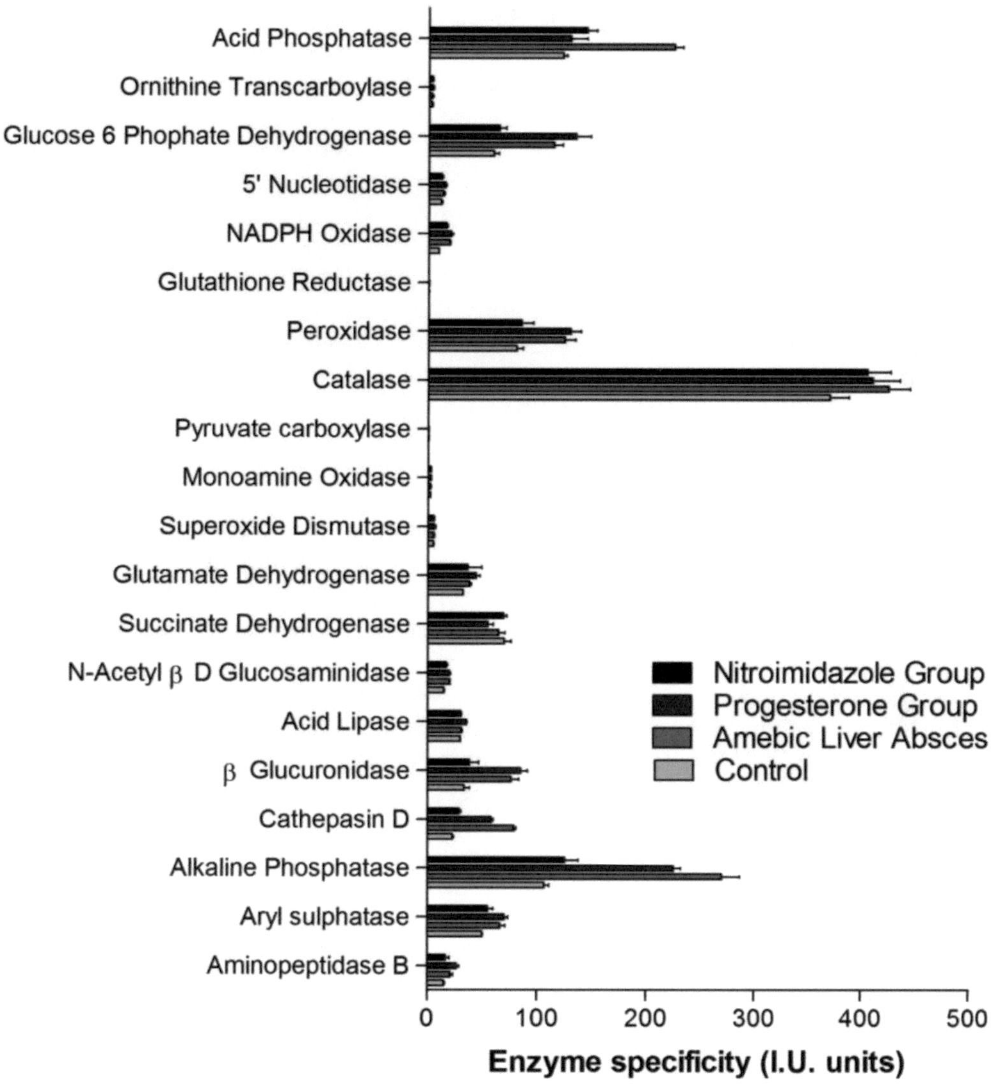

Fig. 16. Lysosomal and microsomal enzymes in isolated nonparenchymal liver cells are shown in control, nitroimidazole, progesterone, amoebic liver abscess recruited livers. Note inhibitory action of nitroimidazole on liver cells to bring normal activities in comparison to amoebic liver abcess recruited cells. Ref. Sharma, R. et al. 2009.

- **Enzyme inhibition in phagocytosis**: In our experiments, liver cell cultures were mixed with zymosan (7:1 particle cell ratio); erythrocytes (5:1 particle cell ratio); amebic trophozoits (2:1 particle ratio), latex (20:1 particle cell ratio) and nitroimidazole was added 1 $\mu M/10^6$ cells. Kupffer cell phagocytic activation in isolated Kupffer cells from human livers (added with amoebic trophozoits) was elevated 2-3 fold over normal Kupffer cells. Kupffer cells incubated with zymosan and trophozoits exhibited maximum phagocytic activity while with latex and sheep erythrocytes these exhibited minimum and insignificantly increased activity respectively. Increased phagocytic activity was time dependent which was exhibited by β-glucuronidase, lactate dehydrogenase and plasminogen activator activity. Phagocytosis of zymosan and sheep erythrocytes triggered the immediate release of β glucuronidase, stimulated synthesis of cellular lactate dehydrogenase and induced delayed production with secretion of plasminogen activator. Inhibition of these enzymes as nitroimidazole response is shown in Figures 16-18, in detail. Major findings were: 1. effect of nitroimidazole and latex were minimum. Similar observations were reported in earlier study [Kamps et al. 1999, Bhatnagar et al. 1981, Kuttner et al. 1982]; 2. nitroimidazole was biotransformed into water soluble products and cleared by drug metabolizing enzyme system of Kupffer cells; 3. latex particles being inert confirmed the independence of Kupffer cell activation from intracellular fate of ingested material; 4. amoebic trophozoits with their toxins or specific membrane surface signals stimulate Kupffer cells for phagocytosis resulting early release of lysosomal hydrolases than the release by other particles [Virk et al 1989, Virk et al. 1988]. However, increased lysosomal enzymes in Kupffer cells incubated with amoebic trophozoits were reported as a consequence of phagocytic activation [Froh et al. 2003]. Recently, these phagocytic activation changes due to nitroimidazole have been reported as associated with prostaglandin synthesis and their release [Spolarics et al. 1997].
- The glucuronidase activity showed the enhanced enzyme activity trend in all additives but different enzyme enhancement for different additives. In control cells, the activity enhanced during 24-36 hours in incubation. The erythrocytes showed most vulnerability to Kupffer cells in initial 24 hours. Zymosan beads showed slow response to phagocytosis while latex remained inert to Kupffer cells. The nitroimidazole showed normal enzyme pattern. It was interesting that maximum rate of phagocytosis was observed during 24 and 36 hours except erythrocytes.
- Lactate dehydrogenase enzyme activity showed the enhanced enzyme activity trend in all additives except latex additive but different enzyme enhancement for different additives. In control cells, the activity enhanced during 12-36 hours in incubation. The erythrocytes showed most vulnerability to Kupffer cells in initial 24 hours. Zymosan beads showed maximum response to phagocytosis while latex remained inert to Kupffer cells. The nitroimidazole showed inhibitory enzyme activity in Kupffer cells but after 24 hours the response was stimulatory to enzyme activity (delayed enhanced enzyme stimulation as a result phagocytosis). It was interesting that maximum rate of phagocytosis was observed during 48 hours. However, the leaked out LDH activity represented the probable phagocytic activity of Kupffer cells relationship with time. Nitroimidazole addition in incubation medium perhaps showed the minimizing amoebic virulence [Sharma 2010]. Earlier lysosomal enzyme release with increased LDH cellular levels has been reported [Koppele et al. 1991].

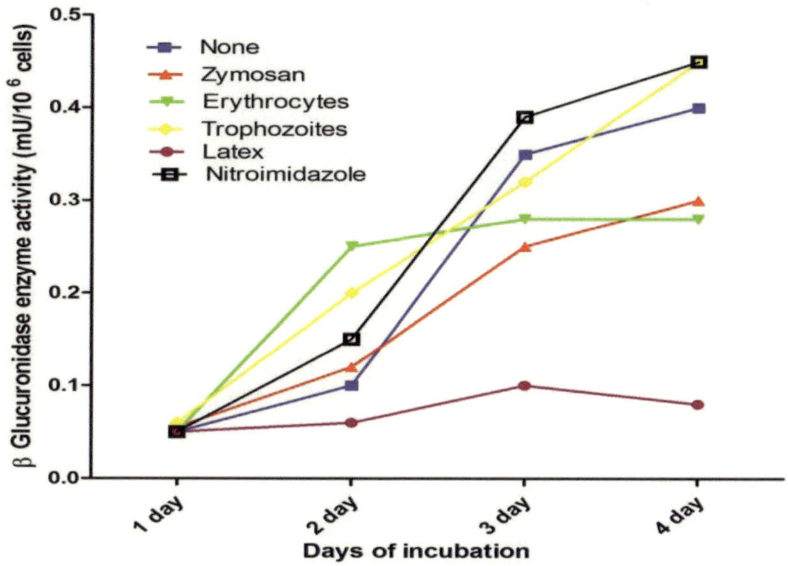

Fig. 17. Phagocytic activity of Kupffer cells is shown as β glucuronidase enzyme activity of cells after adding zymosan, erythrocytes, entamoeba histolytica trophozoits, latex beads additives, and nitroimidazole as drug in cultures of Kupffer cells from liver biopsy samples. The enzyme activity was measured after 1, 2, 3, 4 days of incubation. Each point value represents the average of 3 readings of observations. Ref Sharma,R. et al. 2009.

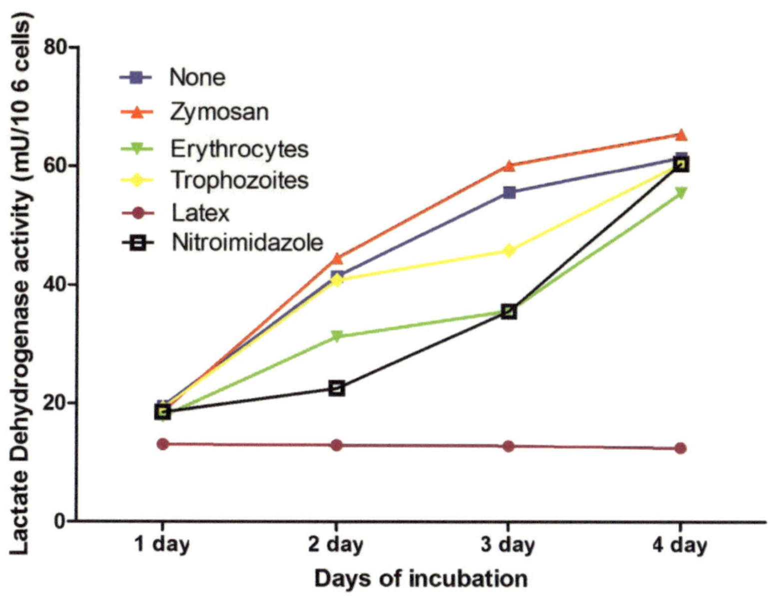

Fig. 18. Phagocytic activity of Kupffer cells is shown as lactate dehydrogenase enzyme activity of cells after adding zymosan, erythrocytes, entamoeba histolytica trophozoits, latex beads additives, and nitroimidazole as drug in cultures of Kupffer cells from liver biopsy samples. The enzyme activity was measured after 1, 2, 3, 4 days of incubation. Each point value represents the average of 3 readings of observations. Ref Sharma,R. et al. 2009.

- The plasminogen activator induced lysis of Kupffer cells showed the enhanced plasminogen activity only after 36 hours. Initially the activity trend in all additives was same and immune to any additive added. In control cells, the plasminogen activity remained unaltered while kupffer cells showed immunity to zymosan and latex additives throughout the incubation period. However Kupffer cells showed sharp enhancement of plasminogen activity after 36 hours during 36-48 hours in incubation. The trophozoits showed maximum effect on plasminogen activator in Kupffer cells after 36 hours. Plasminogen activator induction has been reported dependent on de novo enzyme synthesis in stimulated macrophage by endotoxins and asbestos [Nagaoka et al. 2003].

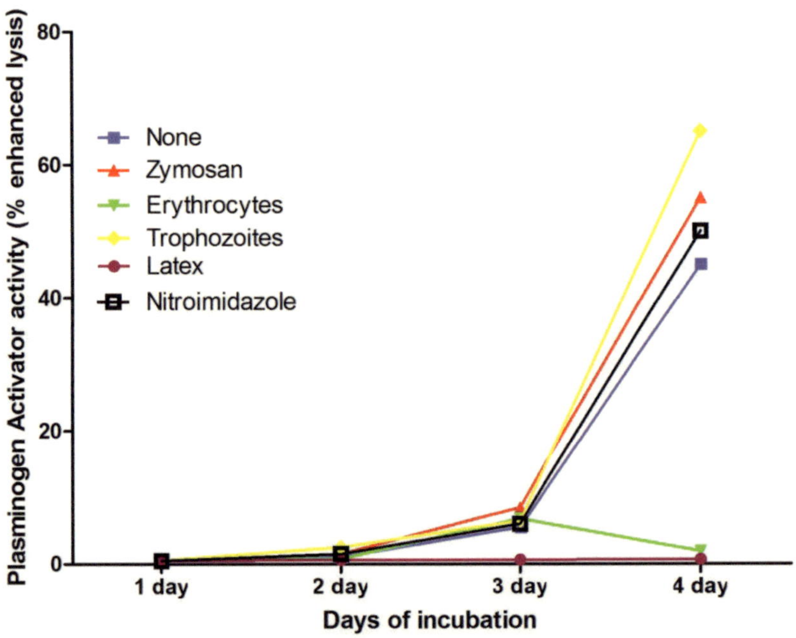

Fig. 19. Phagocytic activity of Kupffer cells is shown as plasminogen activator activity of cells after adding zymosan, erythrocytes, entamoeba histolytica trophozoits, latex beads additives, and nitroimidazole as drug in cultures of Kupffer cells from liver biopsy samples. The enzyme activity was measured after 1, 2, 3, 4 days of incubation. Each point value represents the average of 3 readings of observations. Each point on graph represents the activity as extent of cell lysis in percent increase over basal value of 100. Ref Sharma et al. 2009.

7. DNA synthesis in Kupffer cells

Stimulated DNA synthesis is considered as defense mode. From physiological point of view, it is true that drastic change in enzymatic pattern in liver regeneration permits DNA synthesis. Major enzyme players are DNA polymerase and DNase (allosteric complex enzyme) in mitosis. We present five sequence steps of events: deoxyribonucleotides, ribonucleotides, DNA, RNA and proteins (see Figure 20) [Bresnick et al.1970].

- DNase exposes 3′-OH end groups thus increases capacity of chromosomal DNA in vivo. DNA polymerase plays a passive role in DNA synthesis in presence of high chromosomal DNA and deoxyribonucleotides. Priming DNase action on native and denatured DNA initiates DNA synthesis. Earlier study reported that arginase enzyme acts strong DNA synthase inhibitor perhaps due to lack of neosynthesized histone (arginine-rich histone) by reducing priming state of chromosomal DNA. DNA sysnthesis by means of additive or replicative DNA polymerase needs 3′-OH roup and polymerization proceeds from 5′-phosphate end to the growing 3′-OH end.

- Increased RNA synthesis also increases ribonucleotide pool, rise in RNA polymerase activity, increase in polyamine synthesis and s-RNA methylation. The temporal relationship between kinetic changes in the DNA channel and the RNA channel seem to indicate that RNA channel is primer of DNA channel. In other words, cyclic AMP level may initiate both DNA and RNA synthesis beginning at the nucleotide level under hormonal regulation. Best example of elevated RNA sysnthesis in liver regeneration is elevated orotate pyrophosphorylase/decarboxylase to keep CTP/UTP pool in presence of UTP/CDP coenzymes for glycogenesis and lipogenesis respectively. UMP stimulated orotate-tRNA labeling around 18s-28s RNA complex step is preceded by an increase in tRNA methylation by tRNA Methylase enzyme. For its action, short chain of monocistronic mRNA on 18s component acts as primer for synthesis of constitutive RNA and export protein synthesis. Nobel Prize was awarded in 2009 on resolving the intricacies of ribosomal subunit interplay and consequent role of DNA/RNA Polymerase enzyme reactions in protein synthesis during drug induced regeneration [Schmeing et al.2009]. However, other enzymes RNase activity, cataionic polyamine play role in stabilization of ribosomal RNA (labeled methionine to spermidine) in protein synthesis. Increased polymine synthesis enhances ornithine decarboxylase to increase amino acid pool in presence of cAMP induced prostaglandins mediated growth hormone. Moreover, growth hormone stimulates thymidine kinase in liver. Author describes DNA synthesis in isolated Kupffer cells during liver regeneration after nitroimidazole treatment in amoebic liver abscess in following section.
- Kupffer cells isolated from human amoebic liver abscess liver-biopsies and progesterone treated Kupffer cell in cultures were induced for DNA synthesis by adding nitroimidazole. This offered a simple model to study the mechanism of cell DNA induction and interdependence between two cell populations. Earlier such interactions between cells were reported regulated by thymidine kinase (thymidine pathway) and ribonucleotide reductase (thymidylate pathway)[Bresnick et al.1970]. Conceptually, immediately after disease in regeneration, liver cell gets confronted a challenge of increased protein synthesis for cell division involving metabolic enzymes leading to DNA synthesis (binding of ribonucleotides with deoxyribonucleotides). Two pathways play role in this process by Thymidine pathway and thymidylate pathway. Thymidylate pathway for DNA synthesis needs dTTP supply, thymidine kinase, TDP kinase to keep TMP pool. Basically, a repressor molecule (thymidine) inhibits transcription of thymidine kinase by allosteric regulation. Thymidine kinase is a TTP insensitive dimeric form (one catalytic site and other allosteric site) means K_m for enzyme decreases with elevated ATP concentration (ATP positive effector and TTP negative effector). dimeric form and shows sigmoid behavior of thymidine kinase [Barroso et al. 2005]. Activation of thymidine kinase also depends on cyclic AMP activated protein kinase. On other side, ribonucleotise reductase reduces CDP in presence of dCMP deaminase to make dUMP (a substrate for thymidylate synthesis) by thymidylate sysnthetase in presence of N^5,N^{10} hydroxymethyl-FH_4. dCMP is good enzyme inhibitor of thymidylate sysnthetase and activator of folate reductase to keep net high TMP pool. Ribonucleotide synthesis is regulated by

ribonucleotide reductase. Increase in CTP build up dCTP, purine deoxyribonucleotides and TTP pools[Cheung et al.1970].

- Several types of growth factors have been reported as regulatory substances participating in liver regeneration process. Stimulation of DNA synthesis in Kupffer cells by amoeba trophozoites is shown here as response induced by colony stimulating factor. Since measured increase of radio label uptake by increasing amoeba trophozoite concentration in medium reflected an increase of cell number entering into S phase rather than higher rate of [^{3}H] TdR incorporation seen in all the cells, it was possible that different Kupffer cells normal or pretreated with amoeba (from these different liver cell populations) may vary in susceptibility to colony stimulating factor (CSF). High concentrations of amoeba caused a reduction of thymidine uptake and remarkably reduction was measured on whole cover slip scintillation counts. It reflected lower incorporation of [^{3}H] TdR into each DNA synthesizing cells. This phenomenon may be due to reduced capacity of Kupffer cells uptake as a result of DNA synthesis but it requires further inhibitor study or presence of CSF. Thus higher concentration of amoebae trophozoites caused a reduced Kupffer cell uptake capacity during CSF production.

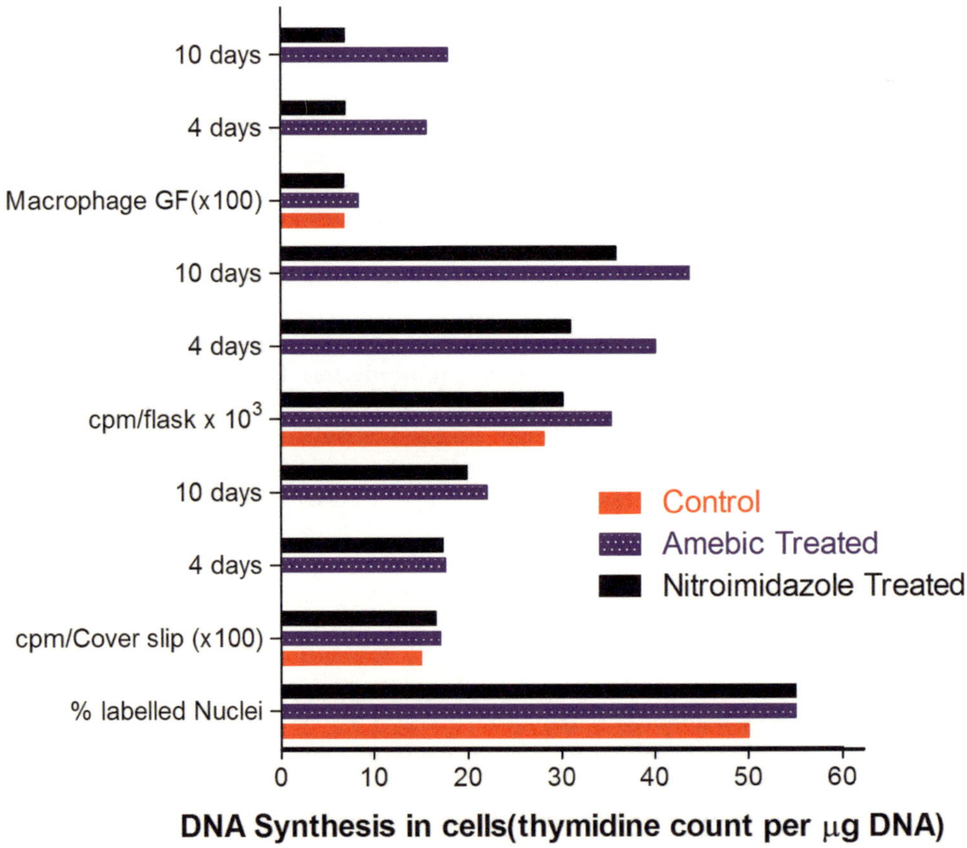

Fig. 20. DNA synthesis in liver cells in culture medium is shown in presence of macrophage growth factor measured as cpm/flask and cpm/cover slip during 4-10 days. Notice the nitroimidazole induced inhibition of DNA synthesis in amebic treated cells.Source: Sharma, R.(1990) Ph.D dissertation.

8. Inhibition of liver lysosomal enzymes in protein metaboilism

Isolated lysosomal preparations from L[1-^{14}C] leucine (25 μCi per 100 gm wt) intraperitonially injected human Kupffer cell cultures exhibited percent degradation at pH 3.0, 4.0,5.0, 6.0,7.0, 8.0 and 9.0 at temperatures 10 °C, 20 °C, 30 °C, 40 °C, 50 °C and 60 °C, in presence of amoebic trophozoits after 15 minutes(short lived) and 15 hours(long-lived) addition as shown in Figures 21. Major players in this liver regeneration process are ribosomes (sitting on endoplasmic reticulum), GTP (ribotide pool) and coenzymes for ribotide polymerase enzymes (a balance in thymidine kinase and thymidylate synthetase). Other enzyme inhibition is LDH-M type gets converted to LDH-H type or lactate → pyruvate (accelerated glycolysis). In simultaneous other set of experiment, Kupffer cell lysosomes in presence of leupeptin, pepstatin inhibitors and nitroimidazole in 100 μg, 100 μg and 50 mM per ml concentration respectively showed aryl sulphatase enzyme activity changes (see Figures 21,22) in the presence of 0.3 M sucrose at 37 °C for 60 minutes at pH 5.5 in incubation mixture of Kupffer cell lysosomes and additives. Proteolysis rate and degradation of short and long lived proteins in isolated lysosomes from the Kupffer cell

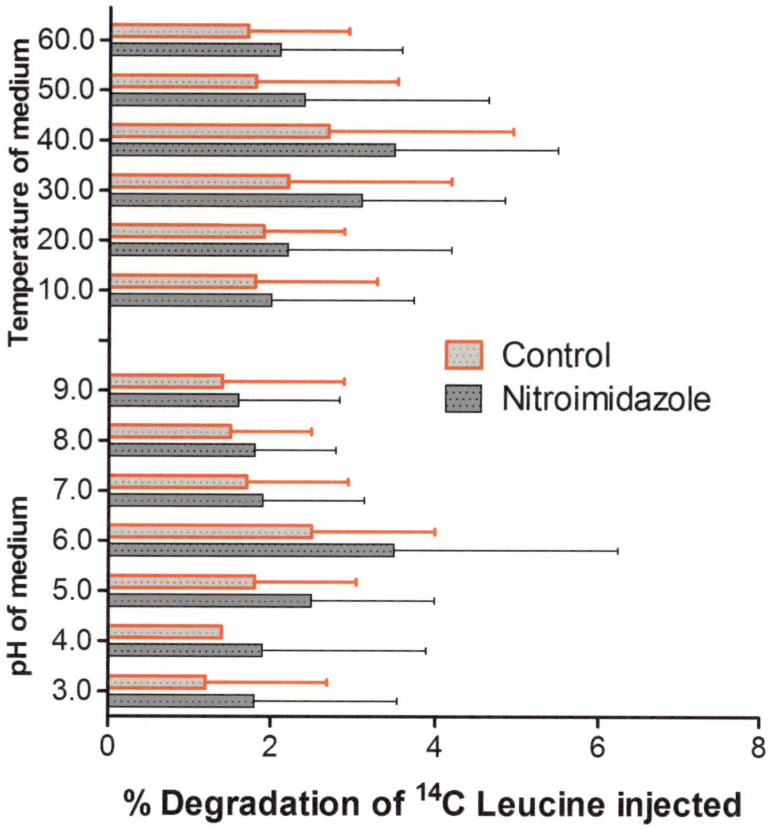

Fig. 21. Figure shows the effect of of radiolabeled [^{14}C]-Leucine in Kupffer cultures represented by aryl sulphatase enzyme activity in normal and ALA recruited cells at different pH and temperatures. Notice the low enzyme activity in normal cells while maximum leucine degradation was at pH 6 and 40 °C.

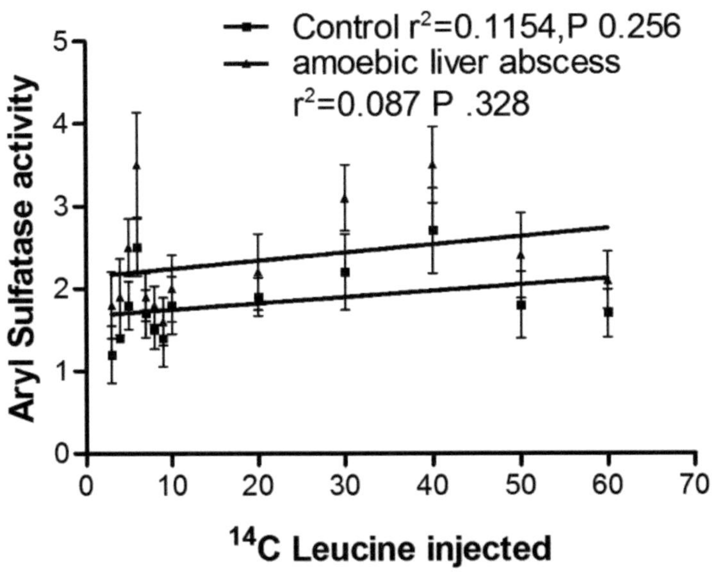

Fig. 22. Figure shows the effect of radiolabeled [^{14}C]-Leucine in Kupffer cultures represented by aryl sulphatase enzyme activity in normal and ALA recruited cells. Notice the threshold of leucine turnover at 40 IU in terms of enzyme activity. Control cells showed slow % protein degradation and low enzyme concentration. Source. Sharma 1990. Ph.D dissertation.

cultures added with trophozoits, nitroimidazole, exhibited increased protein degradation. Lysosome as proteolytic component was shown to contribute for protein degradation as earlier various reports indicated protein size dependence [Neff et al.1990]. Proteolytic rate in isolated lysosomes increased several fold with decreased intralysosomal pH. However, leakage of hydrolytic enzymes (elevated orotate phosphorylase/decarboxylase) may cause degradation of extra lysosomal orotate substrates. Thus incubation time should be short. Degradation of short and long-lived proteins in isolated lysosomes demonstrates that after 15 minutes of [^{14}C] leucine labeling, these lysosomes were active in degradation of short time labeled proteins. Since transit time for lysosomal enzymes from endoplasmic reticulum to lysosomes was around 2-3 hours [Zhu et al. 1994]. So after, 15 minutes some other proteins may be thought to enter lysosomes. The degradation rate in isolated lysosomes was sometimes faster for short time labeling because of short lived proteins as cytosolic or secretory in nature. On the other hand, membrane proteins had half lives in range of few days. So these long labeled proteins were degraded at different rates within lysosomes. Moreover short lived proteins were reported more susceptible to digestion by lysosomal proteases. The effect of proteolytic inhibitors on protein breakdown in isolated lysosomes for distinguishing lysosomal from extra lysosomal protein degradation was dependent upon cell type, culture medium, labeling schedule and inhibitors used. Inhibitors of lysosomal enzymes impede their supposed target of action viz. the lysosomes. But complete inhibition of proteolysis could not be possible due to lysosomal and nonlysosomal proteolysis. Pepstatin inhibited only selective aspartic proteinase cathepsin D, leupeptin inhibited thiol proteinase. Similarly nitroimidazole either intralysosomal pH and restored proteinases. Moreover, cathepsin D inhibition by pepstatin, cathepsin B, H and L inhibition by leupeptin are reported commonly [Katunuma, 2011]. Thus lysosomes were active in degradation of both short and long lived proteins during basal state.

Infected Kupffer cell lysosomes exhibited more pronounced elevated lysosomal degradation at pH 5.0 at temperature range 30 °C-40 °C as shown in Figure 23.

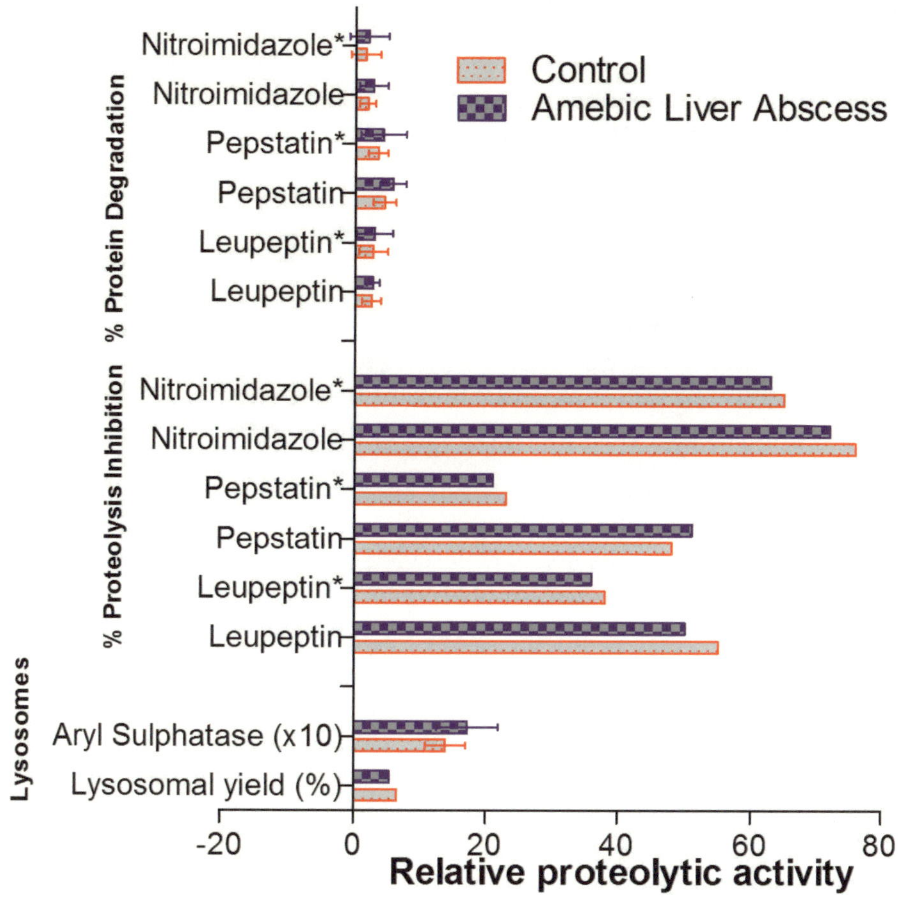

Fig. 23. Figure shows a comparison of % proteolysis inhibition and % protein degradation in presence of nitroimidazole, pepstatin and leupeptin. Proteolysis is represented in terms of arylsulphatase enzyme activity. Notice the size of peptide length makes difference shown by * for long duration and short duration of reaction in each additive. Control cells showed slow % protein degradation and low enzyme concentration. Source. Sharma R 1990,Ph.D dissertation.

9. Liver regeneration and role of lysosomal enzymes

During metabolic integrity loss and resultant hypoxia state, other factors in liver cell signal the alarm to produce nitric oxide, release of cytokines (IFN-α with IL-1β or TNF-α). These changes also trigger the sequence of Kupffer cell stimulation and lysosomal action simultaneously. As a result, lysosomal enzymes, growth factor, plasminogen stimulating factor consistently participate or stimulate the Kupffer cell suicidal phagocytic action (hepatic necrosis) by nitroimidazole analogs and previously reviewed [Michalopoulos, 1990; Thurman, 2000]. In drug induced hepatitis, liver lysosomal enzymes have been reported elevated as stimulation of hepatocellular defense apart from initiating respiratory burst and chemotaxis in liver cells

[Vician et al.2009]. Our study showed the associated electron microscopic observations that Kupffer cells accumulate around the site of hypoxia and exhibited hyperplasia condition showing degenerated nucleus, enlarged lysosomal vesicles [Sharma, 2009]. The enlarged lysosomal vesicles were further correlated by elevated lysosomal enzyme levels in serum. In liver biopsies, these lysosomal enzymes were significantly enhanced. Acid phosphatase, leucine aminopeptidase enzymes catalyzing phosphorylation and amino-peptidization, seem to be secreted more and suggestive of active protein degradation. β-glucuronidase and aryl sulphatase high enzyme secretion was suggestive of continuous breakdown of aryl substituted and β-glucuronidation reactions during cytolysis and proteolysis [Knook et al. 1981]. No previous data is available on liver cell enzymes estimated in liver cells isolated from liver biopsy of nitroimidazole treated livers in hypoxia. Our observations of enzyme inhibition in hypoxic liver cells are very valuable to design a nitroimidazole based radiosensitizer or drug for parasites, myocardial ischemia, tuberculosis, neurotoxicosis etc. Liver cell enzymes may extend the better biochemical explanation of complex hypoxia state at molecular level. However, several scattered studies on serum choline esterase, alkaline phosphatase, glucose-6P-dehydrogenase, ornithine carbamoyl transferase, cyclooxygenase-2, lactate dehydrogenase enzymes have been reported significant in hepatic hypoxia or hepatic damage evaluation with addition of new members of enzymes [Carlson et al. 1976; Koppele et al. 1991]. Enzyme inhibition in liver cells is significant information to design new liver cell metabolic drug stimulators in detoxification, evaluation of hypoxia imaging probes, and cultivation of type of monoaxenic liver cell cultures (hepatocytes and Kupffer cells) with specific enzyme profile.

There appear two main reasons of enzyme level hepatic cell recovery by nitroimidazole. First, altered enzyme changes were recovered by nitroimidazole therapy as drug inhibits the enzymes participating in energy status in hepatocytes and reduces Kupffer cell macrophagal activity with reduced hepatic enzyme secretion to bring back enzyme levels to normal. Second, regenerating hepatocytes may also regulate the normal cell recovery process and initialize the signaling Kupffer cells to store enough lysosomal enzymes.

10. Role of enzyme inhibition in liver transplantation and tissue engineering

Liver regeneration after the enzyme inhibition or loss of hepatic tissue is a fundamental parameter of liver response to injury. Enzyme inhibition phenomenon is recognized as specific external stimuli involving sequential changes in gene expression, growth factor production, and morphologic structure. Author sums up an account on different enzymes in following description:

- Angiotensin Converting enzyme inhibition by lisinopril degrades bradykinin and subsequently induces liver regeneration in rats. During this enzyme inhibition process (liver regeneration) induces 3.4 S DNA polymerase activity in both cytoplasmic fraction and nuclear fractions as constant while 6 S to 8 S DNA polymerase activities in the cytoplasmic fraction increased 6- to 7-fold [Ramalho et al.2001; Chang et al.1972]. The activity of histidine decarboxylase was markedly increased in regenerating liver tissues put. Ornithine decarboxylase is an enzyme catalyze in polyamine synthesis. Inhibition of ornithine decarboxylase, histidine decarboxylase, and other amino acid decarboxylases, was reported in regenerating rat liver and several rat tumors.[Russell, et al.1969]. Amine synthesis in rapidly growing tissues: ornithine decarboxylase activity in regenerating rat liver and various tumors. Apart from enzyme inhibition, many growth factors and cytokines such as most notably hepatocyte growth factor, epidermal growth factor, transforming growth factor-α, interleukin-6, tumor necrosis factor-α, insulin, and norepinephrine, appear to play important roles in this enzyme inhibition and liver response. Many reviews attempted to integrate in last three decades and looked toward clues to the nature of triggering mechanisms in liver and cellular response. [Michalopoulos et al. 1997]. The major events are signaling cascades involving growth factors, cytokines, matrix remodeling, and several feedbacks of stimulation and inhibition of growth related signals. Liver manages to restore any lost mass and adjust its size to that of the organism, while at the same time providing full support for body homeostasis during the entire regenerative process [Michalopoulos et al. 2007].

- Recently, role of tissue engineering concept showed several advantages in liver transplantation. However, several factors are significant in regeneration: matrices design, cell mass up to whole-organ equivalents, optimized microarchitecture, cell-adhesion peptides, growth factors and extracellular matrix molecules to the polymeric scaffold, hepatic enzymes in microenvironment [Mooney et al. 1992]. The use of enzyme characterization is very helpful in hepatocyte culture to test biocompatibility required for the retention of tissue-specific gene expression [Kim et al. 2000; Voytik-Harbin et al. 1997]. Liver cell transplantation into polymeric matrices (cell-matrix-constructs) is emerging as routine in hospitals to meet absolute requirement of the transplanted hepatocytes for hepatotrophic factors that liver constantly receives through its portal circulation [Starzl et al. 1973]. Apart from cellular therapies, two experimental approaches are worth mentioning for successful clinical translation. The first "cell sheet" technology developed by Sasagawa et al.,2007. Ott et al. reported perfusion decellularization to generate whole organ scaffolds [Ott et al. 2008].
- In previous report, investigators developed a similar perfusion decellularization method for the liver [Baptista et al.2009] for organ bioengineering. These bioscaffolds preserve their tissue microarchitecture and an intact vascular network that can be readily used as a route for recellularization (intact enzymes of regulatory metabolic reactions) by perfusion of culture medium containing different liver cell populations. Significant enzyme inhibition assays are used as biomarkers of liver cell viability. In following example, a bioengineered liver (the regenerated hepatic tissue) is shown clearly visible by the naked eye after 7 days in the bioreactor (Fig. 24A). Regenerated tissue was highly cellular (Fig. 24B) and displays biliary duct structures positive for cytokeratin 19 (Fig. 24C), as well as clusters of hepatocytes expressing cytochrome P450 3A and albumin (Fig. 24D, E, respectively). Expression of endothelial cell nitric oxide synthase (eNOS) enzyme in seeded human endothelial umbilical vein cells (hUVECs) were also observed coating and spreading from vascular structures (Fig. 24F). In other approach, Uygun et al. 2010 decellularized rat livers and repopulated them with rat primary hepatocytes, showing promising hepatic function and the ability of heterotopically transplanting these bioengineered livers into animals for up to eight hours [Uygun et al. 2010]. Bioengineered livers displayed some functions of a native human liver (albumin and urea secretion, CYP450 enzyme expression, etc) and also exhibited an endothelial vascular network that prevented platelet adhesion and aggregation, critical for blood vessel patency after transplantation. Hence, this liver regeneration technology with proper active enzyme functions has the potential to translate in the future into the bioengineering of human size

livers to offer readily available liver for drug discovery applications and for liver transplantation, overcoming organ shortage.

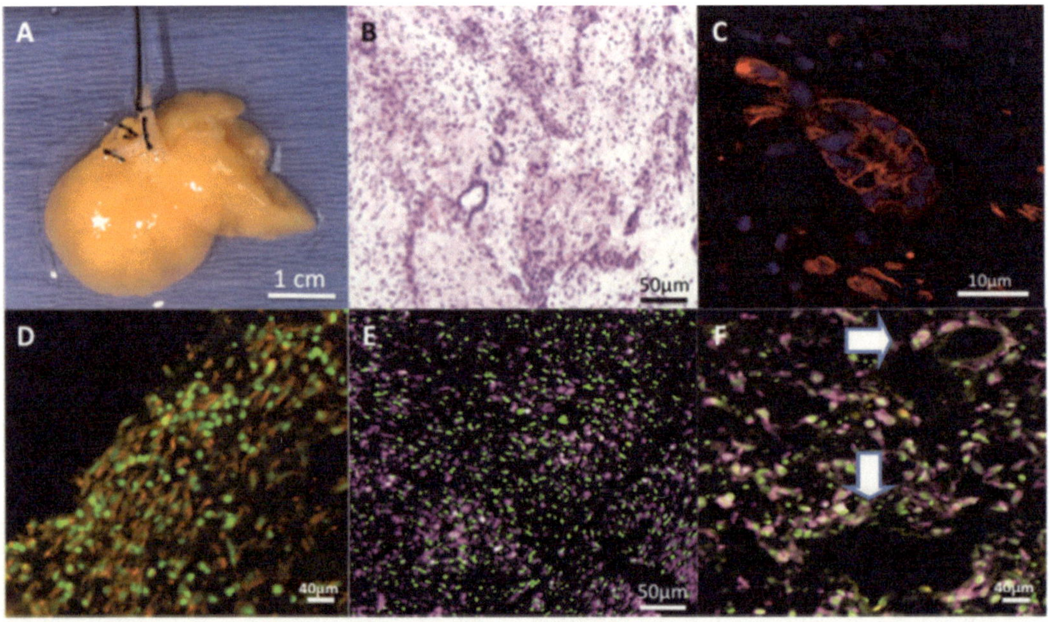

Fig. 24. Human bioengineered livers are highly cellular and display some of the functions observed in native hepatic tissue. (A) Macroscopic appearance of a seeded liver bioscaffold 7 days after seeding with primary human fetal liver and endothelial cells. (B) H&E showing broad recellularization of the bioscaffold with the formation of biliary ductal structures. (C) Immunofluorescence staining for cytokeratin 19 (red) showing a biliary duct formed within the bioscaffold with a visible lumen. (D) Immunofluorescence staining of cytochrome P450 3A (orange) and (E) albumin (purple) showing groups of hepatocytic lineage cells expressing these more mature hepatic markers. (F) Immunofluorescence staining of eNOS (purple) showing hUVECs coating and spreading from vascular structures (arrows). All nuclei stained with YO-PRO1 (green) or DAPI (blue). Reproduced with permission from reference Baptista et al. 2010.

11. Challenges, limitations in enzyme inhibition

There are two major challenges. First challenge was to get enough biopsy to estimate several enzymes. Second challenge was to choose significant enzymes as representative of hepatocellular criteria. The main limitation was that the 'hepatocellular hypoxia criteria' was used in small number of human subjects. The biopsy samples for electron microscopy observations needs more thoroughly controlled experiments. Other limitation was that the enzyme estimations in serum and biopsy samples may not be perfect representative samples and it needs to establish the measurable and actual enzyme activities in cells. The most crucial issue is that hypoxia is a combination of sequence of several metabolic and subphysiological reactions in cells. Moreover, other inflammatory cytokines, apoptosis, nitric oxide production and phagocytosis are associated changes during hypoxia. It further needs tracer technique to track the details of hypoxia in cell.

12. Current developments and future prospects

Nitroimidazoles are rediscovered antiamoebic drugs having great potential of radiosensitizers in diagnosis and treatment of tumors, tuberculosis, myocardial infarction, and hypoxia. Nitroimidazole cytotoxicity was recently capitalized in hypoxic tumor detection and tumor killing. Nitroimidazoles act as smart enzyme inhibitors on one hand and play as enzyme modulators. Liver enzyme inhibition is unique for developing drugs in drug discovery at molecular level. The major challenge is to explore enzyme inhibitor that selectively and specifically catalyzes enzyme inhibition in only abnormal or diseased liver cells leaving safe normal cells in acute stage of liver damage such as hepatic coma, threatening hepatitis, hepatic cancer. The success of enzyme inhibition mainly depends on intracellular physiological conditions such as pH, temperature, concentration of active enzyme or inhibitor, coenzyme, co-factor, enzyme substrate structural-functional conformation state, type of biochemical reaction such as coupled, cascade or individual enzyme reaction and finally regulatory enzyme behavior to control metabolism. Currently, art is developing in liver transplantation using cultured cells to replace diseased or non-functional liver cells from diseased liver. Tissue engineering has opened the new opportunity of using conditioned monoaxenic and monoclonal liver cells in both experimental and pre-clinical or clinical therapeutics in both industries and academics. Recently, gene therapy in liver was a great success and it will be a boom in medical science of detoxification or drug metabolism if and when clinically accepted. Battery of enzymes in hepatocytes and rate limiting enzyme regulatory behavior in presence of additives or inhibitors will certainly enhance better understanding of disease development and new drug pharmaceutical action in drug discovery. Still today, liver cells remain as major research tool to discover drugs in therapeutics and search of transplantable disease resistant liver cells because of liver as sole organ responsible of bile formation, detoxification and drug metabolism. Cirrhosis is a serious health hazard that solely requires functional liver. In future, it remains to see if molecular biology will assist in silencing genes ON or OFF responsible of enzyme stimulation or inhibition in liver as method of disease therapy and cure. The major discoveries will be discovery and better understanding of new growth factors, colony stimulating factors affecting enzyme reactions in hepatocytes and Kupffer cells as major engineered cancer resistant or cirrhosis resistant liver cells. In future, chip technology or silica or polymer coated enzyme estimation techniques may be more reliable in small sample collections or in liver cells with high degree of accurate enzyme estimation.

13. Conclusion

The 2′nitroimidazole is both antiaparasitic drug and radiosensitizer hypoxia marker in tumor therapy. It shows hepatocellular cytotoxicity. The 'hepatocellular hypoxia criterion' distinguishes the nitroimidazole induced hepatocellular oxygen depletion and associated organelle changes by enzyme levels in serum, biopsy samples and cell organelles by electron microscopy. Initially, glucose regulation leads to metabolic integrity loss. Later, it may be oxygen insufficiency and slow cell death. The 2′-nitroimidazole is drug of choice in hepatic infections and its derivatives are emerging choice of tumor treatment. Its action may be evaluated rapidly by enzyme biochemical estimations without time consuming drug monitoring and therapeutic assay techniques. Entamoeba histolytica, an intestinal protozoan parasite can proliferate, lyse and destroy human liver cells. Within the last decade, new animal models of E. histolytica infection showed amoebic trophozoites cause intestinal disease and liver abscess. Recent studies have expanded our understanding of a large number of enzyme inhibition reactions in hepatic regeneration after liver abscess is treated by nitroimidazole as remarkable drug to kill parasites. Still interactions between E. histolytica and human intestine; and between E. histolytica and hepatocytes or Kupffer cells are not fully known. In present chapter, a scheme of enzyme changes in liver cells during development of amoebic liver abscess and enzyme inhibition is explored in detail as rehabilitative action of nitroimidazole to regenerate liver cells. Characteristics of enzyme regulatory behavior highlight the metabolic integrity loss, energy depletion, oxidative phosphorylation, fatty deposition, phagocytosis and slow cell death as main culprits in development of liver abscess and hypoxia. Nitroimidazole inhibits several rate limiting enzymes to control oxygen (hypoxia) levels and keeps balance of NADPH, ATP, fatty acids, amino acids and proteins. In hepatocytes, enzyme inhibition and regulatory behavior of apoptosis (caspases, cysteine protease); carbohydrate metabolizing (glucokinase, pyruvate kinase, phosphofructokinase); energy metabolism (NADPH cytochrome P450 Reductase, Phosphodiesterase); DNA synthesis (Thymidylate Synthetase) describe the process of liver regeneration. In Kupffer cells, enzyme inhibition of drug metabolizing enzymes (oxidoreductases, proteases, esterases, phosphatases) predicts the liver regeneration and recovery from liver abscess. In last decade, several studies reported the use of enzyme inhibition mechanisms in enhancing liver regeneration and recovery as tool of tissue engineering and liver transplantation. In conclusion, enzyme inhibition mechanisms in liver damage and regeneration explain the better information of liver transplantation to maintain complex liver cell environment in balance. Enzyme inhibition mechanisms throw a light on characterization of enzyme regulation in engineered liver cells, amoebic liver abscess development and desired design of targeted nitroimidazole analogues to treat liver abscess.

14. Appendix 1: A protocol of hepatocellular dysfunction

Patients: In 10 control subjects, stool, ELISA, serum and liver biopsies were collected[1]. Only proved 10 subjects free from any hepatic disease were examined for enzyme estimations in isolated liver cells from biopsy specimens. Other ten subjects were treated with nitroimidazole (Tiniba and Zil from Hindustan Lever Ltd, Bombay) therapy 2 x 5 gm one-time dose thrice at 2 days interval. At the end of dose, liver biopsy and serum collection was done.

- All the 10 control subjects had no clinical findings of any hepatic disease and showed stool -ve, ELISA -ve. All 10 showed normal liver and no abscess confirmed by ultrasound and liver scan. The subjects were selected of 35 ± 15 years (mean age ± sd), average monthly income 1750 ± 100 INR, no hepatomegaly, no diarrhea, fever and pain. The subjects after nitroimidazole treatment showed induced hepatomegaly insignificant less than 2 cm by liver scan. The subjects with amoebiasis and amoebic liver abscess are not included and not shown here.

[1]The patients were studied for ongoing other research study on amoebic hepatitis vs amoebic liver abscess by Hepatocellular Dysfunction Criteria at Liver Unit, AIIMS. The results of hepatic damage are shown in Figure 3.

The ELISA was done by method of Prakash et al.[44]. Liver biopsy was taken by Manghini needle [45].

Biochemical assays: In biopsy sonipreps and serum samples the hepatocellular enzymes were estimated as described elsewhere [Sharma, 2009]. All substrates of enzymes were obtained from Sigma Chemical Company, St Louis, USA. The estimations were done on UV spectrophotometer Cecil Inc. England [Sharma, 2009].

Electron microscopy: The biopsy specimens at stored in glutaraldehyde at -20°C. For fixing samples were fixed in dental wax and emersed in pool of 0.1 M phosphate buffer containing 0.1 M Ca^{++}. The samples were cut 1 mm cube by Gillette blade. The cubes were fixed for 2 hours in 0.2 M phosphate buffer in 4 changes at 30 minutes interval in capped vials. The vials were washed for 2 hours in 0.2 M phosphate buffer 4 times at 30 minutes interval in capped vials. The vials were post fixed in buffered 1% osmium tetroxide for 1-2 hours at 4°C, dehydrated in cold 10% ethanol for about 5 minutes followed by dehydrations in cold 50%, 70%, 80%, 95% ethanol for about 5 minutes each and tissues were kept at room temperature. Further tissues were dehydrated at 100% ethanol for 15 minutes. Electron Microscope PA model was used for screening hepatic cell damage by epoxy resin blocks and osmium tetra oxide staining [46].

Art of liver cell isolation and enzyme characterization

The procedures of liver biopsy excision and preparation of liver cells were followed as per ethical committee of institute as described in following section [Sharma, 2009].

Reagents Used

The collagenase enzyme was purchased from Sigma-Aldrich, St Louis, MO and pronase enzyme was purchased from Boerhinger Meinheim GMbH.

Isolation and preparation of Kupffer cells: The liver biopsy samples (6-10 grams) were digested with collagenase enzyme (Sigma-Aldrich, St Louis, MO) 15 mg/ml for 10 minutes and bottom pallet was used for hepatocyte experiments. Other remained supernatant part of digested liver was further digested with 2 mg/ml pronase enzyme (Boerhinger Meinheim GMbH) in 0.2 % Kreb's Hansleit Buffer pH 7.4 containing 5 mm glucose and 3 mM calcium chloride after separation of hepatocytes. The said suspension was incubated for 45 minutes at 37°C stirring 250 revolutions per minute. After incubation, the cell suspension was filtered through nylon sieve 79 × 79 microns and filtrate was centrifuged and washed twice with KHB containing 5 mM glucose and 3 mM calcium chloride pH 7.4. After centrifugation and final washing the filtrate was centrifuged at 350 × g in ice cold centrifuge. The settled pallet at bottom was washed with TC 199 medium. The number of Kupffer cells was counted on Naubourgh counting chamber and visibility was determined by trypan blue.

Kupffer cell fractionation: The isolated Kupffer cells were divided in two parts. One part was used for Kupffer cellular enzymes and other part was used for in vitro experiments. For enzyme estimations Kupffer cells from freshly harvested preparations were sonicated at zero degree temperature in said KHB buffer in soniprep. Later these were fractionated for different cellular fractions as described following by methods [Alabraba et al.2007; Heuff et al. 1994; Kirn et al.1982].

1. In refrigerated centrifuge above soniprep preparations were centrifuged at 1000 × g to take out cell nuclear fraction and again centrifuged at 9000 × g to take out mitochondrial fraction as pallet at bottom.
2. Above supernatant was centrifuged at 23000 × g to isolate microsomal bodies fraction along with fraction rich in peroxisomes.
3. The light turbid supernatant was centrifuged at 100000 × g to isolate lysosomal fraction as clear white pallet at bottom. Sometimes microsomes were contaminated with lysosomes. The lysosomal purity evaluation was used to exclude microsomes free from lysosomes [Wisse et al.1996].

Kupffer cells in cultures: The Kupffer cell cultures were maintained for 48 hours after isolation from pronase perfused liver biopsy samples [Brouwer et al. 1988]. The in vitro cultured cells were screened for trypan blue exclusion [Froh et al.2003]. The Kupffer cells were used for their in vitro phagocytic action upon foreign particles which were sheep erythrocytes, latex beads and zymosan suspended in 10 mM phosphate buffer saline pH 7.4 at 4°C. After harvesting Kupffer cells were added in phosphate buffered saline pH 7.4 containing 4.6 x 10^6 adherent cells and three ml PBS-TC 199 medium in culture flasks containing Kupffer cells in mediukm with particle cell ratio 1:5 for erythrocytes, 1:40 for latex beads and 1:4 for trophozoits. The ratio for zymosan and cell was 20:1. They were kept up to 48 hours or more in carbogen atmosphere (95 % O_2 and 5% CO_2) in desiccator with twice changes of medium after every 24 hours.

15. Acknowledgements

The author acknowledges the assistance provided by Professor Rakesh K Tandon to provide assistant research officer job during this work with grant assistance from ICMR, New Delhi and Dr V.S.Singh for guidance and valuable Council of Scientific and Industrial Research fellowship and research grant from ICMR, New Delhi to do Ph.D.

16. References

Adams, G.E., Ahmed, I., Clarke, E.D., O'Neill, P., Parrick, J., Stratford, I.J., Wallace, R.G., Wardman. P., Watts, M.E. (1980) Structure-activity relationships in the development of hypoxic cell radiosensitizers. III. Effects of basic substituents in nitroimidazole sidechains. Int J Radiat Biol Relat Stud Phys Chem Med. Vol 38, No 6, pp 613-26.

Abuabara, S.F., Barrett, J.A., Hau, T., Jonasson, O.(1982) Amebic liver abscess. Arch Surg. Vol 117, pp 239–44.

Adams, E.B., MacLeod, I.N. (1977) Invasive amebiasis: amebic liver abscess and its complications. Medicine (Baltimore) Vol 56, pp 325–34.

Akgun, Y., Tacyildiz, I.H., Celik, Y. (1999) Amebic liver abscess: changing trends over 20 years. World J Surg. Vol. 23, pp102–06.

Allison, A.C., Ferluga, J., Prydz, H., Schorlemmer, H.U. (1978) The role of macrophage activation in chronic inflammation. Agents Actions. Vol 8, No 1-2, pp 27-35

Ankri, S., Stolarsky, T., Bracha, R., Padilla-Vaca, F., Mirelman, D. (1999) Antisense inhibition of expression of cysteine proteinases affects Entamoeba histolytica-induced formation of liver abscess in hamsters. Infect. Immun. Vol 67, pp 421–422.

Ankri, S. Stolarsky, T., Mirelman, D. (1998) Antisense inhibition of expression of cysteine proteinases does not affect Entamoeba histolytica cytopathic or haemolytic activity but inhibits phagocytosis. Mol. Microbiol. Vol 28, pp 777–785

Aoshiba, K., Rennard, S. I., Spurzem, J. R. (1997) Fibronectin supports bronchial epithelial cell adhesion and survival in the absence of growth factors. Am J Physiol Vol. 273, No 3 Pt 1, pp L684-93.

Ballinger, J.R., Kee, J.W., Rauth, A.M. (1996) In vitro and in vivo evaluation of a technetium-99m-labeled 2-nitroimidazole (BMS181321) as a marker of tumor hypoxia. J Nucl Med. Vol. 37, No 6, pp 1023-31.

Ballinger, J.R. (2001) Imaging hypoxia in tumors. Semin Nucl Med. Vol. 31, No 4, pp 321-9.

Baptista, P.M., Vyas, D., Soker, S. (2011) Liver regeneration: Role of Bioengineering. Chapter 11. In: Progress in Molecular and Environmental Bioengineering– From Analysis and Modeling to Technology Applications. Editor: Angelo Carpi. ISBN 978-953-307-268-5 Intech Open Access Publishers Company.pp257-268.

Baptista, P.M., Orlando, G., Mirmalek-Sani, S.H., Siddiqui, M., Atala, A., Soker, S.(2009) Whole organ decellularization - a tool for bioscaffold fabrication and organ bioengineering. Conf Proc IEEE Eng Med Biol Soc 2009, pp 6526-6529.

Barnes, P.F., DeCock, K.M., Reynolds, T.N., Ralls, P.W. (1987) A comparison of amebic and pyogenic abscess of the liver. Medicine (Baltimore) Vol. 66, pp 472–83.

Baron, R., Neff, L., Brown, W., Louvard, D., Courtoy, P.J. (1990) Selective internalization of the apical plasma membrane and rapid redistribution of lysosomal enzymes and mannose 6-phosphate receptors during osteoclast inactivation by calcitonin. J Cell Sci. Vol. 97, No 3, pp 439-47.

Barroso, J.F., Carvalho, R.N., Flatmark, T. (2005) Kinetic analysis and ligand-induced conformational changes in dimeric and tetrameric forms of human thymidine kinase 2. Biochemistry. Vol 44, No 12, pp 4886-96.

Berninghausen, O. and Leippe, M. (1997) Necrosis versus apoptosis as the mechanism of target cell death induced by Entamoeba histolytica. Infect.Immun. Vol 65, pp 3615–3621.

Berube, L.R., Farah, S., McClelland, R.A., Rauth, A.M. (1992) Depletion of intracellular glutathione by 1-methyl-2-nitrosoimidazole. Int J Radiat Oncol Biol Phys. Vol 22, No 4, pp 817-20..

Bhatnagar, R., Schirmer, R., Ernst, M., Decker, K. (1981) Superoxide release by zymosan stimulated rat kupffer cells in vitro. European J Biochemistry. Vol. 119, No 1, pp 171-175.

Bilzer, M., Roggel, F., Gerbes, A.L. (2006) Role of Kupffer cells in host defense and liver disease. Liver Int. Vol 26, No 10, pp 1175-86.

Bos HJ, Van Den Eijk AA, Steeremberry PA.(1976) Application of ELISA in serodiagnosis of amoebiasis. Trans Roy Soc Trop Med Hyg. Vol 6, pp 440.

Bresnick, E., Mainigi, K.D., Buccino, R., Burleson, S.S. (1970) Studies of deoxythumidine kinase of regenerating rat liver and Eschericia coli. Cancer Res.Vol. 30, pp 2502-2506.

Brezden, C.B., McClelland, R.A., Rauth, A.M. (1997) Apoptosis and 1-methyl-2-nitroimidazole toxicity in CHO cells. Br J Cancer. Vol.76, No 2, pp 180-8.

Brouwer, A., Barelds, R.J., de Leeuw, A.M., Blauw, E., Pla,s A., Yap, S.H., van den Broek, A.M.W.C., Knook, D.L. (1988) Isolation and culture of Kupffer cells from human liver: Ultrastructure, endocytosis and prostaglandin synthesis Journal of Hepatology. Vol 6, No 1, pp 36-49.

Bruchhaus, I., Loftus, B.J., Hall, N., Tannich, E. (2003) The intestinal protozoan parasite Entamoeba histolytica contains 20 cysteine protease genes of which only a small subset is expressed during in vitro cultivation. Eukaryotic Cell, Vol 2, No 3, pp 501-509.

Carlson, R.P., Lefer, A.M. (1976) Hepatic cell integrity in hypodynamic states. Am J Physiol. Vol 231, No 5 Pt. 1, pp 1408-14.

Chang, L.M.S., Bollum, F.J. (1972) Variation of deoxyribonucleic acid polymerase activities during rat liver regeneration. The Journal of Biological Chemistry, Vol 247, pp 7948-7950

Chao, C.F., Subjeck, J.R., Johnson, R.J. (1982) Nitroimidazole inhibition of lactate production in CHO cells. Int J Radiat Oncol Biol Phys. Vol. 8, No 3-4, pp 729-32.

Chu, T., Li, R., Hu, S., Liu, X., Wang, X. (2004) Preparation and biodistribution of technetium-99m-labeled 1-(2-nitroimidazole-1-yl)-propanhydroxyiminoamide (N2IPA) as a tumor hypoxia marker. Nucl Med Biol. Vol.31, No 2, pp 199-203.

Carré, J.E., Orban, J.C., Re, L., Felsmann, K., Iffert, W., Bauer, M., Suliman, H.B., Piantadosi, C.A., Mayhew, T.M., Breen, P., Stotz, M., Singer, M. (2010) Survival in critical illness is associated with early activation of mitochondrial biogenesis. Am J Respir Crit Care Med. Vol.182, No 6, pp 745-51.

Cohen-Tannoudji, J., Vivat, V., Heilmann, J., Legrand, C., Maltier, J. P. (1991). Regulation by progesterone of the high-affinity state of myometrial beta-adrenergic receptor and of adenylate cyclase activity in the pregnant rat. J Mol Endocrinol Vol 6, No 2, pp 137-45.

Cowan, D.S., Matejovic, J.F., McClelland, R.A., Rauth, A.M.(1994) DNA-targeted 2-nitroimidazoles: in vitro and in vivo studies. Br J Cancer. Vol 70, No 6, pp 1067-74.

D'Alessandro, A., Zolla, L. (2011) The SOD yssey: superoxide dismutases from biochemistry, through proteomics, to oxidative stress, aging and nutraceuticals. Expert Rev Proteomics. Vol. 8, No 3, pp 405-21.

Dan, C., Wake, K. (1985) Modes of endocytosis of latex particles in sinusoidal endothelial and Kupffer cells of normal and perfused rat liver. Exp Cell Res. Vol 158, No 1, pp 75-85.

Das, P., Debnath, A., Munoz, M.L. (1999) molecular mechanisms of pathogenesis in amoebiasis. Ind J Gastroenterol Vol. 18, No 4, pp 161-166.

Ding, A., Nathan, C. (1988) Analysis of the nonfunctional respiratory burst in murine kupffer cells. J Exp Med. Vol.167, pp 1154-1170.

Duan, L., Aoyagi, M., Tamaki, M., Yoshino, Y., Morimoto, T., Wakimoto, H., Nagasaka, Y., Hirakawa, K., Ohno, K., Yamamoto, K. (2004) Impairment of both apoptotic and cytoprotective signalings in glioma cells resistant to the combined use of cisplatin and tumor necrosis factor alpha. Clin Cancer Res. Vol. 10, No 1 Pt 1, pp 234-43.

Dusting, G.J., Selemidis, S., Jiang, F. (2005) Mechanisms of suppressing NADPH oxidase in the vascular wall. Mem Inst Oswaldo Cruz, Rio de Janeiro, Vol. 100, No suppl. 1, pp 97-103

Edwards, D.I. (1981) Mechanisms of cytotoxicity of nitroimidazole drugs. Prog Med Chem. Vol 18, pp 87-116.

Edwards, D.I., Knox, R.J., Knight, R.C. (1982) Structure-cytotoxicity relationships of nitroimidazoles in an in vitro system. Int J Radiat Oncol Biol Phys. Vol 8, No 3-4, pp 791-3.

Edward, B., Alabraba, E.B., Curbishley, S.M., Wai, K., Lai, W.K., Wigmore, S.J., Adams, D.H., Afford, S.C. (2007)A new approach to isolation and culture of human Kupffer cells Journal of Immunological Methods, Vol 326, No 1-2, pp 139-144.

Emmelot, P., Bos, C.J., Benedetti, E.L., Rümke, P.H., Sato, O.(1964) Studies on plasma membranes I. Chemical composition and enzyme content of plasma membranes isolated from rat liver. Biochimica et Biophysica Acta BBA Vol. 90, No 1, pp 126-145

Ersoz, G., Karasu, Z., Akarca, U.S., Gunsar, F., Yuce, G., Batur, Y.(2001) Nitroimidazole-induced chronic hepatitis. Eur J Gastroenterol Hepatol. Vol 13, No 8, pp 963-6.

Eschmann, S.M., Paulsen, F., Reimold, M., Dittmann, H., Welz, S., Reischl, G., Machulla, H.J., Bares, R.(2005) Prognostic impact of hypoxia imaging with 18F-misonidazole PET in non-small cell lung cancer and head and neck cancer before radiotherapy. J Nucl Med. Vol. 46, No 2, pp 253-60.

Forman, H.J., Torres, M. (2002) Reactive oxygen species and cell signaling. Am J Resp Clin Care Med.Vol.166, pp S4-S8.

Froh, M., Konno, A., Thurman, R.G. (2003) Isolation of Liver Kupffer Cells. Current Protocols in Toxicology. Vol 14, pp 14.4.1–14.4.12.

Gandhi, B.M., Irshad, M., Chawla, T.C., Tandon, B.N.(1987) Enzyme linked protein-A: an ELISA for detection of amoebic antibody. Transactions of the Royal Society of Tropical Medicine and Hygiene. Vol 81, No 2, pp 183-185.

Gilman, A. G. (1984). Guanine nucleotide-binding regulatory proteins and dual control of adenylate cyclase. J Clin Invest Vol. 73, No 1, pp 1-4.

Grönroos, T., Bentzen, L., Marjamäki, P., Murata, R., Horsman, M.R., Keiding, S., Eskola, O., Haaparanta, M., Minn, H., Solin, O. (2004) Comparison of the biodistribution of two hypoxia markers [18F]FETNIM and [18F]FMISO in an experimental mammary carcinoma. Eur J Nucl Med Mol Imaging. Vol.31, No 4, pp 513-20.

Guengerich, F.P., Martin, M.V., Sohl, C.D., Cheng, Q. (2009)Measurement of cytochrome P450 and NADPH–cytochrome P450 reductase. Nature Protocols. Vol. 4, pp 1245 - 1251.

Guerrant, R.L., Brush J., Ravdin J.I., Sullivan J.A., Mandell, G.L.. (1981) Interaction between Entamoeba histolytica and human polymorphonuclear neutrophils. J. Infect. Dis. Vol 143, pp 83–93.

Hamann, L. Nickel, R., Tannich, E. (1995) Transfection and continuous expression of heterologous genes in the protozoan parasite Entamoeba histolytica. Proc. Natl. Acad. Sci. U. S. A. Vol. 92, pp 8975–8979.

Heimbrook, D.C., Shyam, K., Sartorelli, A.C. (1988) Novel 1-haloalkyl-2-nitroimidazole bioreductive alkylating agents. Anticancer Drug Des. Vol. 2, No 4, pp 339-50.

Hendriks, H.F., Brouwer, A., Knook, D.L. (1990) Isolation, purification, and characterization of liver cell types. Methods Enzymol. Vol.190, pp 49-58.

Heuff, G., Meyer, S., Beelen, R.H.J. (1994) Isolation of rat and human Kupffer cells by a modified enzymatic assay. Journal of Immunological Methods, Vol 174, No 1-2, pp 61-65.

Hidi, R., Timmermans, S., Liu, E., Schudt, C., Dent, G., Holgate, S.T., Djukanovic, R. (2000) Phosphodiesterase and cyclic adenosine monophosphate-dependent inhibition of T-lymphocyte chemotaxis. Eur Respir J. Vol.15, No 2, pp 342-9.

Hodgkiss, R.J., Webster, L., Wilson, G.D. (1997) Measurement of hypoxia in vivo using a 2-nitroimidazole (NITP). Adv Exp Med Biol. Vol.428, pp 61-7.

Horlen, C.K., Seifert, C.F., Malouf, C.S. (2000) Toxic metronidazole-induced MRI changes. Ann Pharmacother. Vol.34, No 11, pp 1273-5.

Jankovic, B., Aquino-Parsons, C., Raleigh, J.A., Stanbridge, E.J., Durand, R.E., Banath, J.P., MacPhail, S.H., Olive, P.L. (2006) Comparison between pimonidazole binding, oxygen electrode measurements, and expression of endogenous hypoxia markers in cancer of the uterine cervix. Cytometry B Clin Cytom. Vol. 70, No 2, pp 45-55.

Jarumilinta, R. and Kradolfer, F. (1964) The toxic effect of Entamoeba histolytica on leucocytes. Ann. Trop. Med. Parasitol. Vol 58, pp 375–381.

Kamps, J.A., Morselt, H.W., Scherphof, G.L.(1999) Uptake of liposomes containing phosphatidylserine by liver cells in vivo and by sinusoidal liver cells in primary culture: in vivo-in vitro differences. Biochem Biophys Res Commun. Vol 256, No 1, pp 57-62.

Katanuma, N. (2011) Structure-based development of specific inhibitors for individual cathepsins and their medical applications.Proc Jpn Acad Ser B Phys Biol Sci. Vol. 87, No 2, pp 29-39.

Kausalya, S., Kaur, S., Malla, N., Ganguly, N.K., Mahajan, R.C. (1996) Microbicidal mechanisms of liver macrophages in experimental visceral leishmaniasis. APMIS. Vol. 104, No 3, pp 171-5.

Kim, B.S., Baez, C.E., Atala, A. (2000) Biomaterials for tissue engineering. World J Urol Vol 18, pp 2-9.

Keeling, P.L., Smith, L.L., Aldridge, W.N. (1982) The formation of mixed disulphides in rat lung following paraquat administration. Correlation with changes in intermediary metabolism. Biochem Biophys Acta. Vol.716, pp 249-257.

Keene, W.E. et al. (1986) The major neutral proteinase of Entamoeba histolytica. J. Exp. Med. Vol. 163, pp 536–549.

Kirn, A., Bingen, A., Steffan, A.M., Wild, M.T., Keller, F., Cinqualbre, J. (1982) Endocytic Capacities of Kupffer Cells Isolated from the Human Adult Liver. Hepatology. Vol 2, No 2, pp 216S–222S.

Knook, D.L., Sleyster, E.C., Teutsch, H.F. (1980) High actiity of glucose-6-phosphate dehydrogenase in Kupffer cells isolated from rat liver. Histochemistry. Vol. 69, No 2, pp 211-6.

Knook, D.L., Barkway, C., Sleyster, E.C. (1981) Lysosomal enzyme content of Kupffer and endothelial liver cells isolated from germfree and clean conventional rats. Infect Immun. Vol 33, No 2, pp 620-2.

Ko, C., In, Y. H., Park-Sarge. O. K. (1999) Role of progesterone receptor activation in pituitary adenylate cyclase activating polypeptide gene expression in rat ovary. Endocrinology Vol. 140, No 11, pp 5185-94.

Koch, C.J., Evans, S.M. (2003) Non-invasive PET and SPECT imaging of tissue hypoxia using isotopically labeled 2-nitroimidazoles. Adv Exp Med Biol. Vol. 510, pp 285-92.

Koppele, J.M., Keller, B.J., Caldwell-Kenkel, J.C., Lemasters, J.J., Thurman, R.G. (1991) Effect of hepatotoxic chemicals and hypoxia on hepatic nonparenchymal cells:

impairment of phagocytosis by Kupffer cells and disruption of the endothelium in rat livers perfused with colloidal carbon. Toxicol Appl Pharmacol. Vol 110, No 1, pp 20-30.

Kumar, P., Pai, K., Pandey, H., Sundar, S.(2002) NADH oxidase, NADPH oxidase and myeloperoxidase activity of visceral leishmaniasis patients. J. Med. Microbiol. Vol. 51, pp 832-836.

Kuttner, R.E., Schumer, W., Apantaku, F.O. (1982) Effect of endotoxin and glucocorticoid pretreatment on hexose monophosphate shunt activity in rat liver. Circ Shock Vol 9, No 1, pp 37-45..

Kwon S, Sharma R. (2009) Role of Immunoregulators in Nitric Oxide Production in Hypoxia-Induced Apoptosis in Alveolar Cells.*Int J Cancer Prevention*. Vol 3, No 3, pp 99-110.

Kwon S, Sharma, R.(2009) Role of Effectors on Hypoxia Due to Nitric Oxide Production in Human Alveolar Epithelial Cells and Oxygen Depletion in Human Hepatocytes. *International Journal of Medical and Biological Frontiers*. Vol. 15, No 5, pp 425-442.

Larrey, D.(2000) Drug-induced liver diseases. J Hepatol. Vol.32, No. 1 Suppl, pp 77-88.

Lasserre, R., Jaroonvesama, N., Kurathong, S., Soh, C-T. (1983) Single-day drug treatment of amebic liver abscess. Am J Trop Med Hyg Vol. 32, pp 723–26.

Mahy, P., De Bast, M., Leveque, P.H., Gillart, J., Labar, D., Marchand, J., Gregoire, V.(2004) Preclinical validation of the hypoxia tracer 2-(2-nitroimidazol-1-yl)- N-(3,3,3-[(18)F]trifluoropropyl)acetamide, [(18)F]EF3. Eur J Nucl Med Mol Imaging. Vol. 31, No 9, pp 1263-72.

Martina, S.D., Ismail, M.S., Vesta, K.S. (2006) Cilomilast: orally active selective phosphodiesterase-4 inhibitor for treatment of chronic obstructive pulmonary disease. Ann Pharmacother. Vol.40, No 10, pp 1822-8.

Martin, B. R., Farndale, R. W. Wong, S. K. (1987) The role of Gs in activation of adenylate cyclase. Biochem Soc Trans Vol. 15, No 1, pp 19-21.

McCuskey, R.S., McCuskey, P.A. (1990) Fine structure and function of Kupffer cells. J Electron Microsc Tech. Vol.14, No 3, pp 237-46.

Mehghini Giorgie (1959) New Series. Vol 4, No 9, pp 682-692.

Melo, T., Ballinger, J.R., Rauth, A.M. (2000) Role of NADPH:cytochrome P450 reductase in the hypoxic accumulation and metabolism of BRU59-21, a technetium-99m-nitroimidazole for imaging tumor hypoxia. Biochem Pharmacol. Vol. 60, No 5, pp 625-34.

Melo, T., Duncan, J., Ballinger, J.R., Rauth, A.M. (2000) BRU59-21, a second-generation 99mTc-labeled 2-nitroimidazole for imaging hypoxia in tumors. J Nucl Med. Vol. 41, No 1, pp 169-76.

Michalopoulos, G.K.(1990) Liver regeneration: molecular mechanisms of growth control. FASEB J.Vol 4, pp 176-187.

Michalopoulos, G.K, DeFrances, M.C. (1997) Liver regeneration. Science. Vol. 276, pp 60-66.

Michalopoulos, G.K.(2007) Liver regeneration. *J Cell Physiol*. Vol 213, No 2, pp 286–300.

Moller, P., Wallin, H., Vogel, U., Autrup, H., Risom, L., Hald, M.T., Daneshvar, B., Dragsted, L.O., Poulsen, H.E., Loft, S. (2002) Mutagenicity of 2-amino-3-methylimidazo[4,5-f]quinoline in colon and liver of Big Blue rats: role of DNA adducts, strand breaks, DNA repair and oxidative stress. Carcinogenesis. Vol 8, pp 1379-85.

Mooney, D., Hansen, L., Vacanti, J., Langer, R., Farmer, S., Ingber, D. (1992) Switching from differentiation to growth in hepatocytes: control by extracellular matrix. J Cell Physiol Vol 151, pp 497-505.

Moselen, J.W., Hay, M.P., Denny, W.A., Wilson, W.R.(1995) N-[2-(2-methyl-5-nitroimidazolyl)ethyl]-4-(2-nitroimidazolyl)butanamide (NSC 639862), a bisnitroimidazole with enhanced selectivity as a bioreductive drug. Cancer Res. Vol. 55, No 3, pp 574-80

Mulcahy, R.T., Gipp, J.J., Carminati, A., Barascut, J.L., Imbach, J.L.(1989) Chemosensitization at reduced nitroimidazole concentrations by mixed-function compounds combining 2-nitroimidazole and chloroethylnitrosourea. Eur J Cancer Clin Oncol. Vol 25, No 7, pp 1099-104.

Nagaoka, M.R., Kouyoumdjian, M., Borges, D.R. (2003) Hepatic clearance of tissue-type plasminogen activator and plasma kallikrein in experimental liver fibrosis. Liver Int. Vol. 23, No 6, pp 476-83.

Neaud, V., Dubuisson, L., Balabaud, C., Bioulac-Sage, P. (1995) Ultrastructure of human Kupffer cells maintained in culture. J Submicrosc Cytol Pathol. Vol.27, No 2, pp 161-70.

Niu, J., Profirovic, J., Pan, H., Vaiskunaite, R., Voyno-Yasenetskaya, T. (2003) G Protein betagamma subunits stimulate p114RhoGEF, a guanine nucleotide exchange factor for RhoA and Rac1: regulation of cell shape and reactive oxygen species production. Circ Res Vol. 93, No 9, pp 848-56.

Noss, M.B., Panicucci, R., McClelland, R.A., Rauth, A.M. (1988) Preparation, toxicity and mutagenicity of 1-methyl-2-nitrosoimidazole. A toxic 2-nitroimidazole reduction product. Biochem Pharmacol. Vol.37, No 13, pp 2585-93.

Noss, M.B., Panicucci, R., McClelland, R.A., Rauth, A.M. (1989) 1-Methyl-2-nitrosoimidazole: cytotoxic and glutathione depleting capabilities. Int J Radiat Oncol Biol Phys. Vol 16, No 4, pp 1015-9.

Ott, H.C., Matthiesen, T.S., Goh, S.K., Black, L.D., Kren, S.M., Netoff, T.I., Taylor, D.A. (2008) Perfusion decellularized matrix: using nature's platform to engineer a bioartificial heart. Nat Med Vol 14, pp 213-221.

Papadopoulou, M.V., Ji, M., Rao, M.K., Bloomer, W.D. (2003) Reductive metabolism of the nitroimidazole-based hypoxia-selective cytotoxin NLCQ-1 (NSC 709257). Oncol Res. Vol.14, No 1, pp 21-9.

Pilkis, S., Schlumpf, J., Pilkis, J., Claus, T.H. (1979) Regulation of phosphofructokinase activity by glucagon in isolated rat hepatocytes. Biochem Biophys Res Commun. Vol.88, No 3, pp 960-7.

Powell, S.J., Wilmot, A.J., Elsdon-Dew, R. (1969) Single and low dosage regimens of metronidazole in amoebic dysentery and amoebic liver abscess. Ann Trop Med Parasitol, Vol.63, pp 139–42.

Prakash

Ragland, B.D., Ashley L.S., Vaux, D.L., Petri, W.A.. (1994) Entamoeba histolytica: target cells killed by trophozoites undergo DNA fragmentation which is not blocked by Bcl-2.Exp. Parasit. Vol 79, pp 460–467.

Ramalho, F.S., Ramalho, L.N.Z., Castro-e-Siva, O., Zucoloto, S., Correa, F.M.A.(2001) Angiotensin-converting enzyme inhibition by lisonopril enhances liver regeneration in rats. Braz. J. Med. Biol Research Vol 34, pp 125-27.

Ramirez-Emiliano, J., Flores-Villavicencio, L.L., Segovia, J., Arias-Negrete, S. (2007) Nitric oxide participation during amoebic liver abscess development. Medicina (B.Aires) Vol. 67, No 2, pp 167-176.

Rauth, A.M., Mohindra, J.K. (1981) Selective toxicity of 5-(3,3-dimethyl-1-triazeno)imidazole-4-carboxamide toward hypoxic mammalian cells. Cancer Res. Vol.41, No 12 Pt 1, pp 4900-5.

Riche, F., d'Hardemare, A.D., Sepe, S., Riou, L., Fagret, D., Vidal, M. (2001) Nitroimidazoles and hypoxia imaging: synthesis of three technetium-99m complexes bearing a nitroimidazole group: biological results. Bioorg Med Chem Lett. Vol. 11, No 1, pp 71-4.

Rumsey, W.L., Patel, B., Kuczynski, B., Narra, R.K., Chan, Y.W., Linder, K.E., Cyr, J., Raju, N., Ramalingam, K., Nunn, A.D. (1994) Potential of nitroimidazoles as markers of hypoxia in heart.Adv Exp Med Biol. Vol. 345, pp 263-70.

Russell, D., Synder, S.H. (1969) Amine synthesis in regenerating rat liver: extremely rapid turnover of ornithine decarboxylase. Mol Pharmacol. Vol 5, No 3, pp 253-62

Sánchez-Ramírez, B.E., Ramírez-Gil, M., Ramos-Martínez, E., Rohana, P.T. (2000) Entamoeba histolytica induces cyclooxygenase-2 expression in macrophages during amebic liver abscess formation. Arch Med Res. Vol.31, No.4 Suppl, pp S122-3.

Sapora, O., Paone, A., Maggi, A., Jenner, T.J., O'Neill, P. (1992) Induction of mutations in V79-4 mammalian cells under hypoxic and aerobic conditions by the cytotoxic 2-nitroimidazole-aziridines, RSU-1069 and RSU-1131. The influence of cellular glutathione. Biochem Pharmacol. Vol 44, No 7, pp 1341-7.

Sasagawa, T., Shimizu, T., Sekiya, S., Haraguchi, Y., Yamato, M., Sawa, Y., Okano, T.(2010) Design of prevascularized three-dimensional cell-dense tissues using a cell sheet stacking manipulation technology. Biomaterials. Vol 31, No 7, pp 1646-54.

Scholze, H. and Schulte, W. (1988) On the specificity of a cysteine proteinase from Entamoeba histolytica. Biomed. Biochim. Acta, Vol 47, pp 115–123.

Schmeing, T.M., Ramakrishnan, V. (2009) What recent ribosome structures have revealed about the mechanism of translation. Nature. Vol. 461, No 7268, pp 1234-42.

Schorlemmer, H.U., Kurrle, R, Schleyerbach, R., Bartlett, R.R. (1999) Generation of O_2-radicals in macrophages can be inhibited in vitro and in vivo by derivatives of leflunomide's primary metabolite. Inflamm Res. Vol.48, No Suppl 2, pp S117-8

Schulte-Hermann, R. Bursch, W. Grasl-Kraupp, B. (1994) Active cell death (apoptosis) in liver biology and disease. Prog. Liver Dis. Vol 13, pp 1–35.

Seydel, K.B., Stanley, S.L., Jr (1998) Entamoeba histolytica induces host cell death in amebic liver abscess by a non-Fas-dependent, nontumor necrosis factor α-dependent pathway of apoptosis. Infect. Immun. Vol 66, pp 2980–2983.

Seydel, K.B., Li, E., Swanson, P.E., Stenley, S.L.(1997) Human intestinal epithelial cells produce pro-inflammatory cytokines in response to infection in a SCID mouse-human intestinal xenograft model of amebiasis. Infect. Immun. Vol 65, pp 1631–1639.

Shahhosseini, S., Guttikonda, S., Bhatnagar, P., Suresh, M.R. (2006) Production and characterization of monoclonal antibodies against shope fibroma virus superoxide dismutase and glutathione-s-transferase.J Pharm Pharm Sci. Vol. No 2, pp 165-8.

Shandera, W.X., Bollam, P., Hashmey, R.H. Jr, et al. (1998) Hepatic amebiasis among patients in a public teaching hospital. South Med J Vol 91, pp 829–37.

Sharma, R. (2011a) Pathophysiology of amoebiasis. TRENDS in Parasitology, Vol 17, No 6, pp 280-285.

Sharma, R.(2011b) Nitroimidazole Radiopharmaceuticals in Bioimaging: 1: Synthesis and Imaging Applications. Curr Radiopharma. Vol 4, No 4, pp 361-378..

Sharma R., Tandon, R.K. (2010) The Hepatocellular Dysfunction Criteria: The Hepatocyte Carbohydrate Metabolizing Enzymes and Kupffer Cell Lysosomal Enzymes in Development of Amoebic Liver Abscess (Electron Microscopic - Enzyme Approach). International J Biological Frontiers.Vol 16, No 7-8, pp 747-762.

Sharma, R. (2009) The Hepatocellular Dysfunction Criteria: Hepatocyte Carbohydrate Metabolizing Enzymes and Kupffer Cell Lysosomal Enzymes in 2'Nitroimidazole Effect on Amoebic Liver Abscess (Electron Microscopic - Enzyme Approach). Chapter 7, in Parasitology Research Trends (Nova). Editors: Olivier De Bruyn,S. Peeters. pp 143-159.

Sharma, R., Kwon, S., Chen, C.J. (2007) Role of immunoregulators as possibility of tumor Hypoxia induced apoptosis in alveolar cells. *International Journal of Cancer Prevention. Vol* 3, No 3, pp 99-110.

Sharma R, Singh VS (2009) Hepatocyte Cellular Enzymes in Guinea Pigs Induced with Amoebic liver Abscess and 2'nitroimidazole Treatment. *Nanotech Res J*, Vol. 3, No 2, pp 131-144.

Sharma R, Singh VS (2009) Kupffer Cell Enzymes as Biomarkers in Guinea Pigs Induced with Amoebic liver Abscess and 2'Nitroimidazole Treatment. *Nanotech Res J*, Vol 3, No 2, pp 119-130.

Sharma, R., Singh, V.S. (2010) Phagocytic Activity and Hexose Monophosphate Shunt Activity of Cultured Human Kupffer Cells Upon Zymosan, Erythrocytes, Amoebic Trophozoits, Latex Beads and Nitroimidazole. *Int JMed Biol Front* Vol. 17, No 6,pp 521-532.

Sharma, R. (1995) The effect of tinidazole on isolated liver cells during development of amoebic liver abscess. Ph.D dissertation submitted to Indian Institute of Technology Delhi and CCS University.

Singh, D. R., Nair, C. K., Pradhan, D. S. (1991) Chemical radiosensitization by misonidazole: production and repair of DNA single-strand breaks in Yoshida ascites tumor cells. Indian J Exp Biol Vol. 29, No 7, pp 601-4.

Small, K. M., Forbes, S. L., Rahman, F. F., Liggett, S. B. (2000) Fusion of beta 2-adrenergic receptor to G alpha s in mammalian cells: identification of a specific signal transduction species not characteristic of constitutive activation or precoupling. Biochemistry. Vol. 39, No 10, pp 2815-21.

Spolarics, Z., Wu, J.X. (1997) Tumor necrosis factor alpha augments the expression of glucose-6-phosphate dehydrogenase in rat hepatic endothelial and Kupffer cells. Life Sci. Vol. 60, No 8, pp 565-71

Spolarics, Z., Bautista, A.P., Spitzer, J.J. (1991) Metabolic response of isolated liver cells to in vivo phagocytic challenge.Hepatology,Vol.13, No 2, pp 277-281.

Stalla, G. K., Stalla,J., von Werder, K., Muller,, O. A., Gerzer, R., Hollt, V., Jakobs, K. H. (1989) Nitroimidazole derivatives inhibit anterior pituitary cell function apparently by a direct effect on the catalytic subunit of the adenylate cyclase holoenzyme. Endocrinology Vol. 125, No 2, pp 699-706.

Stanley S.L. (2003) Amoebiasis. The Lancet, Vol 361, No 9362, pp 1025-1034.

Stanley, S.L., Zhang, T., Rubin, D., Li, E. (1995) Role of the Entamoeba histolytica cysteine proteinase in amebic liver abscess formation in severe combined immunodeficient (SCID) mice. Infect. Immun. Vol 63, pp 1587–1590.

Stanley, S.L.Jr.(2001) Pathophysiology of amoebiasis.Trends in Parasitology. Vol 17, No 6, pp 280-285.

Starzl, T.E., Francavilla, A., Halgrimson, C.G., Francavilla, F.R., Porter, K.A., Brown, T.H., Putnam, C.W. (1973) The origin, hormonal nature, and action of hepatotrophic substances in portal venous blood. Surg Gynecol Obstet Vol 137, pp 179-199.

Stratford, I.J., Williamson, C., Hardy, C. (1981) Cytotoxic properties of a 4-nitroimidazole (NSC 38087): a radiosensitizer of hypoxic cells in vitro. Br J Cancer. Vol 44, No 1, pp 109-16.

Su, Z.F., Zhang, X., Ballinger, J.R., Rauth, A.M., Pollak, A., Thornback, J.R. (1999) Synthesis and evaluation of two technetium-99m-labeled peptidic 2-nitroimidazoles for imaging hypoxia. Bioconjug Chem. Vol 10, No 5, pp 897-904.

Sugiyama, M., Tsuzuki, K., Haramaki, N. (1993) DNA single-strand breaks and cytotoxicity induced by sodium chromate(VI) in hydrogen peroxide-resistant cell lines. Mutat Res. Vol 299, No 2, pp 95-102.

Sunahara, R. K., Tesmer, J.J., Gilman, A.G., Sprang, S. R. (1997). Crystal structure of the adenylyl cyclase activator Gsalpha. Science Vol. 278, No 5345, pp 1943-7.

Tabak, F., Ozaras, R., Erzin, Y., Celik, A.F., Ozbay, G., Senturk, H. (2003) Ornidazole-induced liver damage: report of three cases and review of the literature. Liver Int. Vol. 23, No 5, pp 351-4.

Tandon BN, Tandon HD, Puri BK. (1975) An electron microscopic study of liver in hepatomegaly presumably caused by amoebiasis. Exp Mol Pathol Vol 22, pp 118-132.

Thompson, J.E. Jr, Forlenza, S., Verma, R.(1985) Amebic liver abscess: a therapeutic approach. Rev Infect Dis Vol 7, pp 171–79.

Thurman, R.G. (2000) Sex-related liver injury due to alcohol involves activation of Kupffer cells by endotoxin. Can J Gastroenterol. Vol 14, Suppl D, pp 129D-135D.

Uygun, B.E., Soto-Gutierrez, A., Yagi, H., Izamis, M.L., Guzzardi, M.A., Shulman, C., Milwid, J., et al. (2010) Organ reengineering through development of a transplantable recellularized liver graft using decellularized liver matrix. Nat Med 2010.

Ventura-Juárez, J., Jarillo-Luna, R.A., Fuentes-Aguilar, E., Pineda-Vázquez, A., Muñoz-Fernández, L., Madrid-Reyes, J.I., Campos-Rodríguez, R. (2003) Human amoebic hepatic abscess: in situ interactions between trophozoites, macrophages, neutrophils and T cells. Parasite Immunol. Vol.25, No 10, pp 503-11.

Verala A.T., Anabela, P.R., Carlos, P.M.(2011) Fatty Liver and Ischemia/Reperfusion: Are there Drugs Able to Mitigate Injury?. Curr Med Chem. Vol. 18(32), pp 4987-5002.

Vician, M., Olejnik, J., Michalka, P., Jakubovsky, J., Brychta, I., Gerge,l M. (2009) Warm liver ischemia in experiment and lysosomal markers. Bratisl Lek Listy. Vol 110, No 10, pp 587-91.

Vines, R.R., Purdy J.E., Ragland, B.D., Samuelson, J., Mann, B.J., Petri, W.A. (1995) Stable episomal transfection of Entamoeba histolytica. Mol. Biochem. Parasitol. Vol 71, pp 265–267.

Virk, K.J., Ganguly, N.K., Dilawari, J.B., Mahajan, R.C. (1989) Role of lysosomal enzymes in tissue damage during hepatic amoebiasis. An experimental study. Liver. Vol. 9 No. 6, pp 338-45.

Virk, K.J., Mahajan, R.C., Dilawari, J.B., Ganguly, N.K. (1988) Role of beta-glucuronidase, a lysosomal enzyme, in the pathogenesis of intestinal amoebiasis: an experimental study. Trans R Soc Trop Med Hyg. Vol. 82, No 3, pp 422-5.

Voytik-Harbin, S.L., Brightman, A.O., Kraine, M.R., Waisner, B., Badylak, S.F. (1997) Identification of extractable growth factors from small intestinal submucosa. J Cell Biochem Vol. 67, pp 478-491.

Whitmore, G.F., Varghese, A.J., Gulyas, S. (1986) Reaction of 2-nitroimidazole metabolites with guanine and possible biological consequences. IARC Sci Publ. Vol. 70, pp 185-96.

Widel, M., Watras, J., Suwinski, J., Salwinska, E. (1982) Radiosensitizing ability and cytotoxicity of some 5(4)-substituted 2-methyl-4(5)-nitroimidazoles. Neoplasma. Vol 29, No 4, pp 407-15.

Wisse, E., Braet, F., Luo, D., Zanger, R.D., Jans, D., Crabbe, E., Vermoesen, A.M. (1996) Structure and Function of Sinusoidal Lining Cells in the Liver. Toxicol Pathol January Vol 24, No 1, pp 100-111.

Whitaker, S. J., McMillan, T. J. (1992) Oxygen effect for DNA double-strand break induction determined by pulsed-field gel electrophoresis. Int J Radiat Biol Vol. 61, No 1, pp 29-41.

Worthington Enzyme Manual. (1993) Worthington Biochemical Corporation, 730 Vassar Avenue, Lakewood, New Jersey 08701, Editor: Worthington, V. pp 1-1947.

Yun-Cheung, K., In-Fai, L.(1970) Biochemistry of rat liver regeneration. The Chung Chi Journal. Pp 96-109.

Zhang, Z., Yan, L., Wang, L., Seydel, K.B., Li, E., Ankri, S., Mirelman, D., Stanley, S.L. (2000) Entamoeba histolytica cysteine proteinases with interleukin-1□ converting enzyme (ICE) activity cause intestinal inflammation and tissue damage in amoebiasis. Mol. Microb. Vol 37, pp 542–548.

Zheng, H., Olive, P. L.(1997) Influence of oxygen on radiation-induced DNA damage in testicular cells of C3H mice. Int J Radiat Biol Vol. 71, No 3, pp 275-82.

Zhong, Z., Lemasters, J.J. (2004)Role of free radicals in failure of fatty liver grafts caused by ethanol.Vol 34, No 1, pp 49-58.

Zhu, Y., Conner, G.E.(1994) Intermolecular association of lysosomal protein precursors during biosynthesis. J Biol Chem. Vol. 269, No 5, pp 3846-51.

Ziemer, L.S., Evans, S.M., Kachur, A.V., Shuman, A.L., Cardi, C.A., Jenkins, W.T., Karp, J.S., Alavi, A., Dolbier, W.R. Jr, Koch, C.J.(2003) Noninvasive imaging of tumor hypoxia in rats using the 2-nitroimidazole 18F-EF5. Eur J Nucl Med Mol Imaging. Vol. 30, No 2, pp 259-66

8

Reversible Inhibition of Tyrosine Protein Phosphatases by Redox Reactions

Daniela Cosentino-Gomes and José Roberto Meyer-Fernandes
Universidade Federal do Rio de Janeiro (UFRJ),
Instituto de Bioquímica Médica (IbqM)
Brazil

1. Introduction

Protein phosphorylation/dephosphorylation is considered a widespread mechanism of protein regulation. The covalent addition of a phosphate group to tyrosine residues in cellular proteins may occur as an appropriate response to a series of morphological and biochemical processes, with particular importance in complex functions such as growth, proliferation, differentiation, adhesion and motility. Protein tyrosine phosphorylation is regulated in the cell by the opposing activities of two classes of enzymes: protein tyrosine kinases, which phosphorylate specific tyrosine residues in protein using ATP as the phosphate source, and protein tyrosine phosphatases, which hydrolyze the phosphotyrosines yielding the restored amino acid residue and inorganic phosphate as products (Kolmodin & Åqvist, 2001; Mayer, 2008). The number of active protein phosphatases with the ability to dephosphorylate phosphotyrosine are very similar to the number of active tyrosine phosphatases. Both types of enzymes also display comparable tissue distribution patterns (Alonso et al., 2004). Abnormalities in tyrosine phosphorylation play a role in the pathogenesis of numerous inherited and acquired human diseases from cancer to immune deficiencies (Alonso et al., 2004)

Phosphorylation, acetylation, ubiquitinylation and glycosylation are among the most well-known post-translational modifications. The concept of protein oxidation as a post-translational modification has only gained acceptance more recently. Protein oxidation occurs as an outcome of a chemical attack by reactive oxygen species (ROS) or reactive nitrogen species (RNS) on susceptible amino acids such as tyrosine, tryptophan, histidine, lysine, methionine, and cysteine (Spickett et al., 2006). Indeed, it is important to note that signal transduction must occur in a coordinated manner in response to a previous stimuli. The key elements of a signaling response is reversibility and specificity (Tonks, 2005). In this way, an oxidative-dependent chain reaction may be short and employ low concentration of oxidants to avoid irreversible damage to cellular components.

In this chapter, we will discuss the following: the structural mechanism of action of ROS in the enzymatic activity of tyrosine phosphatases and how it interacts with their target molecules; the reversible regulation of this enzyme by oxidants and antioxidants; and the major consequences of this tightly controlled mechanism on cell signaling.

2. Protein tyrosine phosphatases family

The protein tyrosine phosphatase (PTP) superfamily is encoded by approximately 100 genes in the human genome. They all possess an overall structure with a core catalytic domain composed of four parallel β-strands surrounded on both sides by α-helices. The active site sequence $Cys(X)_5Arg$ defines the PTP family, and this sequence is referred to as the "PTP signature motif". Moreover, this conserved motif creates a very similar structural motif, termed the PTP loop, which connects a central β -strand to an α -helix at the center of the catalytic site. The diversity within the families is conferred by the regulatory domains (Salmeen & Barford, 2005; Tiganis & Bennett, 2007). All protein tyrosine phosphatases are characterized by: a) their sensitivity to vanadate; b) an ability to hydrolyze the artificial substrate *p*-nitrophenyl phosphate; c) an insensitivity to okadaic acid; and d) a lack of metal ion dependence for catalysis (Chiarugi & Buricchi, 2007; Denu & Dixon, 1998). PTPs have the capacity to function both positively and negatively in the regulation of signal transduction, being able to stimulate or inhibit protein functions through their specific dephosphorylation activity (Tonks, 2006).

The protein tyrosine phosphatase superfamily can be grouped into two subfamilies, the tyrosine-specific phosphatases that comprise 38 different enzymes and the dual-specific phosphatases with 61 components in the human genome. The common classification of the protein tyrosine phosphatase family is depicted in table 1. Despite different subcellular localizations and substrate specificities, the protein tyrosine phosphatase classes employ a similar chemical mechanism for phosphate hydrolysis involving a transient cysteinyl-phosphate intermediate (Chiarugi et al., 2005; Östman et al., 2011).

2.1 Protein tyrosine-specific phosphatases

The subfamily of tyrosine-specific phosphatases is constituted by the well-known classical protein tyrosine phosphatase (PTP), which acts strictly on phosphotyrosine-containing proteins. The active site of this group is deeper than the catalytic site of dual-specific protein phosphatase, which makes these enzymes much more selective for p-Tyr-containing substrates than the dual-specific protein phosphatase. The classical PTPs can also be divided into transmembrane, receptor-like enzymes (RPTPs) and the cytosolic, nonreceptor PTPs (NRPTPs) (Chiarugi & Buricchi, 2007).

2.1.1 The receptor-like protein tyrosine phosphatases

The receptor-like PTPs such as RPTPα and CD45 are cell-surface receptors and generally have an extracellular ligand-binding domain, a single transmembrane region, and one or two cytoplasmic catalytic domains. Receptor-like PTPs are mainly negatively regulated through dimerization, a known regulatory mechanism for many signal transduction molecules, such as receptor proteins (Chiarugi & Buricchi, 2007). The number of PTP catalytic domains encoded by the human genome is greater than the number of PTP genes because many of the receptor-type proteins have tandem catalytic domains (Alonso et al., 2004). Receptor-like tyrosine phosphatases have the potential to regulate signaling through ligand-controlled protein tyrosine dephosphorylation. Many of these tyrosine phosphatases display features of cell-adhesion molecules in their extracellular segment and can be implicated in processes that involve cell-cell and cell-matrix contact (Tonks, 2006).

2.1.2 The intracellular protein tyrosine phosphatases

The intracellular protein tyrosine phosphatases contain a single catalytic domain and various amino or carboxyl terminal extensions. They may contain SH2 (Src homology domain) domains that have targeting or regulatory functions, PEST (Pro-Glu-Ser-Thr domain) domains for proteolytic control, phosphorylation sites for protein/protein interactions, and enzymatic activity regulation. Cytosolic tyrosine phosphatases are characterized by regulatory sequences that flank the catalytic domain and control activity either directly, by interactions at the active site that modulate activity or by controlling substrate specificity. This class of enzymes is also activated through its own tyrosine phosphorylation, being essentially inactive under normal basal conditions. Among the members of this group are the prototype PTP1B, the most studied tyrosine phosphatase, the low molecular weight-PTP (LMW-PTP), and the Shp2 phosphatase (Alonso et al., 2004; Chiarugi & Buricchi, 2007).

Tyrosine phosphorylation is common to all cytosolic enzymes, including those that do not contain an SH2 domain, such as PTP1B or LMW-PTP. Both phosphatases have been reported to be activated through autophosphorylation. At the same time, autodephosphorylation has been reported for several tyrosine phosphatases, including Shp2, RPTPα, Shp1, and LMW-PTP (den Hertog et al., 1994; Giannoni et al., 2005; Meng et al., 2002; Zhao et al., 1994). PTP1B action encompass numerous substrates, including different PTK receptors, such as the EGFR (epidermal growth factor receptor), PDGFR (platelet-derived growth factor receptor), CSF-1 (colony-stimulating factor 1), IR (insulin receptor) and IGF-1 (insulin-like growth factor-1) receptors, and cytoplasmic PTKs members, such as c-Src and JAK2 (Janus protein tyrosine kinase 2). The non-transmembrane Shp1 and Shp2 (protein tyrosine phosphatase Src homology 1/2 (SH2) containing domain) perform opposing tasks in signaling pathways. Shp-1, predominantly expressed in hematopoietic cells, is a negative regulator in the signaling pathways mediated by the chemokine and cytokine receptors and by other receptor tyrosine kinases. Unlike other tyrosine phosphatases that negatively regulate signaling, Shp2 phosphatases can positively regulate cell signaling through the dephosphorylation of substrates that are negatively regulated by tyrosine phosphorylation and can participate in the propagation of the ERK (extracellular-signal-regulated kinase) and PI3K (phosphoinositide 3-kinase)/Akt (protein kinase B) pathways in which multiple receptor PTKs play a role (C. Y. Chen et al., 2009; Tiganis & Bennett, 2007).

2.2 The dual-specific tyrosine protein phosphatases

The dual-specific protein phosphatases (DSPs) can dephosphorylate a variety of substrates, including phosphotyrosine-containing proteins, phosphothreonine, phosphoserine residues and phospholipids, and is the most diverse group in terms of substrate specificity because of this feature. The dual-specific protein phosphatases are much more diverse than tyrosine-specific phosphatases and can be divided into several subgroups. Eleven of the 61 dual-specific PTPs encoded by the human genome are specific for the mitogen-activated protein kinases (MAPK); the other ones can be represented by the well-known cell cycle regulators Cdc25 phosphatases and the tumor suppressor phosphatase (PTEN), which is also able to dephosphorylate lipid substrates (Alonso et al., 2004; Chiarugi & Buricchi, 2007). The dual-specific tyrosine protein phosphatases also include the VH1-like enzymes, which are related to the prototypic VH1 DSP, a 20-kDa protein that is a virulence factor of the vaccinia virus (Tonks, 2005).

The MAPK phosphatases are characterized by dual phosphothreonine and phosphotyrosine specificity and the presence of a CH2 region and other MAPK-targeting motifs. They can dephosphorylate tyrosine and threonine residues within the activation loop of MAPKs. The different specificities towards the various MAPKs mean that MAPK phosphatase plays a critical negative regulation on the MAPK-mediated signaling process (Tiganis & Bennett, 2007). Cdc25 phosphatases are involved in the dephosphorylation of cyclin-dependent kinases at their inhibitory, dually phosphorylated N-terminal Thr-Tyr-motifs. The reaction is required for activation of these kinases to drive progression of cell cycle (Alonso et al., 2004). Another subgroup of dual-specific protein phosphatases, the atypical dual-specific protein phosphatases, includes a number of poorly characterized enzymes that lack specific MAPK-targeting motifs and tend to be much smaller enzymes. This is the case for PIR (phosphatase interacting with RNA/RNP), which dephosphorylates mRNA. Cdc14 is involved in the inactivation of cyclin-dependent kinases and in exit from mitosis, while the slingshots and PRLs (phosphatase of regenerating liver) are poorly understood. PTEN dephosphorylates phosphatidylinositol-3,4,5-trisphosphate at the plasma membrane, while the myotubularins primarily dephosphorylate phosphatidylinositol-3-phosphate on internal surface of cell membranes (Alonso et al., 2004).

Regarding substrate specificity, dual-specific protein phosphatases may present a preference for either tyrosine or serine/threonine residues. For cyclin-dependent kinase (CDK)-associated phosphatase (KAP), the phosphatase preferentially dephosphorylates the threonine residue in the activation loop of CDKs, while for the VH1-related dual-specific phosphatases, tyrosine is the preferred residue in MAPKs (Tonks, 2006).

<table>
<tr><th colspan="3">Classification</th><th>PTPs</th><th>Ligand-inducing oxidation</th></tr>
<tr><td rowspan="12">Protein Tyorosine Phosphatase $C(X)_5R$</td><td rowspan="7">Tyrosine specific PTPs</td><td rowspan="3">Receptor PTPs</td><td>PTPα</td><td></td></tr>
<tr><td>PTPκ</td><td></td></tr>
<tr><td>CD45</td><td>B-cell receptor stimulation</td></tr>
<tr><td rowspan="4">Cytosolic PTPs</td><td>Shp1</td><td>B-cell receptor stimulation</td></tr>
<tr><td>Shp2</td><td>PDGF, T-cell receptor stimulation, endothelin 1</td></tr>
<tr><td>PTP1B</td><td>EGF, insulin</td></tr>
<tr><td>TCPTP</td><td>Insulin</td></tr>
<tr><td colspan="2" rowspan="5">Dual specific PTPs</td><td>VHR</td><td></td></tr>
<tr><td>MKPs</td><td>TNFα</td></tr>
<tr><td>LMW PTP</td><td>PDGF, integrin engagement</td></tr>
<tr><td>PTEN</td><td>PDGF, EGF, insulin</td></tr>
<tr><td>Cdc25</td><td></td></tr>
</table>

Table 1. Classification of the protein tyrosine phosphatase family and some of its principal members and ligand-inducing oxidation, according to Tonks in 2006. (PDGF, platelet-derived growth factor; EGF, epidermal growth factor; TNFα, tumor-necrosis factor-α; TCPTP, T-cell protein tyrosine phosphatase; MKP, mitogen-activated protein kinase phosphatase; LMW-PTP, low molecular weight protein tyrosine phosphatase).

3. Structural characteristics of the cysteine-based protein tyrosine phosphatase catalytic domain

Protein tyrosine phosphatases are a family of enzymes with high structural multiplicity. Cysteine-based phosphatases share the common function of hydrolyzing phosphorester bonds in proteins and/or lipids via a conserved cysteine-based mechanism. The most significant feature of the protein tyrosine phosphatase superfamily is the presence of a signature motif $Cys(X)_5Arg$, where X is any amino acid. This well-conserved motif forms the phosphate-binding loop in the active site, known as the P-loop or PTP-loop, and is responsible for structural features required for phosphate recognition and phosphorester hydrolysis. The structural arrangement ensures that the catalytic cysteine functions as the nucleophile in catalysis, which specifically binds negatively charged substrates, and the arginine, which is involved in phosphate binding, remains in close proximity to form a cradle to hold the phosphate group of the substrate in the correct place for nucleophilic attack (Chiarugi & Buricchi, 2007; Salmeen & Barford, 2005; Tabernero et al., 2008). Figure 1 shows the crystal structure of PTP1B complexed with phosphotyrosine.

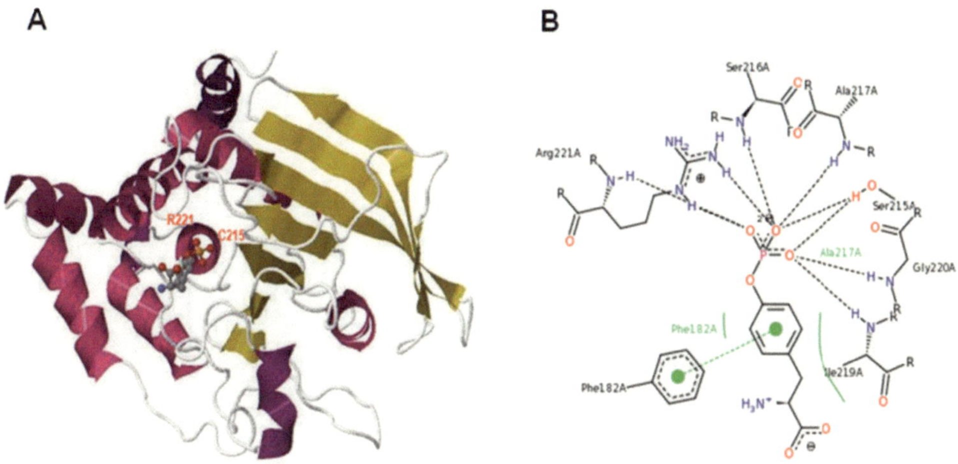

Fig. 1. (A) Crystal structure of PTP1B complexed with phosphotyrosine (PDB entry 1PTV) (Jia et al., 1995). A substrate positioned in the active site is represented by phosphotyrosine (ball and stick). The cysteine (C215) represented in the active site is the nucleophile that attacks the substrate phosphorous atom, forming the cysteinyl-phosphate intermediate. The arginine (R221) is involved in the intermediate stabilization and substrate binding (Tabernero et al., 2008). (B) Binding of phosphotyrosine to the catalytic site. Here the nucleophile cysteine 215 is mutated to serine 215 (Jia et al., 1995).

Specific protein tyrosine phosphatases utilize a two-step reaction for phosphate monoester hydrolysis (Fig. 2). The first step is initiated by a nucleophilic attack by the active-site cysteine (C215) on the phosphorus atom of the bound substrate that leads to the formation of a cysteinyl-phosphate intermediate. Besides substrate binding, the arginine is also involved in the stabilization of this intermediate. During the formation of the transition-state intermediate, the only aspartic residue in the WPD (Trp-Pro-Asp) loop acts as a general acid by donating its proton to the phenolic oxygen of the tyrosyl-leaving group, thus cleaving the phosphate of tyrosine to form the cysteine-phosphate intermediate. This first substitution reaction allows the phosphate group to be covalently attached to the nucleophile via a thioester linkage. In the second step of catalysis, the same aspartic residue functions as a general base by accepting a proton from water during hydrolysis of the cysteinyl phosphate intermediate, thus restoring the free enzyme to its normal basal Cys-SH conformation (Kolmodin & Åqvist 2001; Tabernero et al., 2008; Tiganis & Bennett, 2007). Figure 2 illustrates the general two-step catalysis of protein tyrosine phosphatases.

Most of the cysteine residues within proteins present a normal pKa of 8.3, and this characteristic ensures that this amino acid has a relatively good nucleophile attribute. Because of the unique environment of the tyrosine phosphatase active site, the catalytic cysteine presents an unusually low pKa value. Unlike the free cysteines, which are shown to be protonated at physiological pH, the catalytic cysteines of the tyrosine-specific PTPs are more acidic and are therefore proposed to be deprotonated in the free enzymes at a physiological pH (or even at the pH optimum of 5-6), existing as a thiolate anion (Cys-S$^-$). This property enables them to act as nucleophiles in the first step of catalysis but also makes them highly susceptible to oxidations by different reactive oxygen species and, to a lesser degree, with reactive nitrogen species. Exposure of thiolate anions to reactive nitrogen species causes S-nitrosothiol formation, whereas treatment with peroxynitrite yields S-nitrothiol formation (Brandes et al., 2009; Chiarugi et al., 2005; Salmeen & Barford, 2005).

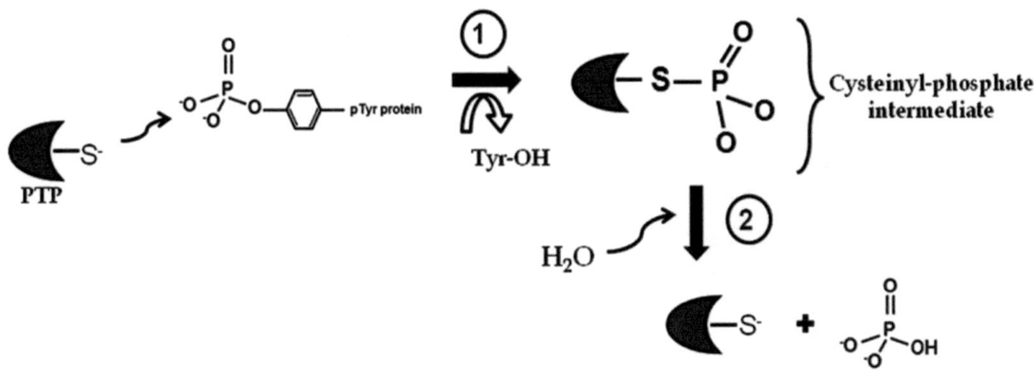

Fig. 2. Schematic mechanism of two-step catalysis of protein tyrosine phosphatases. The first step is initiated by a nucleophilic attack of the active site cysteine, which is in its thiolate anion conformation (Cys-S$^-$), on the phosphorus atom of the bound substrate (1). The dephosphorylated protein tyrosine is released from the active site of the phosphatase while a cysteinyl-phosphate intermediate is formed. In the second step of catalysis, the aspartic residue in the P-loop accepts a proton from water during hydrolysis of the cysteinyl phosphate intermediate, thus restoring the free enzyme to its basal Cys-S$^-$ conformation (2).

After oxidation, the modified cysteine can no longer function as a phosphate acceptor in the first step of catalysis, which abrogates the catalytic function of tyrosine phosphatase (Chiarugi et al., 2005; Salmeen & Barford, 2005). A network of hydrogen bonds stabilizes the negative charges in the active site. Among other interactions in the PTPs, it is found that the hydroxyl group of the serine or threonine residue in the signature motif is important for stabilizing the thiolate form of the active site cysteine, which results in the characteristically lower pKa. The P-loop is found at the N-terminus of an α-helix and should also contribute to the thiolate stabilization, as well as, the presence of neighbouring acid (Asp or Glu) and basic (Arg, His or Lys) amino acids near the catalytic cysteine (Hess et al., 2005; Kolmodin and Åqvist, 2001).

The structure of cytoplasmatic tyrosine phosphatases is highly conserved with only minor differences in the main structural core. The catalytic domain is formed by about 280 amino acids, which is responsible for the specific folding of the enzyme and other differing characteristics. The catalytic domain often exhibits relatively broad substrate specificities *in vitro;* however, under *in vivo* conditions, full-length tyrosine phosphatases have more stringent specificities. This is in part due to the presence of additional regulatory domains that direct their subcellular localization and interactions with specific substrates (Tabernero et al., 2008).

4. Sources of oxidizing species in biological systems

4.1 Reactive oxygen species

Reactive oxygen species (ROS) are a group of molecules produced in cells when oxygen is metabolized and are more reactive than molecular oxygen (Groeger et al., 2009). In aerobic organisms, ROS can originate from different sources, including mitochondrial electron transport chain, xanthine oxidase, myeloperoxidase, nicotinamide adenine dinucleotide phosphate (NADPH) oxidases (Nox enzymes), and lipoxygenase (Groeger et al., 2009; Östman et al., 2011; Salmeen & Barford, 2005). The last two are responsible for the production of ROS in response to hormones, growth factors and cytokines (Bae et al., 1997; Salmeen & Barford, 2005; Sundaresan et al., 1995). Among these oxygen metabolites are superoxide anions ($O_2^{\bullet -}$), hydrogen peroxide (H_2O_2) and hydroxyl radicals ($^{\bullet}OH$) (Chiarugi & Buricchi, 2007). These species are not all equally reactive regarding to prospective targets. Many of them have very short half-lives, leading to little relevance in terms of signaling. For example, the hydroxyl radical is the most unstable radical, reflecting its limited ability to transmit signals across any significant distance. At the same time, superoxide and hydrogen peroxide can be considered more stable species and because of these, may be the most favorable ROS to operate as a signaling molecule (K. Chen et al., 2009).

After being considered important components of the biological process by McCord and Fridovich in 1969, free radicals emerged in the subsequent years as a dangerous byproduct of cellular metabolism by bringing deleterious consequences to DNA, lipids and proteins. The first research on reactive oxygen species has focused primarily on their chemical ability to react with and cause toxic and mutagenic effects on cellular components. There are exceptions, such as phagocytes, that deliberately generate reactive oxygen species as a main defense mechanism (Lambeth, 2002). Almost ten years later in 1977, Mittal and Murad showed the first evidence of the potential for free radicals to act in cellular signaling in metabolism. Now, it is largely accepted that living organisms have not only adapted to cope with oxidizing species but also have developed mechanisms for the advantageous use of free radicals (Dröge, 2002). ROS can act as second messengers that are required for the downstream signaling events (Cross & Templeton, 2006) However, mechanisms by which proteins are able to sense ROS and translate signaling into an effective response are not totally clear.

Superoxide anions are derived by the univalent reduction of triplet-state molecular oxygen ($^3O_2$). This process is mediated by enzymes, such as xanthine oxidase, NADPH oxidases (Nox) and dual oxidases (Duox) (also known as Nox enzymes), or nonenzymatically by redox reactive compounds, such as the semi-ubiquinone of mitochondria. Within mitochondria, superoxide is continually generated as a result of electron leakage from the electron transport chain mainly at complex I (NADH/ubiquinone oxidoreductase) and complex III (ubiquinol/cytochrome c oxido-reductase) (Cross & Templeton, 2006; Harrison, 2002).

Nox Members	Expressing cells	Main characteristics
Nox 1	Colon, vascular smooth muscle, prostate	Similar to phagocytic Nox 2 because of their structure and enzymatic activity.
Nox 3	Fetal liver, fetal lung, fetal spleen, fetal kidney	
Nox 4	Kidney, osteoclasts, pancreas, placenta, skeletal muscle, ovary, astrocytes	
Nox 5	Spleen, sperm, lymphnode, ovary, placenta, pancreas	Presents a calmodulin-like domain and can be activated by calcium.
Duox 1	Thyroid, lung	Presents a calmodulin-like domain, and an amino-terminal peroxidase domain.
Duox 2	Thyroid, colon	

Table 2. Nox members in nonphagocytic cells.

Xanthine oxidase catalyzes the oxidation of xanthine to uric acid and $O_2^{\bullet-}$. Further $O_2^{\bullet-}$ can be dismuted into hydrogen peroxide and molecular oxygen. The enzyme is expressed mainly in the liver and epithelia (Harrison, 2002).

Regarding the Nox family, seven mammalian members are known. The seven members are Nox1-5 and dual oxidases (Duoxs) 1 and 2. Nox proteins are integral membrane proteins with six transmembrane domains, which together are thought to form a channel to allow the successive transfer of electrons. They operate by transferring electrons from NADPH (converting it to $NADP^-$) to FAD, then to heme and finally to oxygen to make $O_2^{\bullet-}$. Duox1 and 2 both have a peroxidase-like domain in their N-terminal region, which means that these two members of the family produce hydrogen peroxide and not $O_2^{\bullet-}$ (Groeger et al., 2009). The NADPH oxidase complexes were firstly thought to be specific to macrophages, but now these enzymes are recognized and expressed almost universally in nonphagocytic cell types, and are also able to produce ROS via a similar mechanism (Chiarugi & Buricchi, 2007; Cross & Templeton, 2006). The wide-ranging distribution of Nox members, suggests that ROS production may have a broad implications for the different cellular phenotypes (K. Chen et al., 2009). Besides the NADPH oxidase from phagocytic cells (Nox 2), Nox family can be divided into three groups that present distinct regulatory patterns from those of Nox 2. The three groups of nonphagocytic Nox members are summarized in table 2. The production of reactive species by NADPH oxidases occurs during inflammatory conditions in cells of the immune system or by external stimuli, such as growth factors in most mammalian cells, and their relevance is related to intracellular signaling (Dorgë, 2002; Lambeth, 2002).

Lipoxygenase is a mixed function oxidase involved in the synthesis of leukotrienes from arachidonic acids. Which are released from membrane phospholipids via the activity of cytosolic phospholipase A_2, and are also responsible for the production of superoxide (Chiarugi & Buricchi, 2007; Kim et al., 2008).

Superoxide is not highly reactive and cannot cross cellular membranes; its diffusion is dependent of anion channels in the membranes. In most cases, the potential effects of superoxides are restricted to a single intracellular compartment that can be directed by some organelles like mitochondria, endosomes, and phagosomes (K. Chen et al., 2009). However, superoxide can readily be converted to hydrogen peroxide by superoxide dismutase, which finally penetrates membranes (Cross & Templeton, 2006).

Hydrogen peroxide is a two-electron oxidant that acts as an electrophile and can react with protein thiol moieties to produce a variety of sulfur oxidation states, as will be discussed later. Most of the times, this reactivity affords reversible posttranslational modification of proteins that would be important for cellular signaling. Moreover, the relatively longer half-life of hydrogen peroxide compared with other ROS in biologic systems affords better activity as second messengers (K. Chen et al., 2009). Both, superoxide and hydrogen peroxide are involved in the inhibitory oxidation of the catalytic sites of the protein tyrosine phosphatases.

4.2 Reactive nitrogen species

The reactive nitrogen species ($^{\bullet}NO$) is produced in higher organisms by the oxidation of one of the terminal guanidino-nitrogen atoms of L-arginine. This process is catalyzed by the enzyme nitric oxide synthase (NOS). Nitric oxide (NO) can be generated by the reduction of NO_x species that are derived from exogenous or endogenous sources (Hess et al., 2005). Depending on the microenvironment, NO can be converted to various other reactive nitrogen species (RNS), such as nitrosonium cation (NO^+), nitroxyl anion (NO^-) or peroxynitrite ($ONOO^-$)(Dröge, 2002). Separate genes code for different NOS isoforms, for example, neuronal NOS (nNOS or NOS1) and for endothelial NOS (eNOS or NOS3), which are named after the tissues in which they were discovered, as well as, for inducible/Ca^{2+}-independent NOS (iNOS or NOS2). (Hess et al., 2005). Nitric oxide is produced in large amounts, ranging in nanomoles concentration, for several hours. In contrast, the constitutive NOS found in some tissues remains active for relatively short periods of time (Do et al., 1996). The attachment of the NO moiety to a nucleophilic group or a transition metal is called nitrosyl. Ingeneral, the NO moiety can be provided by NO itself, nitrite, other NOx species (higher NO oxides), metal-NO complexes. Nitric oxide has high reactivity for thiol groups that are present in proteins, reduced glutathione, or cysteine residues. The chemical addition of NO moiety to a reactive cysteine thiol group (nitrosation) is called S-nitrosylation (SNO). S-Nitrosylation-induced conformation can lead to the formation of S-nitrosoproteins, S-nitrosoglutathione, or S-nitrosocysteine. These nitrosylated molecules can also serve as source of NO moieties. (Do et al., 1996; Janssen-Heininger et al., 2008). Accumulation of protein SNOs is the only mechanism of NO toxicity that is able to induce nitrosative stress. Nitrosative stress that results from increased or deregulated production of reactive nitrogen species is countered by several cellular mechanisms triggered by nitrosative modifications of proteins that mobilize adaptive and protective responses (Hess et al., 2005).

4.3 Antioxidant mechanisms

4.3.1 Antioxidants of reactive oxygen species

The transitory fluctuations in ROS provide important regulatory functions, but when present in high amounts, they can induce severe damage to cellular components. To avoid further deleterious outcomes, cells have evolved with the ability to handle increases in ROS production. These include various nonenzymatic and low-molecular-weight antioxidants, for example, reduced glutathione, vitamins A, C, and E, flavonoids, and enzymatic scavengers (Chiarugi & Buricchi, 2007). Reduced glutathione is a tripeptide synthesized in cells by the sequential addition of cysteine to glutamate followed by the addition of glycine, giving rise to the molecular γ-L-glutamyl-L-cysteinyl-glycine. The sulfhydryl group (-SH) of the cysteine can be involved in reduction reaction, removing peroxides, or it can be associated in conjugation reactions with enzymes such as glutathione peroxidases (GPx) and the isoforme 6 of peroxiredoxins (Psdx 6). These enzymes catalyze the reduction of hydrogen peroxide by GSH into H_2O and oxidized glutathione (GSSG). In mammalian cells, GSH is present in millimolar concentrations and the conversion of GSH to its oxidized form GSSG by enzymatic activities serves as a redox buffer system within cells. The GSSG can then be reduced back originating two GSH molecules by the action of glutathione reductase enzymes, at the expense of NADH or NADPH (Cross & Templeton, 2006; Forman et al., 2009). Glutathione also is important for maintaining ascorbate (vitamin c), itself a potent intracellular antioxidant, in a reduced state (K. Chen et al., 2009).

In cells, superoxide can be converted nonenzymatically to hydrogen peroxide and singlet oxygen ($^1O_2$), or through the enzymatic action of superoxide dismutases (SOD), superoxide can be then converted into H_2O_2, which is a less reactive specie. In mammals, three isoforms of superoxide dismutase are recognized: the cytoplasmatic SOD, which is a copper/zinc-dependent dismutase (SOD1); the mitochondrial manganese SOD (SOD 2); and the extracellular Cu/Zn-dependent SOD (SOD3).

In the presence of reduced transition metals, such as ferrous or cuprous ions, hydrogen peroxide can be converted into a highly reactive hydroxyl radical (HO^-). This radical is a less stable radical species and acts in a nonselective, oxidative manner on organic molecules. There are several families of enzymes, *e.g.*, catalases, glutathione peroxidases (GPx), and peroxiredoxins (Prx), that break down hydrogen peroxide directly (Dröge, 2002; Salmeen & Barford, 2005). Catalases are localized entirely in peroxisomes and efficiently converts hydrogen peroxide to water and molecular oxygen, while glutathione peroxidases are largely restricted to the cytosol and mitochondria. Glutathione peroxidases reduce peroxides by transferring electrons from GSH with the generation of oxidized glutathione. Peroxiredoxins exhibit a higher affinity toward hydrogen peroxide and are also abundant in the cytosol. Peroxiredoxins remove hydrogen peroxide to yield water, while form an intermolecular disulfide bond, which then can be reduced by thioredoxins (Trx). Consequently, thioredoxins are not directly involved in the removal of ROS, but indirectly through a supportive role to the peroxiredoxins. Besides their function as ROS scavengers, these enzymes can actually protect proteins from oxidations or regenerate oxidized proteins (K. Chen et al., 2009; Chiarugi & Buricchi, 2007). The overall catalysis mechanism of these antioxidant enzymes is summarized below:

$$O_2^{\bullet -} + O_2^{\bullet -} + 2H^+ \xrightarrow{\text{SOD}} H_2O_2 + O_2 \quad (1)$$

$$H_2O_2 + Fe^{2+} \rightarrow Fe^{3+} + {}^{\bullet}OH + OH^- \quad (2)$$

$$2H_2O_2 \xrightarrow[\text{GPx or Prx}]{\text{Catalase}} 2H_2O + O_2 \quad (3)$$

4.3.2. Antioxidants of Nitrosylated Molecules

As mentioned above nitrosative stress results from increased or deregulated production of reactive nitrogen species (Hess et al., 2005). The formation of S-nitrosoglutathione is regulated by the coordinately action of the enzyme S-nitrosoglutathione reductase (GSNO reductase), also known as glutathione-dependent formaldehyde dehydrogenase. This enzyme converts GSNO to GSNHOH, ultimately forming ammonia and oxidized glutathione. GSNO reductase is responsible for the regulation of the steady-state levels of S-nitrosylated proteins, that are in equilibrium with GSH. Denitrosylation can also be achieved by specific metal ions, such as Cu^+, the low-molecular-weight antioxidant ascorbate, or thioredoxins enzymes (Janssen-Heininger et al., 2008).

5. Redox inhibition of cysteine-based protein tyrosine phosphatases

Reversible oxidation of regulatory proteins by hydrogen peroxide or NO/SNO has been implicated in broad cellular signaling mechanisms. Tyrosine phosphatases can be differentially regulated by phosphorylation, intra- and intermolecular interactions, alternative splicing, transcription, and translation, but recently, it has become apparent that this family of enzymes may also be regulated by reversible oxidation. In general, the redox regulation of proteins occurs primarily when cysteine residues within the protein or bound to transition metals react with ROS, such as H_2O_2 or superoxide, or with reactive nitrogen species, such as nitric oxide, *S*-nitrosothiols, or peroxynitrite. Reactions of cysteines with mild oxidizing or nitrosylating reagents, such as low concentrations of H_2O_2 or $^{\bullet}NO$, can lead to reversible modulations of protein activity. Oxidation or nitrosylation of the catalytic cysteine of protein tyrosine phosphatases renders the residue unable to act as a nucleophile, and the enzyme loses its phosphatase activity (Chiarugi et al., 2005). Protein phosphatases are among the earliest characterized targets of inhibition by thiol group modification. Thiol group modifications are thought to play a central role in cellular signaling process through ROS modifications (Cross & Templeton, 2006).

The first evidence of protein regulation by the redox modulation of a cysteine was described in 1967 by Pontremoli and colleagues, which showed that disulfide formation with cysteamine stimulated fructose-1,6-bisphosphatase activity. The redox regulation of tyrosine phosphatases depends on different oxidative stress conditions and is tissue specific (Chiarugi et al., 2005).

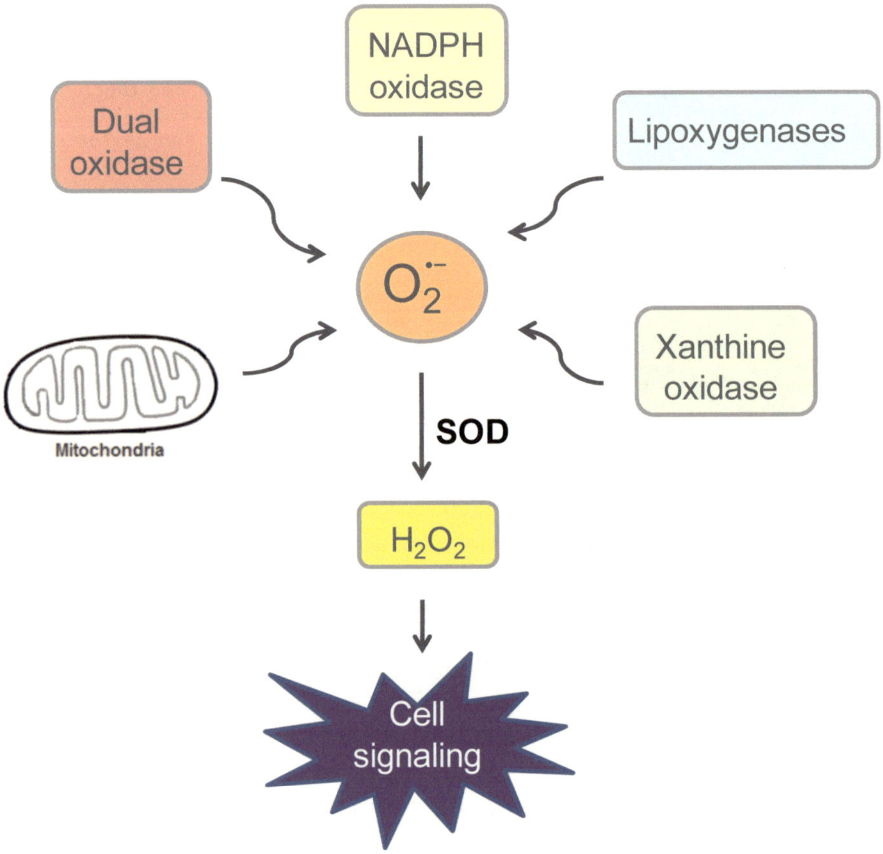

Fig. 3. Sources of reactive oxygen species. Superoxide ($O_2^{\bullet-}$) can be produced from different sources as within the mitochondrial electron transport chain, the activity of xanthine oxidases, Nox members like NADPH oxidase or Dual oxidase, and lipoxygenases. Later, superoxide may be dismuted to hydrogen peroxide (H_2O_2) through the action of superoxide dismutase (SOD). H_2O_2 can then act as second messenger in cell signaling.

For tyrosine phosphatases, the typical element of redox regulation occurs within the catalytic site of the enzyme. Two specific amino acid residues are essential to the catalysis of tyrosine phosphatases, Cys215 and Arg221. Oxidation of the catalytic cysteine of the protein tyrosine phosphatases occurs *in vivo* in response to intracellular ROS production, which can occur in response to a variety of stimuli, such as platelet-derivative growth factor (PDGF), epidermal growth factor (EGF), insulin, several other cytokines, and integrins (Rhee et al., 2003; Tonks, 2005). Cysteine residues are also affected by extracellularly added ROS or by an imbalance of the redox potential within the cell. Therefore, hydrogen peroxide, S-nitrosylation, or other radical species may oxidize cysteine residues in proteins to form cysteine sulfenic acid (Cys-SOH), which can be further stabilized by the formation of inter- or intramolecular disulfide (S-S) or sulfenyl-amide bonds. The last conformation is characterized by a five-membered ring that is formed by the covalent linkage of an atom of the catalytic cysteine to an amide nitrogen in a neighboring residue (Chiarugi & Buricchi, 2007; Denu & Tanner, 1998; Janssen-Heininger et al., 2008).

Different cysteine residues can take part in the disulfide bridge, such as what may occur with the catalytic cysteine of LMW-PTP or PTEN with another cysteine residue or between the classical protein tyrosine phosphatases such as PTP1B and a cysteine from GSH (Barrett et al., 1999; Chiarugi et al., 2001; den Hertog et al., 2005; Kwon et al., 2004; Lee et al., 1998; Salmeen & Barford, 2005). The tripeptide GSH can also induce alterations of proteins function by the formation of mixed disulfide bonds, a reversible process referred to as glutathionylation. The last conformation can result in a temporarily inhibition of PTP activity that protects enzyme from irreversible oxidation. Glutathionylation can be easily reversed by the action of glutaredoxins or thioredoxins, restoring normal enzyme function (Cross & Templeton, 2006).

It is assumed that both disulfide bonds or sulfenyl-amide conformations protect the reactive cysteine from a higher-order, irreversible oxidation or expose the oxidized cysteine to cellular reducing agents, which facilitates the reactivation of the oxidized cysteine, because sulfenyl-amide rings or disulfide bonds represent reversibly oxidized states of cysteine and can be readily reduced (Chiarugi et al., 2005; Salmeen & Barford, 2005; Tonks, 2005; Tabernero et al., 2008). Moreover, these two conformational changes may represent a signal that the given protein tyrosine phosphatase is in a temporarily inactive state (Tabernero et al., 2008), which may be important to stabilize associations within protein complexes, to modify structures, to create, destroy or modulate functional sites, or to regulate enzymatic or transcriptional activities of proteins (Poole & Nelson, 2008).

Dual-specific phosphatases, such as Cdc25, PTEN, and LMW-PTP, and classical protein tyrosine phosphatases are also a target of ROS oxidation. Nevertheless, these enzymes contain a second cysteine residue within the active site. Following oxidation of the cysteine in the signature motif, a disulfide bond is formed with the vicinal cysteine, which protects the enzyme from higher oxidation and irreversible inactivation. Intramolecular disulfide bond formation involving the catalytic cysteine of classical PTPs has not been observed (den Hertog et al., 2005; Lee et al., 2002; Tonks, 2005).

Different phosphatases present distinct sensitivities to oxidation. Regulation of the oxidized states of the cysteine may vary with the structural distances between the cysteines involved in the formation of disulfide bridge and the availability of thiolate ions. This may represent a degree of specificity for protein tyrosine phosphatase redox regulation and lead to a different extent of oxidation (Chiarugi & Buricchi, 2007).

Inactivation of the protein tyrosine phosphatases by hydrogen peroxide could be reversed by thiol-reducing agents such as dithiothreitol (DTT), β-mercaptoethanol, reduced glutathione, cysteine or by the action of some enzymes in which their catalytic cycle involves either a mono- or dithiol mechanism, such as glutaredoxins, thioredoxins, and protein disulfide-isomerase that have a CXXC motif in the active site. Thioredoxins are able to reduce sulfenic acids and disulfide bonds, while glutaredoxins are involved in the reversibility of mixed disulfides with GSH and S-nitrosothiols (Chiarugi & Buricchi, 2007).

The oxidative modification of protein tyrosine phosphatases can be stable over prolonged periods (Salmeen and Barford, 2005). In some cases, the sulfhydryl group is open to further irreversible oxidation if no cysteine derivatives or compounds containing thiol groups are close enough to facilitate the formation of a disulfide bridge or a sulfenyl-amide ring. The addition of one or two oxygen molecules results in the formation of sulfinic acid ($Cys\text{-}SO_2$)

or sulfonic acid (Cys-SO_3), respectively. These oxidative modifications are irreversible, and the phosphatase will be unable to become active again, even in a reducing environment (Groeger et al., 2009). However, it has been shown recently that sulfinic acids can be reduced by the action of a specific enzyme called sulfiredoxin, which has only been described in mammals and yeast so far (Biteau et al., 2003; Woo et al., 2003).

The redox balance of the cells can affect the ability of ROS to oxidize a given protein. At normal cellular conditions, the cellular redox potential may vary from –260 mV to –150 mV. In addition, protein oxidation will depend on the distance between the site of ROS production and the target protein and on the concentration of scavenger enzymes available. The length of time that the protein will be inactivated will also depend on the proximity to enzymes such as glutaredoxin and thioredoxin, which catalyze thiol reduction (Salmeen & Barford, 2005). Regarding protein tyrosine phosphatases, the differences in pKa values of catalytic cysteines, the availability of thiolate ions and the structural distance between the two cysteines involved in the disulfide bridge may also strongly determine the ability of the phosphatases to be rapidly regulated by changes in the intracellular redox conditions, thereby representing a degree of specificity for tyrosine phosphatase regulation (Chiarugi et al., 2005).

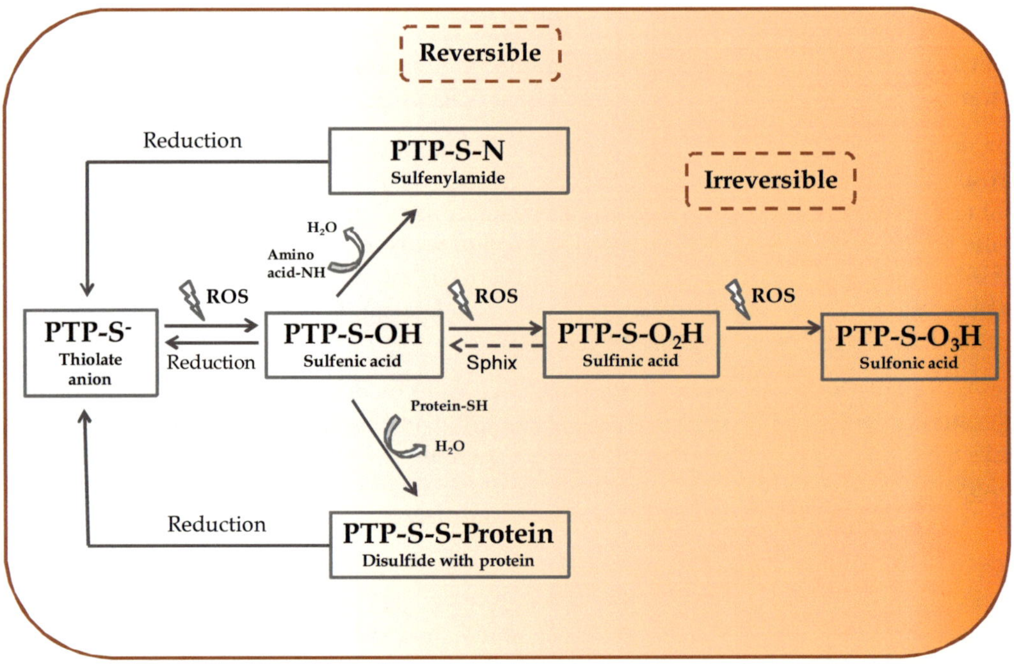

Fig. 4. The redox modulation of a protein tyrosine phosphatase thiolate anion. Because of its low pKa value, the cysteine residue in the active site of the tyrosine phosphatase is present in its thiolate anion form. This conformation facilitates the inhibition of the enzyme by oxidation, which would become inactivated as the sulfenic acid form. Furthermore, sulfenic acid is converted to a more stable conformation, such as a disulfide bridge or sulfenyl-amide ring. Both conformations are stable and reversible through the action of reductant agents. However, if oxidation persists for a long period, stronger and higher oxidation could lead to the formation of sulfinic and sulfonic acids, which are irreversible conformations. In some cases, sulfinic acid can be reversed enzymatically through the action of sulfiredoxin (Sphix) enzyme.

6. Cellular signaling through oxidative inhibition of the protein tyrosine phosphatases

6.1 Receptor tyrosine kinases

Cellular signaling processes are controlled by the coordinated activity of protein phosphatases and kinases. Protein tyrosine kinases (PTKs) are usually divided into two families, the transmembrane receptor family and the nonreceptor family (Blume-Jensen & Hunter, 2001). Receptor tyrosine kinases (RTKs) are cell-surface, transmembrane receptors carrying a multidomain extracellular segment that binds polypeptide ligands, a single-pass transmembrane helix, and a cytoplasmic segment containing a tyrosine kinase domain together with several regulatory sequences located both N- and C-terminal to the kinase domain. The receptor tyrosine kinase family includes epidermal growth factor receptor (PDGFR), fibroblast growth factor receptor (FGFR), and the insulin receptor kinase (IRK), among others (Chiarugi & Buricchi, 2007).

The kinase activities of protein tyrosine kinases are closely regulated through autoregulatory mechanisms and by the action of tyrosine phosphatases. In general, most tyrosine kinases are maintained in a low activity state through a variety of these autoregulatory mechanisms and by avoiding the most favorable configuration of the kinase active site. Activation of receptor tyrosine kinases is achieved through a ligand binding to the extracellular domain, which stabilizes a dimeric receptor arrangement and facilitates *trans*-phosphorylation in the cytoplasmic domain. Autophosphorylation of one or two conserved tyrosines within the juxtamembrane domain of these receptors is required for complete kinase activation (Chiarugi & Buricchi, 2007).

6.2 Hydrogen peroxide as a second messenger in cellular signaling

The classic receptor-mediated signaling involves a ligand engagement, followed by the production of a diffusible second messenger that interacts with a target to affect the signal. This arrangement supports both signal transduction over space and signal amplification, because most second messengers are produced via an enzymatic process. Typically, second messengers are small, diffusible molecules that rapidly activate effector proteins such as protein kinases, phosphatases or ion channels. These molecules can act through binding or chemical modification, thus promoting signal transduction (K. Chen et al., 2009).

It is now widely accepted that hydrogen peroxide may function as a second messenger within cells by being able to induce tyrosine phosphorylation of cellular proteins, which is strongly potentiated by a combination treatment with vanadate. The mechanism was believed to be attributed to the inhibition of tyrosine phosphatase, the activation of tyrosine kinases, or both (Chiarugi & Buricchi, 2007). Furthermore, the addition of exogenous hydrogen peroxide to cells can mimic the effects of growth factors or hormones and lead to hyperphosphorylation of receptor tyrosine kinases. Mitogen-activated protein kinases have also been shown to be activated by extracellular hydrogen peroxide through a specific kinase cascade (Torres, 2003).

Little is known about how hydrogen peroxide is actually distributed to the cytoplasm. Most of the time, Nox family members are responsible for producing hydrogen peroxide near the receptor at the plasma membrane; in addition, hydrogen peroxide, which is produced in response to stimulation by a ligand, is released outside of the cell to then be diffused locally into the cell. Also, NADPH oxidase and lipoxygenase, which can be localized within the endoplasmic reticulum and the nuclear membrane, are able to release hydrogen peroxide into the luminal spaces of the specific organelles (Chiarugi & Buricchi, 2007; Tonks, 2005).

Until recently, it was thought that hydrogen peroxide would cross biological membranes freely; however, many membranes were shown to be almost impermeable to hydrogen peroxide as well as to water (Antunes & Cadenas, 2000; Makino et al., 2004; Seaver & Imlay, 2001). Thus, for transport, hydrogen peroxide must be carried by an appropriate transporter that has not yet been well characterized (Chiarugi & Buricchi, 2007).

6.3 Nitric oxide as a second messenger in cellular signaling

Nitric oxide is a gas, which readily diffuses across membranes and does not act at conventional membrane associated receptors (Do et al., 1996). Like hydrogen peroxide, signal transduction through induced-nitric-oxide modifications relies on the system of Cys-based posttranslational modifications. Accordingly, S-nitrosylation of proteins plays an essential role in downstream cascades.

Nitric oxide exerts a ubiquitous influence on cellular signaling in large part by means of S-nitrosylation/denitrosylation of protein cysteine residues. SNO, or higher oxides is considered a post-translational protein modification that is precisely regulated, conferring specificity to NO-derived effects. These NO-dependent modifications influence protein activity, protein-protein interactions, and protein location. S-nitrosylation thus serves as the prototypical redox-based signal (Janssen-Heininger et al., 2008).

S-Nitrosylation has been implicated in transmitting signals downstream of all classes of receptors, including G-protein-coupled receptor (GPCR), receptor tyrosine kinase, tumor necrosis factor, Toll-receptors, and glutaminergic receptors, acting locally within subcellular signaling domains as well conveying signals from the cell surface to intracellular compartments, including the mitochondria and the nucleus (Janssen-Heininger et al., 2008).

Regarding cellular signaling through S-nitrosylation, specific features of this kind of modification can be considered useful tools in signaling transduction According to Janssen-Heininger et al. (2008). These features include:

1. Temporal regulation of response through a rapid and controlled stimulation;
2. The existence of motifs within proteins that provides S-nitrosylation specificity;
3. Colocalization of target proteins with a source of NO;
4. Reversibility of protein S-nitrosylation;
5. Enzymatic control of S-nitrosylation through the action of S-nitrosoglutathione reductase.

6.4 The role of protein tyrosine phosphatases in cellular signaling

Physiological stimuli, such as growth factors or the engagement of antigen receptors, can trigger localized and controlled production of ROS. Protein tyrosine phosphatases, because of their high susceptibility to inactivation by oxidation at physiological pH, are targets of ligand-induced ROS generation. Thus, the ability of tyrosine phosphatases to dephosphorylate receptor tyrosine kinases becomes temporarily compromised because of the oxidation of the catalytic site of the tyrosine phosphatases. These events are involved in the enhancement of the signaling response through an overall increase in tyrosine phosphorylation. Therefore, complete activation of the tyrosine kinase signaling cascades requires both an inhibition of protein tyrosine phosphatases by oxidation and an activation of receptor tyrosine kinases by phosphorylation (Salmeen & Barford, 2005). Redox signaling through the reversible oxidation of tyrosine protein phosphatases must occur through a reaction that is chemically reversible under physiological conditions to ensure the continuity of cellular mechanisms.

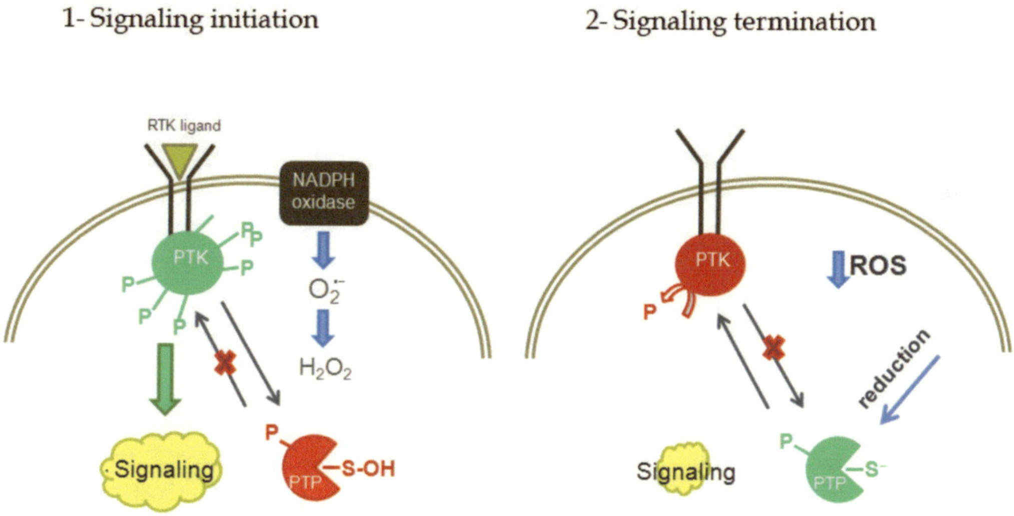

Fig. 5. After ligand engagement, the receptor tyrosine kinase (RTK) needs the first phase in which the tyrosine phosphorylation level must be high to guarantee signal propagation. This high level is granted by a transient inhibition of protein tyrosine phosphatase (PTP) due to oxidation promoted by hydrogen peroxide and by the disrupted association with the receptor substrates. Hydrogen peroxide is produced by NADPH oxidase that also became activated by ligand stimulation of the receptor tyrosine kinase. The concomitant tyrosine phosphorylation of oxidized protein tyrosine phosphatase, which is performed in response to receptor tyrosine kinase activation, is ineffective on the abolished enzymatic activity and preparative for the second phase. In the signaling termination phase, the tyrosine phosphatase recovers its enzymatic activity due to a re-reduction of the catalytic thiol group and immediately becomes hyperactive due to the previous tyrosine phosphorylation, thus promoting the rapid and efficient termination of the signal. The tyrosine phosphatase recurrence after oxidation is followed by a dephosphorylation of the activated receptors, which consequently terminates the signal (Chiarugi et al., 2005).

Besides their regulatory functions, the regulatory domains in tyrosine phosphatases are also involved in the sensitivity to oxidation, which probably contributes to the intrinsic sensitivity differences between various protein tyrosine phosphatases to oxidative inhibition

(Groeger et al., 2009; Östman et al., 2011). The redox regulation of tyrosine phosphatases may depend on different oxidative stress conditions and the extent of oxidation, as it is tissue and differentiation dependent (Chiarugi et al., 2005). As mentioned above, the transient negative regulation of tyrosine phosphatases due to oxidants is dependent on ligand stimulation of the receptor tyrosine kinases.

Hydrogen peroxide is recognized as one of the main ROS capable of acting as a second messenger because of it is relative stability. Cytokines, growth factors and integrins activate the multicomponent NADPH oxidase enzymes as part of the receptor-mediated signaling in cells. Both superoxide and, hence, hydrogen peroxide are generated, with the latter serving a particularly significant role as a second messenger in signaling, which mediates the regulation of survival pathways in cells (Poole & Nelson, 2008; Groeger et al., 2009).

The tyrosine phosphorylation level of a given protein is the result of the collective action of kinases and phosphatases. Signal transduction by ROS through reversible tyrosine phosphatase inhibition represents a widespread and conserved mechanism that is triggered by the ligand stimulation of receptor tyrosine kinases. Conformational changes lead to an increase in ROS production, which causes a transient negative regulation of protein tyrosine phosphatases that are associated with receptor tyrosine phosphatases, and at the same time decreases receptor tyrosine phosphatase dephosphorylation that leads to a high stimulation of its activity and signal propagation. The same PTP that suffered oxidation can be highly phosphorylated by the protein tyrosine kinase that is now activated. The oxidation of the active site of the tyrosine phosphatase protects the enzyme from autodephosphorylation. When the tyrosine phosphatase returns to the reduced form, its highly phosphorylated state influences a hyperactivation of the enzyme's activity, which is essential to terminating the signaling cascade in a fast way (Chiarugi & Buricchi, 2007). This mechanism represents a strategy adopted by cells to promote receptor tyrosine kinase signaling by avoiding its prompt inactivation by tyrosine phosphatases. After the redox potential has been restabilized, tyrosine phosphatase recovers its activity and the signaling triggered by the receptor tyrosine kinases is terminated (Chiarugi et al., 2005; Chiarugi & Buricchi, 2007). Thus, tyrosine kinase activation by oxidation is considered an indirect effect of tyrosine phosphatase inhibition. The overall mechanism is represented in figure 5.

The distinct but complementary function of kinases and phosphatases clarifies the role of both enzymes in controlling cellular signaling. Therefore, it is possible to affirm that kinases are implicated in controlling the amplitude of a signaling response whereas phosphatases are thought to have an important role in controlling the rate and duration of the response (Tonks, 2006).

A key element of a signaling response is reversibility. Regulated reversible phosphorylation of proteins and other cellular molecules plays a ubiquitous role in the control of cellular behavior. For oxidation to illustrate the mechanism for reversible regulation of protein tyrosine phosphatase function, it is essential that the active site cysteine residue not be oxidized to sulfinic or sulfonic acids, which is usually irreversible (Poole et al., 2004; Tonks, 2005). As mentioned elsewhere, thiol groups (SH) do not react at physiologically significant rates with a hydroperoxide, such as hydrogen peroxide, unless the reaction is catalyzed. However, thiolates (S^-) react faster with hydroperoxides depending on their local environment (Forman et al., 2004). The reaction rate of hydrogen peroxide with protein tyrosine phosphatase is about 10^5 times slower than with peroxiredoxin. The question appears on how hydrogen peroxide can react specifically with the active site of PTPs before its metabolization by enzymatic scavengers. Two possible mechanisms can arise:

1. Hydrogen peroxide is generated by NADPH oxidase close to PTP so that the concentration could be sufficient to cause enzyme modification; or
2. PTPs could be compartmentalized to the sites of ROS generation and far from scavenger enzymes (Chiarugi et al., 2005; Forman et al., 2004).

Besides, specificity of ROS to a given PTP may be influenced by contributions from the surrounding amino acids near the catalytic cysteine. The prevalence of basic amino acids near the active site of the enzyme can contribute to the extension formation of thiol ionization giving rise to thiolate anions within active cysteine. Chemical reactivity secondary to the local amino acids environments is likely to contribute to the modification of cysteine residues (Cross & Templeton, 2006). This is the case of disulfide bonds or sulfenyl-amide conformations.

Receptor signaling in which receptor activation induces both the activation of downstream kinases and the oxidative inactivation of inhibitory phosphatases are components required for efficient signaling. The regulated and localized production of ROS can inactivate a subpopulation of phosphatase molecules near the site of ROS generation. The specificity of ROS for tyrosine phosphatase exists through different intrinsic sensitivities to oxidation and a tight control over the production of ROS. This production can be achieved by the controlled action of Nox enzymes, including their requirement for assembly into a multiprotein complex, the phosphorylation of regulatory subunits, the binding of inositol phospholipids and the small GTPase, Rac. In addition, regulation of Duox and NADPH oxidases by Ca^{2+} adds further control of ROS generation in cells (Tonks, 2006; Poole et al., 2004). As an example, Nox4 can regulate the oxidation of PTP1B in response to insulin and localizes with PTP1B on the intracellular membranes. Ecto-phosphatases, which are protein phosphatases that present a catalytic site facing the extracellular medium, are also able to respond to the endogenous production of hydrogen peroxide from mitochondria. The addition of a proton ionophore, which was able to reduce mitochondrial hydrogen peroxide production, was also shown to induce ecto-phosphatase activity due to the reduction of hydrogen peroxide outside of the cell (Cosentino-Gomes et al., 2009).

7. Conclusion

The mechanism by which ROS/RNS exerts a regulatory function in cysteine-based protein tyrosine phosphatases is well known, as well as the role of antioxidants in reverse enzyme oxidation. The reversible inhibition of tyrosine phosphatases by ROS has been intensely investigated in some diseases such as cancer and diabetes, but the complete pathway in which these reactive species are involved is not elucidated yet. Some questions still remain unsolved such as: How exactly ROS/RNS has a specific action on a given tyrosine phosphatase? Is there a compartmentalization of proteins involved in redox pathways? Is there a specific site of ROS/RNS production to induce redox signaling? How much of the total hydrogen peroxide generated in a specific site can be translated into a cellular signaling event? Which molecules would be involved in the redox sensor of the cell? The challenge to solve these questions seems to be very difficult, since we have multiple oxidants species from several sites of the cell. The development of specific probes for a given reactive specie, or for a particular site of ROS/RNS production would be of great contribution to understand the role of redox reactions in the regulation of physiological signaling process.

8. References

Alonso, A., Sasin, J., Bottini, N., Friedberg, I., Friedberg, I., Osterman, A., Godzik, A., Hunter, T., Dixon, J. & Mustelin, T. (2004). Protein tyrosine phosphatases in the human genome. *Cell*, Vol. 117, No 6, pp. 699-711, ISSN 1097-4172

Antunes, F. & Cadenas, E. (2000). Estimation of H2O2 gradients across biomembranes. *FEBS Letter*, Vol. 475, No 2, pp. 121-126, ISSN 1873-3468

Bae, Y.S., Kang, S.W., Seo, M.S., Baines, I.C., Tekle, E., Chock, P.B. & Rhee, S.G. (1997). Epidermal growth factor (EGF)-induced generation of hydrogen peroxide. Role in EGF receptor-mediated tyrosine phosphorylation. *The Journal of Biological Chemistry*, Vol. 272, No 1, pp. 217-221, ISSN 1083-351X

Barrett, W.C., DeGnore, J.P., König, S., Fales, H.M., Keng, Y.F., Zhang, Z.Y., Yim, M.B. & Chock, P.B. (1999). Regulation of PTP1B via glutathionylation of the active site cysteine 215. *Biochemistry*, Vol. 38, No 20, pp. 6699-6705, ISSN 1520-4995

Biteau, B. Labarre, J. & Toledano, M.B. (2003). ATP-dependent reduction of cysteine-sulphinic acid by S. cerevisiae sulphiredoxin. *Nature*, Vol. 425, No 6961, pp. 980-984, ISSN 1476-4687

Blume-Jensen, P. & Hunter, T. (2001). Oncogenic kinase signalling. *Nature*, Vol. 411, No 6835, PP. 355-365, ISSN 1476-4687

Brandes, N., Schmitt, S. & Jakob, U. (2009). Thiol-based redox switches in eukaryotic proteins. *Antioxidants & Redox Signaling*, Vol. 11, No 5, pp. 997-1014, ISSN 1557-7716

Chen, C.Y., Willard, D. & Rudolph, J. (2009). Redox regulation of SH2-domain-containing protein tyrosine phosphatases by two backdoor cysteines. *Biochemistry*, Vol 48, No 6, pp. 1399-1409, ISSN 1520-4995

Chen, K., Craige, S.E. & Keaney Jr., J.F. (2009). Downstream targets and intracellular compartmentalization in Nox Signaling. *Antioxidants & Redox Signaling*, Vol. 11, No 10, pp. 2467-2480, ISSN 1557-7716

Chiarugi, P. & Buricchi, F. (2007). Protein tyrosine phosphorylation and reversible oxidation: two cross-talking posttranslation modifications. *Antioxidants & Redox Signaling*, Vol. 9, No 1, PP.1-24, ISSN 1557-7716

Chiarugi, P., Fiaschi, T., Taddei, M.L., Talini, D., Giannoni, E., Raugei, G. & Ramponi, G. (2001). Two vicinal cysteines confer a peculiar redox regulation to low molecular weight protein tyrosine phosphatase in response to platelet-derived growth factor receptor stimulation. *The Journal of Biological Chemistry*, Vol. 276, No 36, pp. 33478-33487, ISSN 1083-351X

Chiarugi, P., Taddei, M.L. & Ramponi, G. (2005). Oxidation and tyrosine phosphorylation: synergistic or antagonistic cues in protein tyrosine phosphatase. *Cellular and Molecular Life Sciences*, Vol. 62, No 9, PP. 931-936, ISSN 1420-9071

Cosentino-Gomes, D., Russo-Abrahão, T., Fonseca-de-Souza, A.L., Ferreira, C.R., Galina, A. & Meyer-Fernandes, J.R. (2009). Modulation of Trypanosoma rangeli ecto-phosphatase activity by hydrogen peroxide. *Free Radical Biology and Medicine*, Vol. 47, No 2, pp. 152-158, ISSN 1873-4596

Cross, J.V. & Templeton, D.J. (2006). Regulation of signal transduction through protein cysteine oxidation. *Antioxidants & Redox Signaling*, Vol. 8, No 9-10, pp. 1819-1827, ISSN 1557-7716

den Hertog, J., Groen, A. & van der Wijk, T. (2005). Redox regulation of protein-tyrosine phosphatases. *Archives of Biochemistry and Biophysics*, Vol. 434, No 1, pp. 11-15, ISSN 1096-0384

den Hertog, J., Tracy, S. & Hunter, T. (1994). Phosphorylation of receptor protein-tyrosine phosphatase alpha on Tyr789, a binding site for the SH3-SH2-SH3 adaptor protein GRB-2 in vivo. *The EMBO Journal*, Vol. 13, No 13, pp. 3020-3032, ISSN 1460-2075

Denu, J.M. & Dixon, J.E. (1998). Protein tyrosine phosphatases: mechanisms of catalysis and regulation. *Current Opinion in Chemical Biology*, Vol. 2, No 5, pp.633-641, ISSN 1879-0402

Denu, J.M. & Tanner, K.G. (1998). Specific and reversible inactivation of protein tyrosine phosphatases by hydrogen peroxide: evidence for a sulfenic acid intermediate and implications for redox regulation. *Biochemistry*, Vol. 37, No 16, pp. 5633-5642, ISSN 1520-4995

Do, K.Q., Benz, B., Grima, G., Gutteck-Amsler, U., Kluge, I. & Salt, T.E. (1996). Nitric oxide precursor arginine and S-nitrosoglutathione in synaptic and glial function. *Neurochemistry International*, Vol. 29, No 3, pp. 213-224, ISSN 1872-9754

Dröge, W. (2002). Free radicals in the physiological control of cell function. *Physiological Reviews*, Vol. 82, No 1, pp. 47-95, ISSN 1522-1210

Forman, H.J., Zhanq, H. & Rinna, A. (2009). Glutathione: overview of its protective roles, measurement, and biosynthesis. *Molecular Aspects of Medicine*, Vol. 30, No 1-2, pp. 1-12, ISSN 1872-9452

Forman, H.J.m Fukuto, J.M. & Torres, M. (2004). Redox signaling: thiol chemistry defines which reactive oxygen and nitrogen species can act as second messengers. *American Journal of Physiology*, Vol. 287, No 2, pp. 246-256, ISSN 1522-1563

Giannoni, E., Buricchi, F., Raugei, G., Ramponi, G. & Chiarugi, P. (2005). Intracellular reactive oxygen species activate Src tyrosine kinase during cell adhesion and anchorage-dependent cell growth. *Molecular and Cellular Biology*, Vol. 25, No 15, PP. 6391-6403, ISSN 1098-5549

Groeger, G., Quiney, C. & Cotter, T.G. (2009). Hydrogen peroxide as a cell-survival signaling molecule. *Antioxidants & Redox Signaling*, Vol. 11, No 11, PP. 2655-2671, ISSN 1557-7716

Harrison, R. (2002). Structure and function of xanthine oxidoreductase: where are we now? *Free Radical Biology and Medicine*, Vol. 33, No 6, pp. 774-797, ISSN 1873-4596

Hess, D.T., Matsumoto, A., Kim, S.O. Marshall, H.E. & Stamler, J.S. (2005). Protein S-nitrosylation: purview and parameters. *Nature Reviews. Molecular Cell Biology*, Vol. 6, No 2, pp. 150-166, ISSN 1471-0080

Janssen-Heininger, Y.M.W., Mossman, B.T., Heintz, N.H., Forman, H.J., Kalyanaraman, B., Finkel T., Stamler, J.S., Rhee, S.G. & Van der Vliet, A. (2008). Redox-based regulation of signal transduction: principles, pitfalls, and promises. *Free Radical Biology & Medicine*, Vol. 45, No 1, pp. 1-17, ISSN 1873-4596

Jia, Z., Barford, D., Flint, A.J. & Tonks, N.K. (1995). Structural basis for phosphotyrosine peptide recognition by protein tyrosine phosphatase 1B. *Science*, Vol. 268, No 5218, pp. 1754-1758, ISSN 1095-9203

Kim, C., Kim, J.Y. & Kim, J.H. (2008) Cytosolic phospholipase A(2), lipoxygenase metabolites, and reactive oxygen species. *BMB reports*, Vol. 41, No 8, pp. 555-559, ISSN 1976-670X

Kolmodin, K. & Aqvist J. (2001). The catalytic mechanism of protein tyrosine phosphatases revisited. *FEBS Letter*, vol. 498, No.2-3, pp. 208-213, ISSN 1873-3468

Kwon, J., Lee, S.R., Yang, K.S., Ahn, Y., Kim, Y.J., Stadtman, E.R. & Rhee, S.G. (2004). Reversible oxidation and inactivation of the tumor suppressor PTEN in cells stimulated with peptide growth factors. *Proceedings of the National Academy of Sciences of the United States of America*, Vol. 101, No 47, pp. 16419-16424, ISSN 1091-6490

Lambeth, J.D. (2002). Nox/Duox family of nicotinamide adenine dinucleotide (phosphate) oxidases. *Current Opinion in Hematology*, Vol. 9, No 1, pp. 11-17, ISSN 1531-7048

Lee, S.R., Kwon, K.S., Kim, S.R. & Rhee, S.G. (1998). Reversible inactivation of protein-tyrosine phosphatase 1B in A431 cells stimulated with epidermal growth factor. *The Journal of Biological Chemistry*, Vol. 273, No 25, pp. 15366-15372, ISSN 1083-351X

Lee, S.R., Yang, K.S., Kwon, J., Lee, C., Jeong, W. & Rhee, S.G. (2002). Reversible inactivation of the tumor suppressor PTEN by H2O2. *The Journal of Biological Chemistry*, Vol. 277, No 23, pp. 20336-20342, ISSN 1083-351X

Makino, N., Sasaki, K., Hashida, K. & Sakakura, Y. (2004). A metabolic model describing the H2O2 elimination by mammalian cells including H2O2 permeation through cytoplasmic and peroxisomal membranes: comparison with experimental data. *Biochimica et Biophysica Acta*, Vol. 1673, No 3, PP. 149-159, ISSN 0006-3002

Mayer, B.J. (2008). Clues to the evolution of complex signaling machinery. *Proceedings of the National Academy of Sciences of the United States of America*, vol. 105, No 28, pp. 9453-9454, ISSN 1091-6490

McCord, J.M. & Fridovich, I. (1969). The utility of superoxide dismutase in studying free radical reactions. I. Radicals generated by the interaction of sulfite, dimethyl sulfoxide, and oxygen. *The Journal of Biological Chemistry*, Vol. 244, No 22, pp. 6056-6063, ISSN 1083-351X

Meng, T.C., Fukada, T. & Tonks, N.K. (2002). Reversible oxidation and inactivation of protein tyrosine phosphatases in vivo. *Molecular Cell*, Vol. 9, No 2, pp. 387-399, ISSN 1097-4164

Mittal, C.K. & Murad, F. (1977). Properties and oxidative regulation of guanylate cyclase. *Journal of Cyclic Nucleotide Research*, Vol. 3, No 6, pp. 381-391, ISSN 0095-1544

Östman, A., Frijhoff, J., Sandin, A. & Böhmer, F.D. (2011). Regulation of protein tyrosine phosphatases by reversible oxidation. *Journal of Biochemistry*, in press, ISSN 1756-2651

Pontremoli, S., Traniello, S., Enser, M., Shapiro, S. & Horecker, B.L. (1967). Regulation of fructose diphosphatase activity by disulfide exchange. *Proceedings of the National Academy of Sciences of the United States of America*, Vol. 58, No 1, pp. 286-293, ISSN 1091-6490

Poole, L.B., Karplus, P.A. & Claiborne, A. (2004). Protein sulfenic acids in redox signaling. *Annual review of pharmacology and toxicology*, Vol. 44, No 325, pp. 325-347, ISSN 1545-4304

Poole, L.B. & Nelson, K.J. (2008). Discovering mechanisms of signaling-mediated cysteine oxidation. *Current Opinion in Chemical Biology*, Vol. 12, No 1, pp. 18-24, ISSN 1879-0402

Rhee, S.G., Chang, T.S., Bae, Y.S., Lee, S.R. & Kang, S.W. (2003). Cellular regulation by hydrogen peroxide. *Journal of the American Society of Nephrology*, Vol. 14, No 8, pp. 211-215, ISSN 1533-3450

Salmeen, A. & Barford, D. (2005). Functions and mechanisms of redox regulation of cysteine-based phosphatases. *Antioxidants & Redox Signaling*, Vol. 7, No 5-6, PP.560-577, ISSN 1557-7716

Seaver, L.C. & Imlay, J.A. (2001). Hydrogen peroxide fluxes and compartmentalization inside growing Escherichia coli. *Journal of Bacteriology*, Vol. 183, No 24, pp. 7182-7189, ISSN1098-5530

Spickett, C.M., Pitt, A.R., Morrice, N. & Kolch, W. (2006). Proteomic analysis of phosphorylation, oxidation and nitrosylation in signal transduction. *Biochimica et Biophysica Acta*, Vol. 1764, No 12, PP. 1823-1841, ISSN 0006-3002

Sundaresan, M., Yu, Z.X., Ferrans, V.J., Irani, K. & Finkel, T. (1995). Requirement for generation of H2O2 for platelet-derived growth factor signal transduction. *Science*, Vol. 270, No 5234, pp. 296-299, ISSN 1095-9203

Tabernero, L., Aricescu, A.R., Jones, E.Y. & Szedlacsek, S.E. (2008). Protein tyrosine phosphatases: structure-function relationships. *The FEBS Journal*, Vol. 275, No 5, pp. 867-882, ISSN 1742-4658

Tiganis, T. & Bennett, A.M. (2007). Protein tyrosine phosphatase function: the substrate perspective. *The Biochemical Journal*, Vol. 402, No 1, PP.1-15.

Tonks, N.K. (2005). Redox redux: revisiting PTPs and the control of cell signaling. *Cell*, Vol. 121, No 5, pp. 667-670, ISSN 1097-4172

Tonks, N.K. (2006). Protein tyrosine phosphatases: from genes, to function, to disease. *Nature Reviews Molecular Cell Biology*, Vol. 7, No 11, PP. 833-846, ISSN 1471-0080

Torres, M. (2003). Mitogen-activated protein kinase pathways in redox signaling. *Frontiers in Bioscience*, Vol. 1, No 8, pp. 369-391, ISSN 1093-4715

Woo, H.A., Chae, H.Z., Hwang, S.C., Yang, K.S., Kang, S.W., Kim, K. & Rhee, S.G. (2003). Reversing the inactivation of peroxiredoxins caused by cysteine sulfinic acid formation. *Science*, Vol. 300, No 5619, pp. 653-656, ISSN 1095-9203

Zhao, Z., Larocque, R., Ho, W.T., Fischer, E.H. & Shen, S.H. (1994). Purification and characterization of PTP2C, a widely distributed protein tyrosine phosphatase containing two SH2 domains. *The Journal of Biological Chemistry*, Vol. 269, No 12, pp. 8780-8785, ISSN 1083-351X

9

Feasible Novozym 435-Catalyzed Process to Fatty Acid Methyl Ester Production from Waste Frying Oil: Role of Lipase Inhibition

Laura Azócar, Gustavo Ciudad, Robinson Muñoz,
David Jeison, Claudio Toro and Rodrigo Navia
Scientific and Technological Bioresources Nucleous, La Frontera University
Chile

1. Introduction

Fatty acid methyl ester (FAME) or biodiesel is a biofuel conventionally produced from edible oil and methanol, using an alkaline catalyst, through a transesterification reaction. As FAME is mostly produced from edible vegetable oils, crop soils are used for its production, increasing deforestation and producing a fuel more expensive than diesel. In addition, between 70 and 80% of the total FAME production costs correspond to the vegetable oils. Therefore, the use of waste lipids such as waste frying oils (WFO), waste fats and soapstock has been proposed as low-cost alternative to feedstock. Non-edible oils such as jatropha, pongamia and rubber seed oil are also economically attractive. In addition, microalgae, bacteria, yeast and fungi with 20% or higher lipid content are oleaginous microorganisms known as single cell oil and have been proposed as feedstock for FAME production. Alternative feedstocks are characterized by their elevated acid value due to the high level of free fatty acid (FFA) content, causing undesirable saponification reactions when an alkaline catalyst is used in the transesterification reaction. The production of soap consumes the conventional catalyst, diminishing FAME production yield and simultaneously preventing the effective separation of the produced FAME from the glycerin phase. These problems could be solved using biological catalysts, such as lipases or whole cell catalysts, avoiding soap production since the FFAs are esterified to FAME. In addition, by-product glycerol can be easily recovered and the purification of FAME is simplified using biological catalysts.

Lipase-catalyzed processes have been widely investigated for FAME production from alternatives raw material. Although interesting results have been reached up to date, the enzymatic catalysis has not become competitive compared to the conventional chemical process. The main reasons explaining this issue are the long reaction time (until 48 h), the loss of enzymatic activity due to methanol use in the reaction and the high operational costs because the lipases cannot be reused. The present chapter described an investigation to a get a feasible lipase-catalyzed process to FAME production from WFO, avoiding lipase inhibition.

2. Lipase catalyzed process to FAME production from waste frying oil: Improving the yield

In spite of that some investigations have been carried using WFO instead of edible oils in FAME production using lipases, there is not clear the effect in the process of replace the raw material. Using *Rhizopus oryzae* as the biocatalyst, FFA from a synthetic WFO were esterified to produce FAME with an improved reaction yield (Li et al., 2007). In addition, using *Thermomyces lanuginosus* lipases immobilized on a microporous polymer, 97% FAME content from edible sunflower oil was reached, while only 90.2% FAME content was obtained from WFO (Dizge et al., 2009). Watanabe et al. (2001) tested the immobilized the *Candida antartica* lipase immobilized on acrylic resin (Novozym 435) and obtained a 5.5% reduction in FAME conversion yield when using WFO compared to edible oil as the feedstock. They concluded that the oxidized fatty acid compounds in WFO may be responsible for this decrease.

As the effects of using WFO instead of edible oil in lipase-catalyzed processes are not clear, the aim of this study was to elucidate the effect of WFO incorporation in feedstock mixed with rapeseed oil on FAME production yield using Novozym 435 as the catalyst by means of the response surface methodology (RSM). In addition, specific WFO and rapeseed oil chemical characteristics were investigated to identify the components that were responsible for these results. Finally, a preliminary study to establish the optimal time for methanol addition during the reaction was proposed.

Both filtered WFO collected from restaurants and crude rapeseed oil from a local factory from Southern Chile were used as the feedstocks. Novozym 435 from Sigma-Aldrich was used as the catalyst. Methyl heptadecanoate, 1,2,3-butanetriol and 1,2,3-tricaprinoylglycerol were used as internal standards and were chromatographically pure.

RSM was used to analyze and optimize the interaction effects of four variables on FAME production yield (Table 1): the WFO content in the feedstock mixture (% wt), the final methanol-to-oil ratio (mol/mol), temperature (°C) and Novozym 435 dosage (% wt based on oil weight). A central composite matrix with 5 levels was used and 30 runs were carried out in a random order. Each run was performed in triplicate.

All reactions were incubated in flasks containing 1 mL of oil at 200 rpm. The volume of the flasks was selected to maintain perfect agitation of the samples when using both the highest dosage of catalyst and the lowest methanol-to-oil molar ratio. Under these conditions, different combinations of feedstock mixture, dosage of catalyst, methanol-to-oil molar ratio and temperature were used. Methanol was added in two steps to avoid lipase inhibition (Shimada et al., 2002). In the first step, one-third of the total molar ratio was added according to Table 1, while in the second step, the remaining two-thirds of the total molar ratio was added to generate the final methanol-to-oil molar ratio.

To establish the reaction and second methanol addition times, a preliminary study was performed. This experiment was carried out for 48 h using the RSM central point, with 50% (wt) WFO in a mixed feedstock, a methanol/oil molar ratio of 3:1 and 9% (wt) Novozym 435 at 45°C and stirring 200 rpm.

Samples were immediately stored at 4°C to stop the reaction. The upper layer was analyzed by gas chromatography for FAME quantification.

Independent variables	Symbols	Levels				
		-2	-1	0	1	2
WFO in feedstock [wt %]	X_1	0	25	50	75	100
Methanol to oil ratio [mol/mol]	X_2	1.50	2.25	3.00	3.75	4.50
Temperature [°C]	X_3	35	40	45	50	55
Novozym 435 [wt %]	X_4	3	6	9	12	15

Table 1. Variable and levels used in the response surface methodology

The experimental data obtained were fitted to a second-order polynomial equation. This equation describes the relationship between the predicted response variable (FAME production yield) and the independent variables (Table 1). The polynomial model for FAME yield may be written as follows (Eq. 1):

$$\text{FAME yield}=\beta_0+\sum_{i=1}^{4}\beta_i X_i+\sum_{i=1}^{4}\beta_{ii}X_i^2+\sum\sum_{i<j=1}^{4}\beta_{ij}X_iX_j \quad (1)$$

Where β (0 = intercept, i = linear, ii = quadratic and ij = interaction) and Xi, Xj (i = 1, 4; j = 1, 4; i ≠ j represent the coded independent variables) are the model coefficients. With the fitted quadratic polynomial equation, contour plots were developed to analyze the interaction between terms and their effects on FAME production yield.

To identify and quantify the fatty acids in the feedstock a Clarus 600 chromatograph coupled to a Clarus 500T mass spectrometer of Perkin Elmer (GC-MS) was utilized. An Elite-5ms capillary column with a length of 30 m, thickness of 0.1 μm and internal diameter of 0.25 mm was used. The vials were prepared by adding 3 μg of sample to 100 μL methyl heptadecanoate as an internal standard (initial concentration of 1300 mg/L). The following temperature program was used: 50°C for 1 min and then increasing temperature at a rate of 1.1°C/min up to 187°C. Both the injector and detector temperatures were 250°C and He was used as the carrier gas. Before injection into the GC-MS equipment, the WFO and rapeseed oil were methylated according to Araújo (1995). Specific gravity was measured at 20°C using a manual densimeter. Kinematic viscosity was measured at 40°C using a capillary viscosimeter. The acid value was determined by titration with KOH using phenolphthalein as an indicator. The peroxide value was determined by titration with $Na_2S_2O_3$, and the iodine value was determined by the Wijs method (Araújo 1995). The following parameters were measured according to ASTM Standard Methods: water and sediments (ASTM Standard D 1976-97), pour point (ASTM Standard D 97-04), sulfur (ASTM Standard D 7039-04) and flash point (ASTM Standard D 56-02a).

To establish the degree of oil conversion, monoacylglycerols (MG), diacylglycerols (DG), triacylglycerides (TG) and FFA in the oil feedstock were quantified using an HP 6890 series gas chromatography system (GC-MS) with an adaptation of the EN-14214 (The)methodology. TG was determined by mass balance. A 50-m long BPX-5 column with a thickness of 0.5 μm and an internal diameter of 0.32 mm was used. The vials were prepared by combining 10 mg of sample with 0.8 μL of 1,2,3-butanetriol and 10 μL of 1,2,3-tricaprinoylglycerol dissolved in pyridine as internal standards. N-methyl-Ntrimethylsilyltrifluoroacetamide (MSTFA) was added to the vials, which were then shaken and incubated for 15 min to transform the sample into more volatile siliade components. Subsequently, the preparation was dissolved in 0.8 mL of n-heptane. The following temperature program was used: 15°C for 1 min and three consecutive ramps of 15°C/min to 180°C, 7°C/min to 230°C and 10°C/min to 320°C and 320°C for 15 min. The detector temperature was 380°C and a split rate 10 was used for injection of 1 µL of sample.

Methanol addition. Methanol solubility is less than 1.5:1 methanol/oil (mol/mol), but a molar ratio of 3:1 methanol/oil is necessary to complete the transesterification reaction for FAME production (Shimada et al., 2002). However, very high initial concentrations of methanol could lead to the inhibition of the lipase used in this study due to its low solubility in oil (Du et al., 2004; Shimada et al., 2002). Therefore, the experiment was designed so that the methanol was added in two steps (Fig. 1).

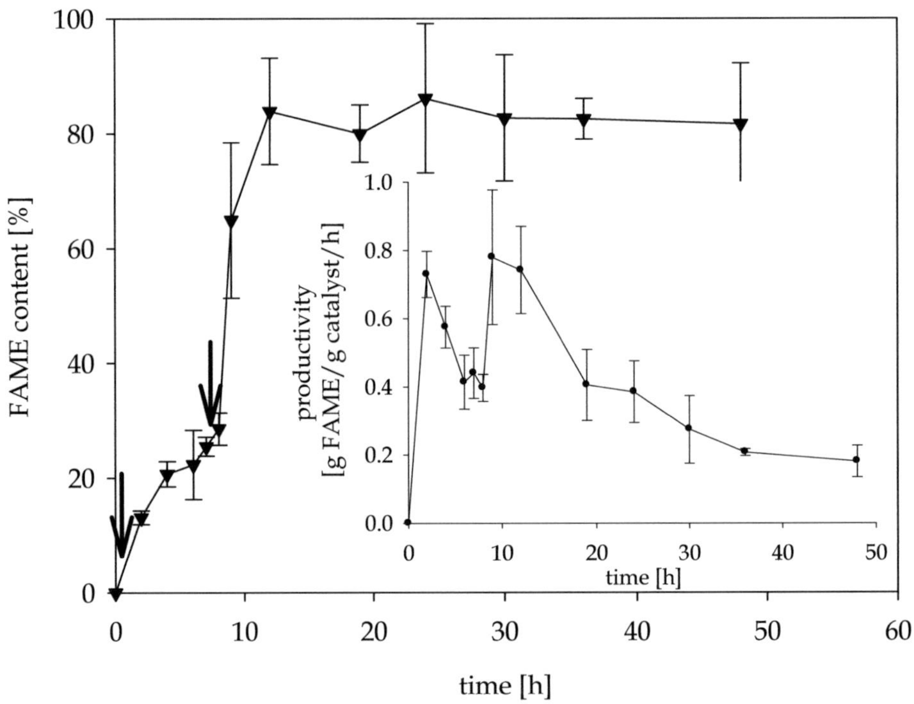

Fig. 1. FAME content and productivity during the reaction time. Operational conditions: 50 wt% WFO in the mixed feedstock, methanol to oil molar ratio of 3:1, 9 wt % Novozym 435 and 45°C, at 200 rpm. Arrows indicate methanol addition.

The experiment was started with an initial concentration of one-third of the necessary moles of methanol. When about one-third of the oil was converted to FAME after 8 hours, the remaining two-thirds of the methanol were added (Fig. 1). The addition of the higher concentration of methanol in the second step was determined based on Shimada et al. (2002), who established that methanol is more soluble in FAME than in TG.

This two-step technique of methanol addition was shown to diminish the reaction time by obviously preventing enzyme inhibition. Li et al. (2009) showed that in spite of the high yield reached using stepwise methanol addition, both the reaction time and rate of FAME conversion were slower. In experiments using a three-step methanol addition and 4% (wt) Novozym 435 as the catalyst, high FAME production yields were reached but only after 50 hours of reaction (Du et al., 2004; Watanabe et al., 2001). Similarly, in another recent study, an reaction time of 50 hours was also used (Ognjanovic et al., 2009). Usually, the second methanol addition starts after one-third (33.3%) of the FAME content has been generated and the initially added methanol has been completely consumed (Du et al., 2004; Shimada et al., 2002). This methodology produces a stationary phase of FAME production yield, diminishing the reaction productivity.

As Novozym 435 was shown to be a robust and stable catalyst compared to other lipases in the presence of short chain alcohols (Hernandez-Martin et al., 2008), all of the experiments in this study were performed using this commercially available immobilized enzyme. When approximately 28.6% of the FAME content was reached after 8 hours of reaction, the second aliquot of methanol was added. This stepwise addition of methanol increased the reaction productivity by avoiding the stationary production phase and led to a plateau in FAME production at 12 hours. Similar reductions in the reaction time by shortening the stationary phase of FAME production have already been reported (Li et al., 2009). Therefore, in the RSM experiments, the first amount of methanol (one-third of the molar ratio shown in Table 1) was added at the start of the reaction, while the second aliquot of methanol (two-thirds of the molar ratio shown in Table 1) was added after 8 hours, and the total reaction time was 12 hours.

Optimization of methanolysis conditions. Four variables previously shown to have significant effects on FAME production yield were investigated. The methanol-to-oil ratio, temperature and Novozym 435 dosage were established based on previous studies in the literature (Du et al., 2004; Fukuda et al., 2001; Kose et al., 2002). According to previous experiments (data not shown), the effect of the WFO content in the feedstock mixture was also tested in the lipase-catalyzed process. The experimental central composite design matrix is presented in Table 1.

To understand and optimize the relationship between the tested variables, the obtained experimental data were analyzed by second-order polynomial equations by means of the RSM. The analysis of variance (ANOVA) of the quadratic polynomial model showed low p-values (0.125) and both high determination coefficients (R^2=95.3%) and high adjustment of the determination coefficients (Adj. R^2=94.5%). The low p-value obtained indicates that the model accurately represented the relationship between response and the variables. The R^2 value obtained indicates that the variation in FAME production yield correlated with 95.3% of the independent variables and the obtained Adj. R^2 value indicates a 94.5% correlation between the independent variables. The lack of fit refers to the fact that a simple linear regression model may not adequately fit the experimental data. The obtained p-value not indicate a significant lack of fit and, therefore, gave no reason to evaluate a more complex model.

The p-values obtained from the regression analysis results showed that all of the coefficients of the linear, interaction and quadratic terms had a significant effect on FAME production yield. All of the linear and quadratic terms, as well as the interaction terms X_1X_3, X_2X_4 and X_3X_4, were significant at the 1% level. The interaction terms X_1X_2, X_1X_4 and X_2X_3 were significant at the 5% level. Therefore, the final response model equation in terms of the variable factors can be written as follows (Eq. 2):

$$\begin{aligned} FAME_{yield} = &-896.52 + 1.90 \cdot X_1 + 91.40 \cdot X_2 + 28.06 \cdot X_3 + 20.00 \cdot X_4 - 0.01 \cdot X_1^2 \\ &- 14.46 \cdot X_2^2 - 0.24 \cdot X_3^2 - 0.85 \cdot X_4^2 - 0.12 \cdot X_1 \cdot X_2 - 0.02 \cdot X_1 \cdot X_3 \\ &- 0.03 \cdot X_1 \cdot X_4 - 0.59 \cdot X_2 \cdot X_3 + 4.13 \cdot X_2 \cdot X_4 - 0.26 \cdot X_3 \cdot X_4 \end{aligned} \tag{2}$$

The main factors that affect the FAME production yield were the linear terms X_1, X_2, X_3 and X_4, the quadratic term X_2^2 and the interaction term X_2X_4. The linear terms with positive coefficients indicate an increase in FAME production yield. As X_1 is a positive term, it is possible to establish that WFO incorporation could increase FAME production yield.

To validate the model obtained, an experimental point was compared to the model prediction (Eq. 2), and an error of less than 8% was obtained. In addition, the optimal methanolysis conditions were obtained through the regression model (Eq. 2) according to the limit criterion of the maximum response of FAME production yield. The obtained optimal conditions were 100% (wt) WFO, a methanol-to-oil ratio of 3.8:1 (mol/mol), 15% (wt) Novozym 435 and 44.5°C, which generated 100% FAME production yield. These predicted conditions show that WFO incorporation increases FAME production yield.

Moreover, using this model, it was also possible to predict several optimal conditions to reach the highest FAME production yields, while simultaneously reducing production costs and enzyme use. For instance, 96% FAME production yield could be reached using 75% (wt) WFO in the mixed feedstock, with a methanol-to-oil ratio of 3.75:1 (mol/mol), 12% (wt) Novozym 435 and 40°C.

To illustrate that several optimal combinations are able to produce the highest FAME production yields, contour plots were generated. Figure 2 and 3 show the effect of variable interaction on FAME production yield in contour plots predicted by the model. The contour plots were generated to show the effect of two variables on the response, while the other two independent variables were held constant.

Figure 2 shows the interaction between two independent variables (WFO in the mixed feedstock and Novozym 435 dosage) and their effects on the response variable FAME production yield, while the other two variables were held at zero. At these conditions, an increase in Novozym 435 dosage led to an enhancement in FAME production yield. This change is additionally increased when WFO is incorporated at up to 80% in the mixed feedstock. WFO seems to be a more available substrate compared to rapeseed oil for Novozym 435 catalysis under the conditions investigated.

FAME production yield was more sensitive to Novozym 435 dosage, since high doses were necessary to obtain high FAME production yields. In this sense, in order to decrease the production costs for future economically sound industrial applications, further experiments should be conducted to investigate the effect of WFO incorporation using another inexpensive biologic catalyst instead of Novozym 435. In addition, because Novozym 435 has shown high residual activity in successive applications (Hernandez-Martin et al., 2008), further experiments with a subsequent recovery protocol and reuse of the immobilized catalyst should be conducted under the optimal conditions established in this work.

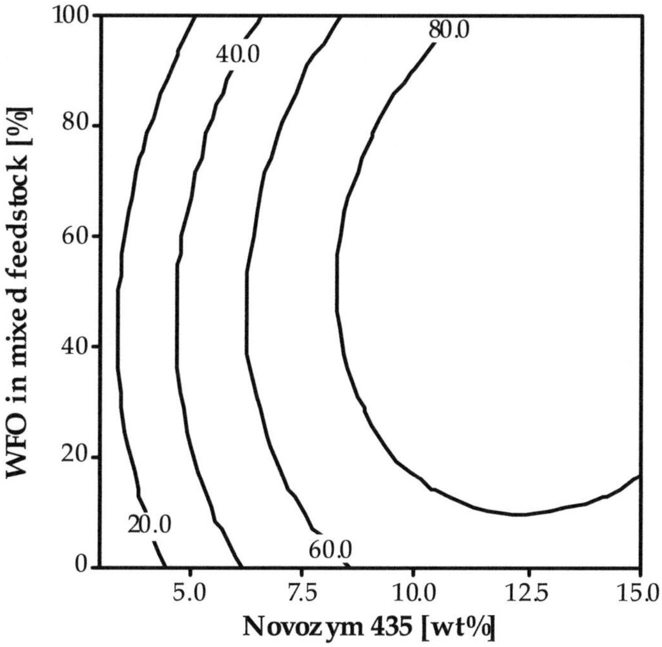

Fig. 2. Contour plot of FAME production yield predicted from the model at 45°C and molar methanol to oil ratio 3:1 mol/mol.

Recently, Issariyakul et al. (2008) investigated the use of mixed WFO and rapeseed oil using an alkaline catalyst. However, lower ester production yields were observed when WFO was incorporated in the reaction mixtures. These results agreed with Fukuda et al. (2001), who established that the FFA in WFO can be completely converted to FAME using lipases as catalysts, whereas soap is produced when an alkaline catalyst is used, which diminishes the production yield. Although the cost of lipases is significantly higher than that of alkaline catalysts, their use may partially solve the main drawbacks of the conventional biodiesel production process: the use of pure vegetable oils (which account for around 80% of the cost of the total process) and the competition with food products (Gui et al., 2008). We have shown that the incorporation of WFO in the feedstock to partially replace rapeseed oil in processes catalyzed by Novozym 435 may diminish the production costs while simultaneously increasing the production yield, which makes it a potential alternative method for FAME production on an industrial scale.

FAME production yield was more sensitive to both the methanol-to-oil ratio and Novozym 435 dosage, compared to temperature (Fig. 3). FAME production yield was enhanced when the temperature increased to 45-50°C; however, the opposite tendency was observed for temperatures higher than 50°C. These results agree with the results obtained by Kose et al. (2002), who established that at about 50°C, FAME production yield decreases as a result of enzyme deactivation at high temperatures (Fig. 3A). An enhancement in FAME production yield was observed when both the methanol-to-oil ratio and Novozym 435 dosage increased (Fig. 3B). When using a high immobilized biocatalyst load, large amounts of alcohol were needed to provide sufficient liquid to maintain a uniform suspension of the biocatalyst (Hernandez-Martin et al., 2008). In addition, as Novozym 435 is a robust biocatalyst in the presence of short chain alcohols, the excess alcohol kept the glycerol in solution, which prevented deactivation of Novozym 435 by glycerol blockage of the entrance to catalyst pores (Hernandez-Martin et al., 2008). Therefore, the levels of Novozym 435 used and the stepwise methanol addition seem to avoid the possible diffusion limitations, which resulted in high production yields.

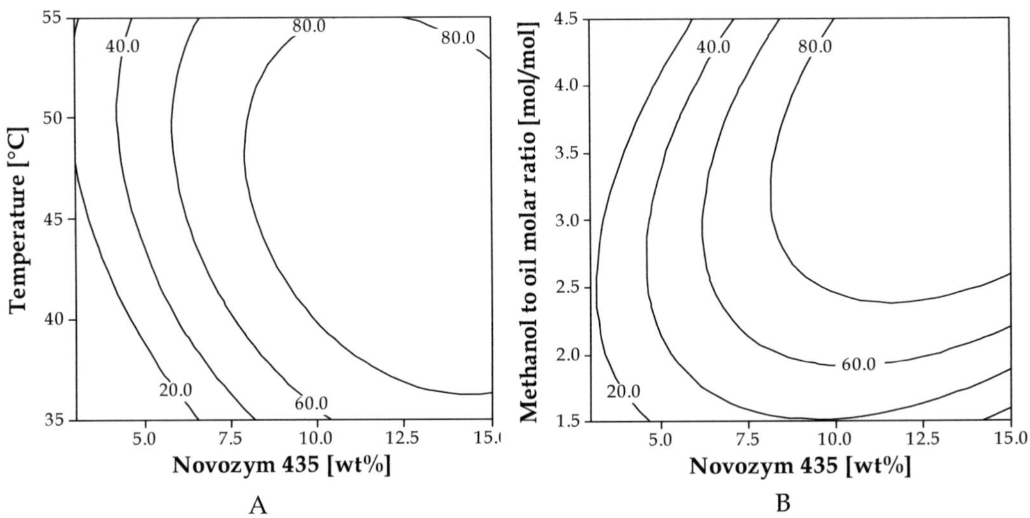

Fig. 3. Contour plots of FAME production yield predicted from the model for 50% (wt) WFO at (A) a methanol-to-oil ratio of 3:1 (mol/mol) and (B) 45°C.

Properties of feedstock that affect methanolysis optimization. The feedstock characteristics were investigated to determine the components responsible for the results obtained in the RSM, particularly the increase in the FAME yield when WFO was incorporated in the process (Table 2).

Differences were found in most of the physical properties of the oil feedstock measured (Table 2). The food frying process produced an increase in both acid and peroxide values from WFO due to the hydrolysis and oxidation reactions, respectively (Araújo 1995). The high peroxide value of the WFO could positively affect the FAME characteristics by increasing the oxygen content, which enhances its burning efficiency (Lin et al., 2007). However, a higher oxygen content in oils may also promote a higher nitrogen oxide concentrations during the burning process (Lin et al., 2007).

The analyzed WFO was originally an edible soybean oil and sunflower oil mixture, which is characterized by a typical iodine value of about 130 g I_2/100 g oil. As the iodine value is diminished by polymerization reactions during the frying process, a similar iodine value compared to rapeseed oil was found for WFO (Table 2). A lower iodine value increases the pour point and may also improve the oxidative stability of the oil (Canakci 2007). However, a high pour point can negatively affect FAME performance in an engine, when used at low temperatures. The kinematic viscosity of WFO was also measured and was slightly higher compared to rapeseed oil, probably as a result of polymerization reactions. As WFO is refined oil, the sulfur content was lower in WFO compared to rapeseed oil, which is an unrefined oil. In addition, a higher FFA content in WFO compared to rapeseed oil was observed, which may enhance FAME production yield using a biological catalyst, such as Novozym 435, through direct FFA esterification (Li et al., 2007).

Physicochemical properties	Feedstock	
	WFO	Rapeseed oil
Specific gravity [kg/m^3 at 20°C]	925	927
Kinematic viscosity at 40°C [cst]	37	34
Acid value [mg KOH/g]	4.5	0.8
Iodine value [g I_2/100 g oil]	111	105
Peroxide value [meq/kg]	15.2	4.1
Water and sediments [%v/v]	1.3	< 0.01
S [mg/L]	2.0	7.0
Pour point [°C]	-6.7	-23.3
Flash point [°C]	>100	>100
Fatty acid composition [wt%]		
Palmitic [C16:0]	12.2	4.2
Stearic [C18:0]	4.1	1.3
Oleic [C18:1]	28.5	66.6
Linoleic [C18:2]	49.5	18.7
Linolenic [C18:3]	3.5	7.7
Arachidic [C20:0]	0.2	0.4
Eicosenic [C20:1]	0.1	1.0
Eicosapentaenoic [C20:5]	0.2	0.1
Erucic [C22:1]	0.0	0.1
Others	1.7	0.0
Acylglycerols and FFA composition [wt%]		
Monoacylglycerol	0.1	0.0
Diacylglycerol	3.3	0.5
Triacylglycerol	93.5	98.7
Free fatty acid	3.1	0.8

Table 2. Properties and composition of oil feedstocks (samples were filtered before to the analysis).

The FFA esterification reaction produces 1 mol of water per 1 mol of FAME produced. In this work, water and sediments were quantified and low levels were found in both rapeseed oil and WFO, with a slightly higher water content for WFO (Table 2). However, as WFO has a high FFA content, incorporation of WFO in the mixed feedstock may increase the water content through FFA esterification reactions. Although lipases need an optimal amount of water to maintain their activity in organic media, a recent work established that Novozym 435 itself appears to contain sufficient water to preserve its catalytic activity without requiring the presence of an oil/water interface (Ognjanovic et al., 2009). It has also been shown that Novozym 435 needs a nearly anhydrous reaction medium to be effective (Salis et al., 2005). According to these previous studies, when WFO was incorporated in a mixed feedstock, the reaction yield diminished as a result of the high water content. Therefore, an improvement in FAME production yield when WFO is incorporated into a feedstock should not be produced due to the presence of water in the transesterification reaction.

Differences between the fatty acid compositions of WFO and rapeseed oil were also detected (Table 2). In order to analyze the role of fatty acids in FAME production, the input and output fatty acid compositions were compared. However, no significant differences between input and output fatty acid composition were detected. (data not shown). Products from oil conversion, i.e., MG, DG, TG and FFA, in the used feedstock were compared using GC-MS. MG, DG and FFA account for about 6.5% of the WFO composition, compared to only 1.3% in the case of rapeseed oil (Table 2). Turkan & Kalay (2006) established that in Novozym 435-catalyzed transesterification, the first step (i.e., conversion of TG to DG) is the rate-determining step, since TG is converted to FAME without significant accumulation of MG and DG. This fact indicates that MG, DG and FFA can be more easily converted to FAME compared to TG. Therefore, WFO is a more available substrate in the process investigated, compared to rapeseed oil, and incorporation of WFO in feedstock increased the FAME production yield.

Finally, is possible to establish that according to the RSM, the optimized predicted combination of conditions to obtain 100% FAME production yield was the use of 100% (wt) WFO, a methanol-to-oil ratio of 3.8:1 (mol/mol), 15% (wt) Novozym 435 and 44.5°C at 200 rpm. Methanol addition in two steps was previously optimized. An earlier second addition of methanol after 8 hours of reaction time was an effective technique to decrease the total reaction time to 12 hours, while preventing enzyme inhibition. The model obtained from the RSM predicted that several optimal conditions were able to reach the highest FAME production yield. According to this model, the addition of WFO increased the FAME production yield, and this effect was mainly attributed to the higher contents of MG, DG and FFA of WFO compared to rapeseed oil, which are more available substrates for enzymatic catalysis. Therefore, a partial replacement of rapeseed oil by WFO in processes catalyzed by Novozym 435 may diminish the cost of biodiesel production by using a less expensive feedstock that simultaneously increases the production yield.

3. Enzymatic biodiesel production in an anhydrous medium with lipase reutilization

A major problem in lipase-catalyzed processes to FAME production is the high acidity of the final product, mainly caused by water presence. In fact, water favors the hydrolysis reaction producing FAME with a high acid value. Therefore, to accomplish with international biofuel standards, additional treatment are needed (Fukuda et al., 2008). In addition, this problem could be enhanced by using alternative feedstock characterized by high acid value, such WFO, jatropha, microalgae and crude palm oil (Azócar et al., 2010b).

To solve these drawbacks the addition of water adsorbents to the lipase-catalyzed reaction has been recently investigated (Li et al., 2009; Wang et al., 2006). Some reported absorbents are blue silica gel, maceo-pored silica gel, fine-pored silica gel, and molecular sieves of 3 Å, 4 Å and 5 Å (Li et al., 2009; Wang et al., 2006). Among the adsorbents examined, blue silica gel has been reported as the best one, increasing FAME yield when immobilized lipase from *Penicillium expansum* was used as catalyst (Li et al., 2009). However, high dosage of silica gel could provoke low biodiesel yield. In fact, as the pore size of silica gel is much larger than the methanol molecule size, silica adsorbs methanol negatively affecting the reaction performance (Wang et al., 2006).

The possible advantages of using water adsorbents in lipase-catalyzed processes to improve FAME production yield may be in contradiction with reports showing that lipases activation may need water presence. In fact, lipases activation involves the active site restructuration, which requires the presence of an oil-water interface (Yu et al., 2010). As a result, small dosages of water in the reaction have been proposed to achieve an effective process (Yu et al., 2010). Novozym 435 has been shown to contain enough water in its support medium to preserve its catalytic activity in an anhydrous medium (Ognjanovic et al., 2009; Tamalampudi et al., 2008). In a response surface methodology study, Organovic et al. (2009) correlated Novozym 435 concentration with water concentration, achieving the highest biodiesel yield of 92% using the medium with the lowest water content level (0%) and the highest enzyme concentration (5% based on oil weight). In another study, Novozym 435 showed low activity in the presence of water and an anhydrous medium for efficient catalyzing was proposed (Tamalampudi et al., 2008).

Novozym 435 has been widely reported as an effective biocatalyst for promoting high FAME production yields (Azócar et al., 2010b). Optimal conditions to reach 100% FAME yield using WFO as feedstock were determined to be the following: Methanol-to-oil molar ratio 3.8:1 (wt), 15% (wt) Novozym 435 and incubation at 44.5°C for 12 h with agitation at 200 rpm (Azócar et al., 2010a). Novozym 435 activity loss and regeneration have been also investigated. It has been established that the immobilization material (acrylic resin) could adsorb polar components such as methanol and glycerol, provoking enzyme inactivation. Therefore, Novozym 435 reutilization by means of washing processes using acetone, soybean oil, *tert*-butanol, isopropanol and 2-butanol has been investigated (Chen et al., 2003; Samukawa et al., 2000). In addition, the incorporation of a co-solvent in the reaction has been recently proposed (Royon et al., 2007; Yu et al., 2010).

According to previous research, Novozym 435 has been proved to be an effective catalyst in FAME production and has been shown to preserve its catalytic activity in an anhydrous medium. However, Novozym 435 has not been investigated yet regarding catalyzed-processes when adsorbent materials are added to the reaction in order to obtain high quality biodiesel with low acid value. In this sense, it is necessary to evaluate the behavior of the enzymatic reaction in anhydrous medium. In addition, it is necessary to determine the effect of water absence in the enzyme activity during consecutive reactions and to evaluate alternatives for enzyme recovery during successive reactions.

The aim of this work was to test an anhydrous medium in Novozym 435 catalyzed-process to produce FAME with a low acid value. In addition, the behavior of Novozym 435 in the anhydrous medium was discussed and enzyme recovery alternatives were proposed.

Filtered WFO collected from restaurants and crude rapeseed oil from a local factory in Southern Chile were used as feedstock. The WFO was characterized by an acid value of 5.61 mg KOH/g oil, 2.8% FFA, 1.3% (v/v) water content, 0.1% (wt) MG, 3.3% (wt) DG and 925 Kg/m^3 of specific gravity. The rapeseed oil was characterized by an acid value of 0.8 mg KOH/g oil, 1.6% FFA and 927 Kg/m^3 of specific gravity. Novozym 435 donated by Novo Industries (Denmark) was used as catalyst. 3 Å molecular sieves (Sigma-Aldrich) were used to generate the anhydrous medium. Chromatographically pure methyl heptadecanoate, 1,2,3-butanetriol and 1,2,3-tricaprinoylglycerol were used as internal standards.

FAME production using Novozym 435 in an anhydrous medium was studied by adding 3 Å molecular sieves to the reaction. All reactions were carried out in flasks containing 1 mL of WFO (0.925 g) and 15% Novozym 435 (% wt based on oil weight) as the biocatalyst. A methanol-to-oil molar ratio of 4:1 was used. Methanol was added in two steps in order to avoid lipase inhibition (according to previous experiments in section 2). The flasks were incubated in a shaker at 35 °C and stirred at 200 rpm. Each flask was managed as a destructive sample at 1, 2, 3, 4, 5, 6, 7, 8, 9, 10, 11, and 12 hours of reaction time and each condition was carried out in triplicate. Controls were carried out under the same conditions but without molecular sieves.

To calculate the dosage of molecular sieves, first the maximal theoretical water content in the reaction was estimated. This value was established by taking the initial water content in the WFO and the amount of water produced during the reaction by esterification of FFA contained in the WFO. The water produced by esterification of FFA was estimated by using the mass balance, considering that 1 mol of esterified FFA produces 1 mol of water. With an FFA content of 2.8% and water content of 1.3% in the WFO described above, a total water content of 4.1% was used to estimate the molecular sieve dosage in the reaction, as shown in Eq. 3.

$$Molecular\ sieves = \frac{V\ oil \bullet \rho\ oil \bullet H_2O}{WAC} \approx \frac{1 \bullet 0.925 \bullet 4.1}{20} = 0.19\ g \qquad (3)$$

Where V_{oil} (mL) is the volume of oil added, ρ_{oil} (g/mL) is the density of the oil, H_2O is the total water content during the reaction (water content in the oil more water produced by esterification of FFA) (% v/v) and WAC (%) is the water absorption capacity of the sieves. At the end of the reaction period, the samples were stored at 4 °C to stop the reaction. The samples were centrifuged and the upper layer was extracted for the analysis of FAME yield and acid value.

Different treatments to recover enzyme activity for further reutilization were investigated. Prior to the enzyme recovery treatments, reactions of FAME production were carried out in flasks which were incubated in a shaker under the same operating conditions. The operational conditions were the following: Methanol-to-oil ratio of 3.8:1 (mol/mol); 15% Novozym 435 (% wt based on oil weight); 44.5 °C; stirring at 200 rpm; and 12 hours of reaction time. Methanol was added in two steps as described previously.

For each reaction, 8 mL of WFO (7.4 g) with an acid value of 5.61 mg KOH/g oil were used. To maintain an anhydrous medium, 0.95 g of 3 Å molecular sieves were added to each reaction flask, which were calculated according to Eq. (3). Each reaction was carried out in triplicate. Samples of 70 μL were taken consecutively at 3, 6, 9 and 12 hours of reaction time. The samples were centrifuged and the upper layer was extracted to analyze the FAME yield.

For enzyme recovery, the molecular sieves were separated with a strainer after the transesterification reaction. A paper filter over a vacuum system was placed underneath the strainer containing the molecular sieves to retain the enzymes. The product of the reaction was passed through the two filters: The molecular sieves captured in the strainer were eliminated, whereas the enzymes captured in the paper filter were placed in a new flask for the recovery treatment.

Three treatments to reutilize the enzymes were performed according to the reference studies (Chen et al., 2003; Samukawa et al., 2000; Yu et al., 2010) and a previous experiments (data not shown). A control experiment reusing the enzymes without treatment was also carried out. The treatments were the following: (A) Acetone washing: The enzymes were washed by adding a small dosage of acetone to the flask containing the enzymes. The flask was shaken and the liquid residue was eliminated. This was repeated successively until the liquid was clear. The total volume used in this process was about 10 mL of acetone per gram of enzyme. Subsequently, a wash with 10 mL of WFO per gram of enzyme was carried out to eliminate residual acetone in the enzymes. (B) Waste frying oil washing: The enzymes were washed in a sufficient quantity of WFO to maintain the enzymes submerged in the flask. The flask was shaken and the liquid residue was eliminated. The washing was repeated 3 times. (C) *tert*-Butanol washing: The washing was carried out by adding a specific dosage of *tert*-butanol to the flask containing the enzymes. Then, the mixture was shaken and the residual liquid was eliminated. This sequence was repeated consecutively until the liquid was clear. The total volume used in this process was about 10 mL of *tert*-butanol per gram of enzyme. After this, a wash with about 10 mL of WFO per gram of enzyme was carried out to eliminate residual *tert*-butanol in the enzymes.

After each washing step, a known dosage of WFO was added to the flasks containing the enzymes, which were incubated in oil overnight at ambient temperature. After about 10 hours, both methanol and 3 Å molecular sieves were added to carry out a new FAME production reaction in the same incubation medium. The control was carried out by transferring the enzymes directly to a new reaction for FAME production. Cycles that consisted of both a reaction for FAME production and a treatment for enzyme recovery were repeated 4 times.

As an alternative to enzyme reuse for FAME production using an anhydrous medium, the addition of *tert*-butanol as a co-solvent in the reaction was investigated. To carry out the experiments, consecutive cycles of enzymatic FAME production using the same enzymes (reutilized) were carried out in an anhydrous medium and in a *tert*-butanol system. Each experiment was carried out in flasks by adding 5 mL of WFO (4.6 g), 15% of Novozym 435 (% wt based on oil weight) and a methanol-to-oil ratio of 3.8:1 (mol/mol). Methanol was added to the flask in two steps, similar to the previous experiment. In addition, both the selected dosage of *tert*-butanol and an amount of 3 Å molecular sieves (estimated according to Eq. 1) were added to the flasks. The *tert*-butanol dosage was established according to previous experiments at 0.75% (V/V) (data not show). Under these conditions, the flasks were incubated at 44.5 °C and 200 rpm during 12 hours of reaction time. All the experiments were carried out in triplicate and samples were taken throughout the reaction time. The upper layer of the centrifuged samples was analyzed to determine the FAME yield, according to the analytical procedures.

The acid value and FFA content in both the feedstock and throughout the reaction time were determined by titration with a KOH solution of known concentration and using phenolphthalein as indicator. Water and sediments were measured according to ASTM Standard Method D 1976-97(American). FAME, MG and DG in feedstock and reaction products were quantified according to methodology previously described in section 2.

Biocatalysis in an anhydrous medium. In order to reduce the acid value of the produced FAME, the alternative of carrying out FAME production reaction in an anhydrous medium was studied. Adsorbent materials such as blue silica gel and molecular sieves of different sizes have been shown to be effective in water removal during biodiesel production (Li et al., 2009). However, in this study, blue silica gel was discarded due to results from previous experiments showing possible methanol adsorption, negatively interfering with the reaction (data not shown). The adsorption of methanol could also occur when molecular sieves with large pore size are used. Thus, 3 Å molecular sieves were chosen to produce an anhydrous medium through water absorption.

The acid value and FAME yield results obtained using an anhydrous medium and a control run are shown in Fig. 4. The acid value obtained from the reaction using the anhydrous medium was maintained in about 1 mg KOH/g oil throughout the entire reaction time, whereas the control showed a value higher than 3 mg KOH/g oil by the end of the reaction (Fig. 4A). The values obtained using the anhydrous medium are close to those established by the biodiesel norm. As FFA present in WFO were esterified producing water and FAME at the start to the reaction (Fig. 4A), water was removed from the reaction by adsorption onto the molecular sieves, avoiding hydrolysis reactions and further FFA production. Therefore, the biocatalysis in an anhydrous medium produces a higher quality product compared to a typical standard reaction using Novozym 435.

FAME yield was also analyzed in an anhydrous medium (Fig. 4B). The results obtained show that water removal during FAME production generated a significant increase in FAME yield using Novozym 435 as the catalyst. These results are in agreement with those obtained by Li et al. (2007) who established that using 3 Å molecular sieves as adsorbent to remove excessive water could significantly increase the FAME yield when 100% FFA is used as raw material for biodiesel production. In this study, WFO containing only 2.8% FFA and 1.3% water was employed as the raw material, but FAME yield increased drastically. In addition, the control run only using molecular sieves indicates that this material did not catalyze the reaction (data not shown). Therefore, the increment in FAME yield is only related to the anhydrous medium. This corroborates the results obtained by Tamalampudi et al. (2008) who suggest that Novozym 435 transesterification activity is inhibited by the presence of added water and that it needs a nearly anhydrous medium for an efficient performance.

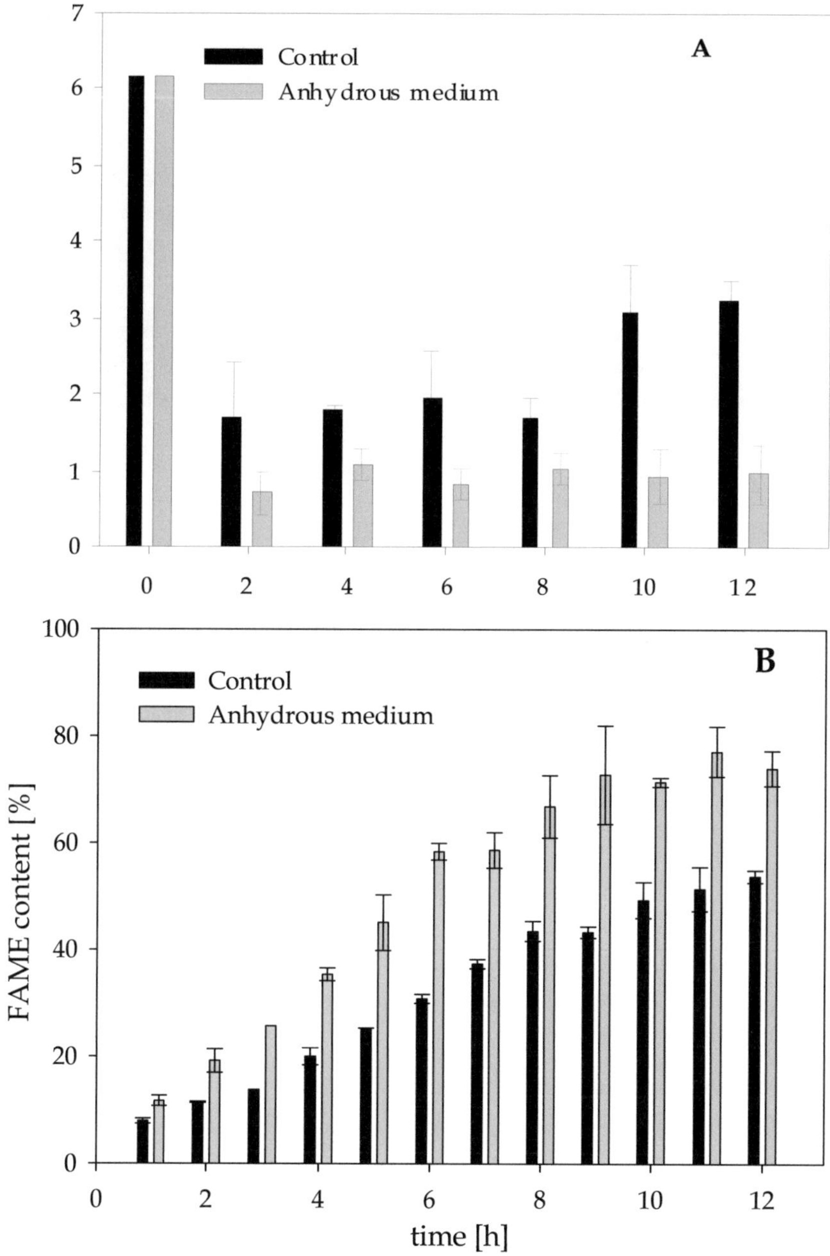

Fig. 4 Acid values (A) and FAME yield (B) during the reaction in an anhydrous medium with a methanol-to-oil ratio of 4:1 (mol/mol), 15% (wt) Novozym 435 (based on the weight of oil), 35 °C, 200 rpm and 12 hours of reaction time.

A reasonable explanation for the high FAME yield obtained may be related to the work of Cabrera et al. (2009). They suggest that lipase may exist in two different structural forms; in one form, the site of the lipase is isolated from the medium whereas in the other form, the active site is exposed to the reaction medium. In a homogeneous aqueous medium the lipase is in equilibrium between these two structures. In the case of Novozym 435, it is immobilized in acrylic resin with hydrophilic properties. Therefore, the higher yield in the reaction may occur because of interactions with the hydrophobic medium, where the lipase shifts towards the open structure form, increasing its activity. These conformational changes enable lipases to be greatly altered by controlled immobilization of the support material properties. This is in accordance with the results obtained by Samukawa et al. (2000) who established that preincubation of the enzyme in oil prior to the reaction improves the yield. This is because water adsorbed during the reaction prevented oil penetration, whereas this did not occur with the preincubated enzyme.

Treatments for enzyme reutilization in an anhydrous medium. The main advantage of using immobilized lipases is that the enzyme can be used repeatedly in a semi-continuous process. However, this objective is not always reached under the optimized conditions as short chain acyl acceptors can produce a loss of enzyme activity in successive reactions (Ognjanovic et al., 2009). In this study, the stability of the enzyme and treatments for its reutilization in an anhydrous medium were examined. Fig. 5 shows the results obtained by different treatments to maintain lipase activity after each FAME production reaction in an anhydrous medium. Fig. 5A shows the reutilization of Novozym 435 without treatment after reaction. According to the results obtained, FAME yield decreased considerably in the second cycle. This tendency was maintained in successive cycles, with values close to 0% of FAME yield in the fourth cycle. These results are in accordance with the findings of Ognjanovic et al. (2009), who reported 0% FAME yield in the fourth cycle using Novozym 435 and methanol as the acyl acceptor in an organic medium. Thus, although the activity of the enzyme increased in the first cycle using an anhydrous medium compared to the control (Fig. 4), the loss of activity seems to be similar in both types of medium when successive cycles are carried out (Fig. 5A). Lipases have shown high synthesis activity and stability in hydrophobic solvents, but alcohol and glycerol are immiscible in those solvents (Halim et al., 2008). This situation could produce poor solubilization of polar compounds in the medium, leading to their adsorption onto the lipases hydrophilic support and therefore provoking a low transesterification rate (Halim et al., 2008). So far, an alternative process is needed to allow the reuse of the enzyme achieving a low-cost process which can feasibly be implemented at an industrial scale.

The results of the first treatment studied for reusing the enzyme are shown in Fig. 5B. Applying an acetone washing step a higher FAME yield was achieved in the second cycle compared to the control (Fig. 5B and 5A, respectively). This result is caused by the fact that acetone is a hydrophilic solvent that could remove the glycerol and the methanol adsorbed onto the hydrophilic support material of the enzyme. However, in the third cycle the FAME yield diminished drastically (Fig. 5B). Acetone is a hydrophilic solvent with a very low log P (partition coefficient in a standard octanol-water two-phase system) value of -0.24. According to Yu et al. (2010) water has higher affinity to these hydrophilic solvents rather than to the enzyme. It has been reported that Novozym 435 can contain a sufficient quantity of water to preserve its catalytic activity in an anhydrous medium (Ognjanovic et al., 2009; Tamalampudi et al., 2008). Therefore, the enzyme activity may be reduced when using an acetone washing step, because the enzyme might lose its flexibility conformation due to the lack of bound water (Yu et al., 2010). The enzyme washing step using WFO followed by overnight incubation in WFO was also studied (Fig. 5C). The activity of Novozym 435 was higher in the third cycle in comparison to the two previous alternatives. Samukawa et al. (2000) reported that Novozym 435 preincubated in methyl oleate and subsequently in soybean oil could enhance the rate of FAME production through impregnation of these compounds into the enzyme support material. In this study, it was assumed that the enzyme could contain methyl ester residues when incubated in WFO and therefore, the treatment should be similar to that carried out by Samukawa et al. (2000).

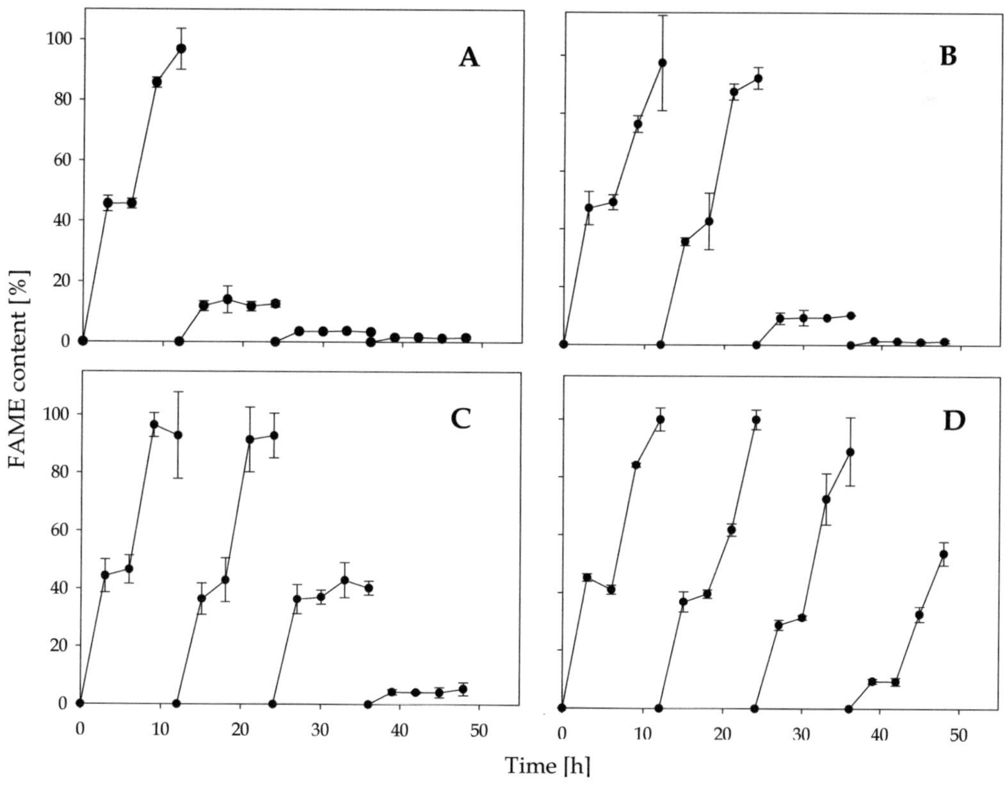

Fig. 5. Comparison of reutilization treatments of Novozym 435 after the reactions to FAME production with a methanol-to-oil ratio of 3.8:1 (mol/mol), 15% (wt) Novozym 435 (based on the weight of oil), 44.5 °C, 200 rpm and 12 hours of reaction time. A) Control, B) Acetone washing, C) Waste frying oil washing D) *tert*-Butanol washing.

This treatment using WFO is advantageous compared to the acetone washing process because it is a cheaper and more environmental friendly process. In addition, the same WFO that was used for the incubation was used subsequently for the reaction of FAME production, and therefore less equipment is needed to carry out this process. In another study, Chen and Wu (2003) achieved a FAME yield five times higher when the enzyme was incubated overnight in soybean oil. However, when they used the same oil in incubation experiments to recover the enzyme after the reaction, the activity began to decay after the fourth reutilization. Although the advantages of using WFO in enzyme recovery, FAME yield also declined in this study, as reported by Chen and Wu (2003). The reason may be the low solubility of alcohol and glycerol in the oil (hydrophobic wash), which impedes the efficient washing of the enzyme by oil.

As the washing process with hydrophobic and hydrophilic solvents did not allow an efficient recovery of the enzyme activity by more than three successive reactions in an anhydrous medium, a moderate polar solvent was also investigated. *Tert*-butanol, a tertiary alcohol with a log P value of 0.35 has been shown to improve enzyme activity more than linear alcohols. This fact could be attributed to the differences in miscibility with triglycerides, as compared to alcohols with the same carbon numbers, branched ones have better miscibility with triglycerides compared to linear isomers (Yu et al., 2010). Therefore, a washing treatment of the enzymes with *tert*-butanol, followed by incubation in WFO, was conducted. The results showed that this enzyme pretreatment achieved the best results, maintaining a higher activity over the time compared to the previous experiments (Fig. 5D). It is likely that *tert*-butanol recovered the enzyme activity because it has the advantages of both hydrophilic and hydrophobic solvents but none of the drawbacks. Therefore, *tert*-butanol should promote the removal of both glycerol and methanol from the lipase support material because of its hydrophilic properties, while its hydrophobic properties should help maintain a high level of lipase activity. This is in accordance to Chen and Wu (2003) who found that *tert*-butanol can be even used to regenerate a deactivated, immobilized enzyme such as Novozym 435. Therefore, *tert*-butanol is a promising alternative for enzyme recovery after reaction in an anhydrous medium, and is also highly stable and less reactive than other butanol isomers. Based on these results, a new experiment was carried out which included *tert*-butanol as a co-solvent in the reaction in an anhydrous medium.

Successive reactions to FAME production in an anhydrous medium using tert-butanol as a co-solvent. A previous study found that *tert*-butanol is inert in the methanolysis system, whereas it is also a potential co-solvent that could maintain enzymatic activity (Li et al., 2006). According to this, the performance of *tert*-butanol as a co-solvent was studied in an anhydrous medium to determine its ability to maintain Novozym 435 activity in successive reactions. In Fig. 6 is shown the successive reactions for FAME production that were carried out in an anhydrous medium under previously optimized operational conditions, using a previously selected dosage of *tert*-butanol (0.75% v/v, data not show). A control without *tert*-butanol as co-solvent was also carried out, showing that FAME yield was drastically reduced in the second cycle (< 20% FAME yield). The best results were obtained in the system using the co-solvent, where FAME yield was maintained over 50% after 17 cycles (Fig. 6). According to similar findings from other studies, the enzyme is inhibited throughout the reaction (Royon et al., 2007). The inhibitory effect at the beginning of the reaction is due to the presence of methanol which has poor miscibility in oil. Subsequently, once methanol concentration decreases, inhibition is caused by a glycerol layer coating the catalyst. Therefore, the positive effect of *tert*-butanol in the reaction could be first related to the increment in the oil-methanol miscibility, preventing direct contact between enzyme and alcohol at the beginning of the reaction and improving the reaction yield. Secondly, due to *tert*-butanol's hydrophilic properties, it is able to dissolve both methanol and glycerol during the reaction, preventing enzyme inhibition throughout the reaction. Thirdly, *tert*-butanol's hydrophobic properties maintained the high lipase activity through the successive reactions.

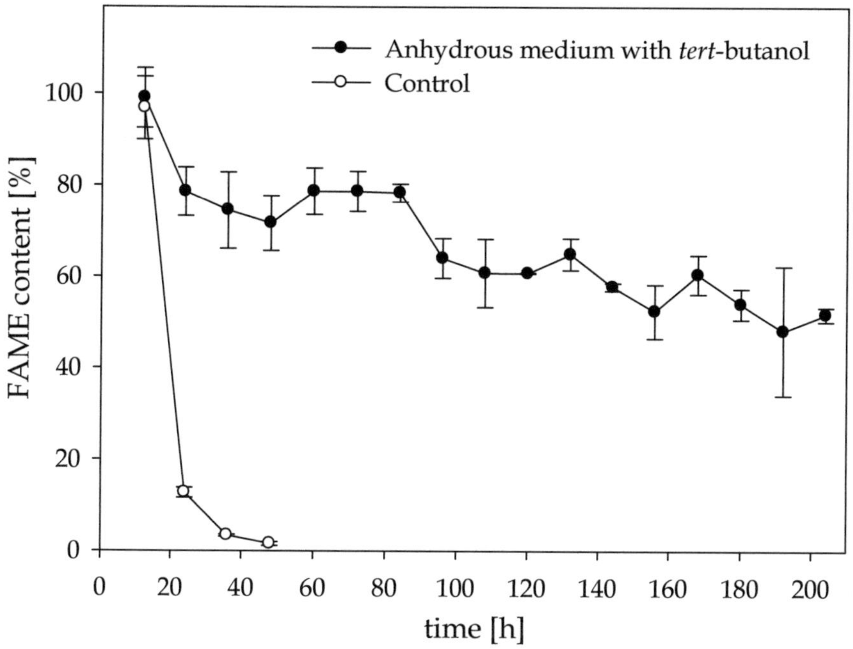

Fig. 6. Time course of FAME yield during 17 batch reaction cycles with a methanol-to-oil ratio of 3.8:1 (mol/mol), 15% (wt) Novozym 435 (based on the weight of oil), 44.5 °C, 0.75% v/v of *tert*-butanol, 0.9 g of molecular sieves, 200 rpm and 12 hours of reaction time. Symbols: in *tert*- butanol system (closed circles), control (open circles).

According to Yu et al. (2010), different properties such as viscosity, dielectric constant, solubility parameters and log P should be considered when choosing a co-solvent to improve enzyme performance in biodiesel production. Another aspect which should be considered is the feasibility of recovering the co-solvent through distillation, which is related to the boiling point. In this sense, *tert*-butanol has advantages compared to others co-solvents reported, such as amyl alcohol, which has very similar properties compared to the *tert*-butanol but a higher boiling point (102 °C) close to the boiling point of the water. As the boiling point of *tert*-butanol is 82 °C and the distillation of methanol occurs at 65 °C, the recovery of *tert*-butanol should not increase the amount of energy spent that much in comparison to the advantages of using this co-solvent. Therefore, employing *tert*-butanol as a co-solvent could allow the use of an anhydrous medium as an industrial alternative for FAME production using Novozym 435 as biocatalyst.

The results of this section showed that the use of an anhydrous medium (resulting from water extraction by using molecular sieves), FAME yield was improved by avoiding hydrolysis and esterification reactions, producing FAME mainly through transesterification of TG. The enzyme activity cannot be recovered successively neither using a hydrophilic washing step with acetone nor by hydrophobic washing with WFO. However, 17 successive cycles of FAME production using *tert*-butanol as a moderate polar co-solvent, show that Novozym 435 can be reused in anhydrous medium. These results also show that the anhydrous medium could enable the implementation of lipase-catalyzed processes on an industrial scale for biodiesel production mainly by transesterification reaction. Different raw materials could be used for this, achieving properties close to the norm and potentially avoiding post-treatments to refine the produced biodiesel.

4. Enzymatic transesterification reaction kinetic in a *tert*-butanol system for biodiesel production

For the design of suitable reactors to FAME production using lipases, kinetic information of the rate of product formation and the effects of changes in system conditions is needed. Investigation about kinetic of FAME production using different lipases has been recently reported. The first researches were focused in the esterification of FFA, where the Ping Pong model with competitive inhibition by methanol was used to describe the reaction (Krishna et al., 2001). After, Al-Zuhair et al. (2007) proposed a kinetic model based in the transesterification of TG. The aim of this work was study the kinetic of the transesterification to FAME production catalyzed by Novozym 435. The investigation was carried out in anhydrous medium, using WFO as raw material, methanol as acyl acceptor and *tert*-butanol as co-solvent. In addition, the set-up of a semi-continuous enzymatic bioreactor was carried out. To realize the study Ping Pong model with competitive inhibition by methanol was used (Eq. 4).

$$\upsilon = \frac{V_{\max}}{1 + \frac{K_W}{[W]}\left[1 + \frac{[M]}{K_{IM}}\right] + \frac{K_M}{[M]}} \tag{4}$$

Where υ (mol/L/min) is the initial reaction rate, $V_{máx}$ is the maximum rate of reaction (mol/L/min), K_W and K_M are the binding constants for the WFO (W) and the methanol (M) (mol/L), [W] is WFO concentration (mol/L), [M] is methanol concentration (mol/L) and K_{IM} is the inhibition constant for the methanol (mol/L).

To determinate the sole effect of methanol in the transesterification, the experiments was run using methanol concentrations in the range of 100-3000 mol/L, equivalent at methanol to oil molar ratio 0.6-15 (mol/mol), at a constant WFO concentration of 300 mol/L. The WFO concentration was chosen to ensure operating outside WFO limitation or inhibition region. To determinate the sole effect of WFO in the transesterification , the experiments was run using WFO concentrations in the range of 200-350 mol/L, at a constant methanol concentration of 1400 mol/L. The methanol concentration was chosen to ensure operating outside alcohol limitation or inhibition region. Michaelis-menten kinetic was used to estimate the kinetic constants Vmáx, K_W y K_M. Subsequently, the values were optimized and the kinetic constant K_{IM} was obtained using Excel solver to find the minimum objective function that compares the measured rate of the reaction with those predicted by the proposed kinetic equation. Each assay were performed in quadruplicate in Erlenmeyer flask of 25mL and incubated in an orbital shaker during 4 hr under the same operational condition: 15% (wt) of Novozym 435, 44°C, 0.75% (V/V) of *tert*-butanol, 0.5 g of molecular sieves and 200 rpm. In order to remove the *tert*-butanol, the samples were centrifuged and heated at 85°C during 30 minutes. The supernatant was used to quantify FAME yield using GC-MS methods.

Using the experimental results the kinetic parameters estimated for the model were: $V_{máx}$= 0.018 mol/L/min, $K_{M,\ metanol}$ = 1030 mol/L, $K_{W,\ WFO}$= 397 mol/L y $K_{IM,\ metanol}$= 1,815 mol/L. The Fig. 7 shows a comparison between the model predictions and the experimental data of the initial rate of reaction at different methanol concentration using Novozym 435.

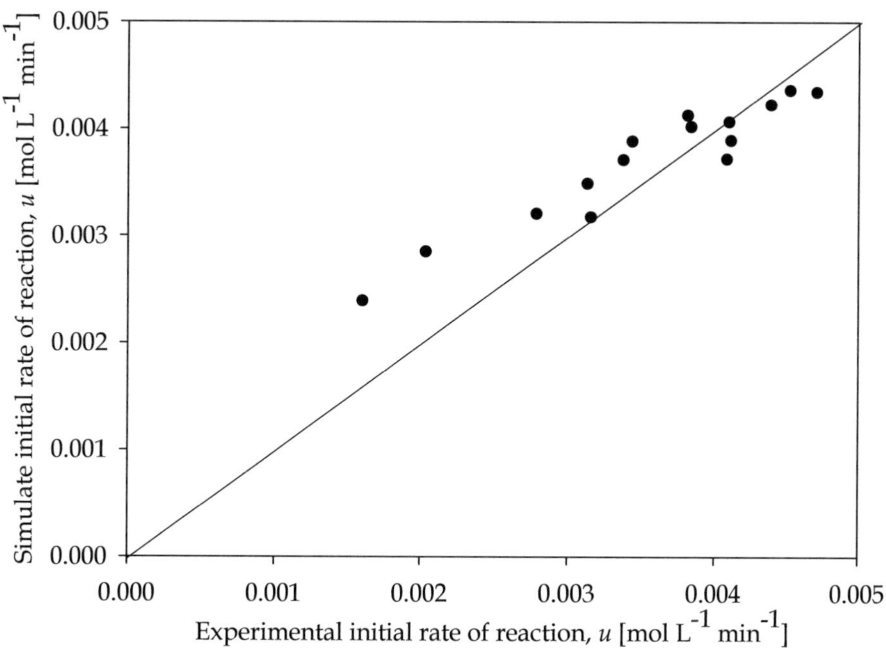

Fig. 7. Comparison between the experimental results and the Ping-Pong kinetic model equation with the estimated constants to different initial methanol concentration and initial WFO concentration of 300 mol L^{-1}.

According to Fig.7 the kinetic model is suitable to predict the behavior of the reaction, indicating that the use Michaelis-Mentel kinectic could simplify the calculation of kinectic constant of complex model such as Ping Pong kinetic model. The model and empirical data showed that Novozym 435 presented a low inhibition by methanol, even at higher concentration maintaining a high initial reaction rate until a methanol to WFO molar ratio of 8/1 (mol/mol). The model predicts a moderate decrease in the initial reaction rate for higher concentration of methanol. This behavior could be associated to the incorporation of the co-solvent in the reaction due its capacity to improve the miscibility of reaction mix. This is an advantage because higher concentration of methanol favor the production formation since the transesterification is an equilibrium reaction, decreasing the operation time of the FAME production using enzymatic catalysts.

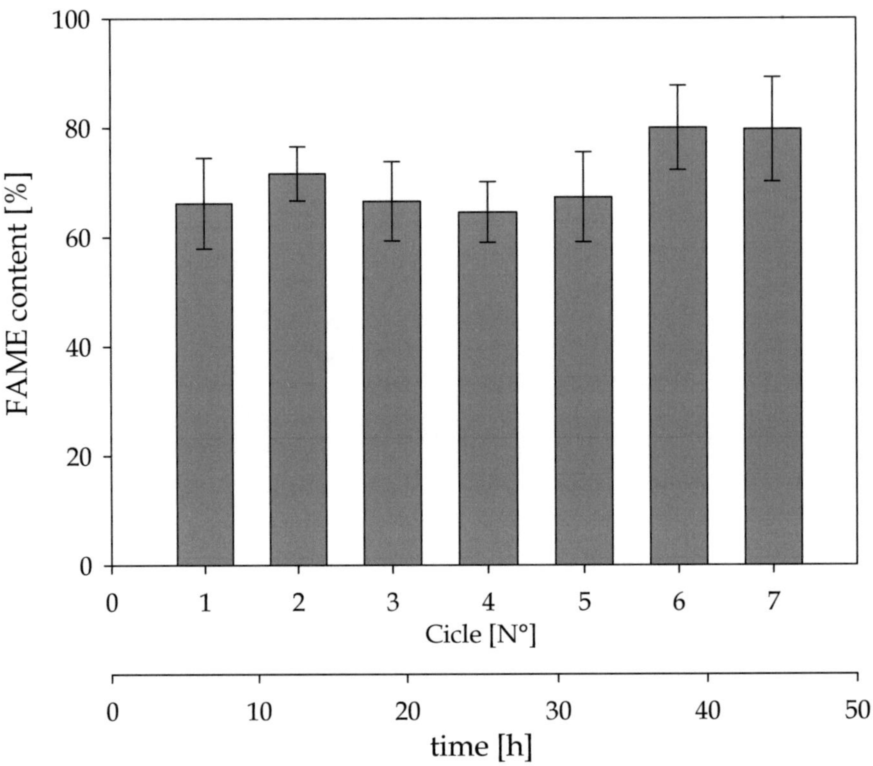

Fig. 8. FAME yield during the set-up of a semi-continuous enzymatic reactor in anhydrous medium with enzymes reutilization. Operational conditions: methanol-to-oil ratio of 8/1 (mol/mol), 15% (wt) Novozym 435 (based on the weight of oil), 44.5 °C, 0.75% v/v of *tert*-butanol, 200 rpm and 4 hours of reaction time.

The results of the kinetic study were applied in the operation of a semi-continuous reactor to enzymatic FAME production. The bioreactor was made of glass with a volume of 0.5 L. The temperature was controller by a thermostat. The agitation of the bioreactor was supply by a magnetic stirrer. In order to extract water constantly from media, a glass column filled with molecular sieves was connected to the bioreactor, where the media was recycling using a peristaltic pump. The glycerol was removed from bottom by a settler connected to the bioreactor. The operational conditions to each reaction cycle were: methanol to oil molar ratio 8/1 (mol/mol), 15 % (wt) Novozym 435, 0.75% (v/v) of *tert*-butanol, 44.5 °C, 200 rpm and 4 hours of reaction time. The enzymes were reused successively remaining into the reactor during all the cycles. In Fig. 9 are showed the results obtained.

The bioreactor was operated several cycles of charge and discharge during 30 h, adding only raw material and changing the molecular sieves to maintain an anhydrous media. Under these conditions was possible to maintain FAME yields over 80% during 7 reaction cycles (Fig. 8). Therefore, the use of both an anhydrous media and *tert*-butanol as co-solvent are effective strategies for the implementation of a semi-continuous enzymatic process that can become economically competitive with traditional chemical catalysts to produce FAME.

5. Conclusions

The short reaction time of the proposed process and the reuse of the enzymes generate a feasible alternative to be implemented at industrial scale. This process has several advantages compared to the chemical process. Although the chemical process requires only 1 hour of reaction time, after the process a washing step to remove catalyst residues is necessary. This washing step is not necessary in the enzymatic process because Novozym 435 is a solid catalyst. In addition, the enzymatic process has the advantage to be flexible allowing the use of alternative and low cost raw materials. Therefore, the results obtained in this work have generated an enzymatic process that could become not only environmentally friendly but also economically competitive with the chemical-catalyzed process.

The kinetic study and the reactor operation showed that Novozym 435 was not inhibited to high methanol concentrations and could be reused without significant loss of activity. These results shown that under the conditions investigated enzymatic FAME production could be a competitive process to be implemented to industrial scale. To reach this objective is necessary to carry on the bioreactor operation in the long time, in order to establish the lifespan of the enzymes. In addition is necessary to establish a molecular sieves recuperation protocol and to looking for inexpensive material such as zeolites to decrease the operational cost.

6. Acknowledgement

Research described was supported by Desert Bioenergy S.A., Chilean project "Inserción de Capital Humano Avanzado en el Sector Productivo Chileno 78110106", Chilean Fondecyt project 3120171, Chilean CONICYT Project 79090009 and FONDECYT iniciación project 11110282.

7. References

Al-Zuhair, S., Wei, F. & Jun, L. (2007). Proposed kinetic mechanism of the production of biodiesel from palm oil using lipase. *Process Biochem*, 42, 951-960.

American Society for Testing and Materials, ASTM (2008): *Annual Book of ASTM Standards: Section 5 - Petroleum Products, Lubricants, and Fossil Fuels*. ASTM International, West Conshohocken.

Araújo, J. (1995). Oxidação de Lipidios, p. 1- 64. *In* Imprensa Universitária (ed.), Química de alimentos: Teoría e prática. Universidad Federal de Viçosa. Viçosa.

Azócar, L., Ciudad, G., Heipieper, H., Muñoz, R. & Navia, R. (2010a). Improving fatty acid methyl ester production yield in a lipase-catalyzed process using waste frying oils as feedstock. *J Biosci Bioeng*, 109, 609-614.

Azócar, L., Ciudad, G., Heipieper, H. & Navia, R. (2010b). Biotechnological processes for biodiesel production using alternative oils. *Appl Microbiol Biotechnol*, 88, 621–636.

Cabrera, Z., Fernandez-Lorente, G., Fernandez-Lafuente, R., Palomo, J. & Guisan, J. (2009). Novozym 435 displays very different selectivity compared to lipase from *Candida antarctica* B adsorbed on other hydrophobic supports. *J Mol Catal B: Enzym*, 57, 171–176.

Canakci, M. (2007). The potential of restaurant waste lipids as biodiesel feedstocks. *Bioresour Technol*, 98, 183–190.

Chen, J. & Wu, W. (2003). Regeneration of immobilized *Candida antarctica* lipase for transesterification. *J Biosci Bioeng*, 95, 466-469.

Dizge, N., Aydiner, C., Imer, D. Y., Bayramoglu, M., Tanriseven, A. & Keskinlera, B. (2009). Biodiesel production from sunflower, soybean, and waste cooking oils by transesterification using lipase immobilized onto a novel microporous polymer. *Bioresour Technol*, 100, 1983-1991.

Du, W., Xu, Y., Liu, D. & Zeng, J. (2004). Comparative study on lipase-catalyzed transformation of soybean oil for biodiesel production with different acyl acceptors. *J Mol Catal B: Enzym*, 30, 125-129.

Fukuda, H., Hama, S., Tamalampudi, S. & Noda, H. (2008). Whole-cell biocatalysts for biodiesel fuel production. *Trends Biotechnol*, 26, 668-673.

Fukuda, H., Kondo, A. & Noda, H. (2001). Review: Biodiesel fuel production by transesterification of oils. *J Biosci Bioeng*, 92, 405-416.

Gui, M., Lee, K. & Bhatia, S. (2008). Feasibility of edible oil vs. non-edible oil vs. waste edible oil as biodiesel feedstock. *Energy*, 33, 1646-1653.

Halim, S. & Kamaruddin, A. (2008). Catalytic studies of lipase on FAME production from waste cooking palm oil in a *tert*-butanol system. *Process Biochem*, 43, 1436–1439.

Hernandez-Martin, E. & Otero, C. (2008). Different enzyme requirements for the synthesis of biodiesel: Novozym (R) 435 and Lipozyme (R) TL IM. *Bioresour Technol*, 99, 277-286.

Issariyakul, T., Kulkarni, M., Meher, L., Dalai, A. & Bakhshi, N. (2008). Biodiesel production from mixtures of canola oil and used cooking oil. *Chem Eng J*, 140, 77-85.

Kose, O., Tuter, M. & Aksoy, H. (2002). Immobilized *Candida antarctica* lipase-catalyzed alcoholysis of cotton seed oil in a solvent-free medium. *Bioresour. Technol.*, 83, 125-129.

Krishna, S. H. & Karanth, N. G. (2001). Lipase-catalyzed synthesis of isoamyl butyrate - A kinetic study. *Biochimica Et Biophysica Acta-Protein Structure and Molecular Enzymology*, 1547, 262-267.

Li, L., Du, W., Liu, D., Wang, L. & Li, Z. (2006). Lipase-catalyzed transesterification of rapeseed oils for biodiesel production with a novel organic solvent as the reaction medium. *J Mol Catal B: Enzym*, 58-62.

Li, N., Zong, M. & Wu, H. (2009). Highly efficient transformation of waste oil to biodiesel by immobilized lipase from *Penicillium expansum*. *Process Biochem*, 44, 685-688.

Li, W., Du, W. & Liu, D. (2007). *Rhizopus oryzae* IFO 4697 whole cell-catalyzed methanolysis of crude and acidified rapeseed oils for biodiesel production in *tert*-butanol system. *Process Biochem*, 42, 1481-1485.

Lin, C. & Lin, H. (2007). Engine performance and emission characteristics of a three-phase emulsion of biodiesel produced by peroxidation. *Fuel Process. Tech.*, 88, 35-41.

Ognjanovic, N., Bezbradica, D. & Knezevic-Jugovic, Z. (2009). Enzymatic conversion of sunflower oil to biodiesel in a solvent-free system: Process optimization and the immobilized system stability. *Bioresour Technol*, 100, 5146-5154.

Royon, D., Daz, M., Ellenrieder, G. & Locatelli, S. (2007). Enzymatic production of biodiesel from cotton seed oil using *t*-butanol as a solvent. *Bioresour Technol*, 98, 648–653.

Salis, A., Pinna, M., Monduzzi, M. & Solinas, V. (2005). Biodiesel production from triolein and short chain alcohols through biocatalysis. *J Biotechnol*, 119, 291-299.

Samukawa, T., Kaieda, M., Matsumoto, T., Ban, K., Kondo, A., Shimada, Y., Noda, H. & Fukuda, H. (2000). Pretreatment of immobilized *Candida antarctica* lipase for biodiesel fuel production from plant oil. *J Biosci Bioeng*, 90, 180-183.

Shimada, Y., Watanabe, Y., Sugihara, A. & Tominaga, Y. (2002). Enzymatic alcoholysis for biodiesel fuel production and application of the reaction to oil processing. *J Mol Catal B: Enzym*, 17, 133-142.

Tamalampudi, S., Talukder, M., Hama, S., Numata, T., Kondo, A. & Fukuda, H. (2008). Enzymatic production of biodiesel from Jatropha oil: A comparative study of immobilized-whole cell and commercial lipases as a biocatalyst. *Biochem Eng J*, 39, 185-189.

The British Standards Institution (2003): BS EN 14214:2003. *Automotive fuels. Fatty acid methylesters (FAME) for diesel engines*. Requirements and test methods, BSI, London.

Turkan, A. & Kalay, S. (2006). Monitoring lipase-catalyzed methanolysis of sunflower oil by reversed-phase high-performance liquid chromatography: Elucidation of the mechanisms of lipases. *J. Chromatogr. A*, 1127, 34-44.

Wang, L., Du, W., Liu, D., Li, L. & Dai, N. (2006). Lipase-catalyzed biodiesel production from soybean oil deodorizer distillate with absorbent present in *tert*-butanol system. *J Mol Catal B: Enzym*, 43, 29-32.

Watanabe, Y., Shimada, Y., Sugihara, A. & Tominaga, Y. (2001). Enzymatic conversion of waste edible oil to biodiesel fuel in a fixed-bed bioreactor. *J Am Oil Chem Soc,* 78, 703-707.

Yu, D., Tian, L., Wu, H., Wang, S., Wang, Y., Ma, D. & Fang, X. (2010). Ultrasonic irradiation with vibration for biodiesel production from soybean oil by Novozym 435. *Process Biochem,* 45, 519–525.

10

Urease Inhibition

Muhammad Raza Shah[1] and Zahid Hussain Soomro[2]
[1]*International Center for Chemical and Biological Sciences, H.E.J. Research Institute of Chemistry, University of Karachi*
[2]*Institute of Materials Science and Research Dawood College of Engineering and Technology Karachi*
Pakistan

1. Introduction

The design, synthesis, characterization and exploring the broad spectrum of potency of biologically active molecular building blocks have attracted the attention of scientific community since last couple of decades. The emergence of pathogenic resistance is a natural phenomenon and investigations toward the advances of new structurally diverse inhibitors have always been at the esteem of pharmaceutical research. This growing field has ever become an active area of research for the synthetic, biological and medicinal perspectives, looking at the molecular level. Enzyme inhibition has attracted great attention of biomedical scientist since last couple of decades. A variety of inhibitors have been discovered and used for the control of various diseases. One of the key features of enzyme is its selectivity towards certain inhibitors; the inhibitor can be a simple organic molecule or complex molecular architecture. The specificity of inhibitor depends on the size, shape and the interactive forces which result in the exact matching of inhibitor and the enzyme. These inhibitions block the activity of the enzyme under the physiological conditions. Urea and urease are the landmark molecules in the early days of synthetic organic chemistry. Urea was the first organic molecule synthesized in laboratory and urease was being first crystallized out from jack bean (Amtul et al., 2002; Hagar et al., 1925; Mobley et al., 1989; Amtul et al., 2006; Zonia et al., 1995). The significance of the urease enzyme is critically investigated due to its capacity to serve as a virulence factor of infections in the urinary and gastrointestinal tracts, maintaining equilibrium in the nitrogen cycles of nitrogenous wastes in the rumens of domestic livestock, and its importance in environmental transformations of nitrogenous compounds, including urea based fertilizers. The physiology of microbial ureases depends on their cellular location, regulation, genetic makeup and the relationships between the microbial urease and the well-characterized potent range of inhibitors. The urease (urea amido hydrolase; EC 3.5.1.5) is a nickel containing enzyme which hydrolyzes urea into ammonia and carbamic acid, the later compound produce ammonia and carbon dioxide on further decomposition. The hydrolysis of the urea yields high concentrations of ammonia which accompanying in pH elevation. Urease has been regarded as the important enzyme for the manipulation of urea and other nitrogenous ingredients in the biological, agricultural and environmental fields. Urease occurs in many plants, selected fungi, and a wide variety of prokaryotes. This enzyme is having important negative implications in medicine, agriculture and environment. The urease inhibitors can play a vital role to counter effect the negative role of urease in living organisms. Urease inhibitors are effective against several serious infections caused by the secretion of urease by Helicobactor pylori which includes gastric tract syndromes, proteus related species in the urinary tract, struvite urolithiasis mainly in dogs and cats. The urease inhibitors have also received great attention from scientists in the soil sciences whereas the nutrient miss-management is associated to the excessive use of synthetic fertilizer and excess of urea products. Research on urease inhibitions yielded several vital therapeutic drugs (Amtul et al., 2002). Hagar and Magath have reported in 1925 that urease is the primary source for the biochemical formation of stone in urinary tract (Hagar et al., 1925). Urease serves as virulence factor in the pathogens responsible for kidney stones entailed in urolithiasis that contributes toward the acute pyelonephritis with other urinary tract infection which induced arthritis and gastric intestinal infections and ultimately the urease imbalance lead to peptic ulcers (Mobley et al., 1989). Urease controls the nitrogen contents in the physiological systems, while it provides a protection mechanism against predators and phytphathogenic organisms (Amtul et al., 2006; Zonia et al., 1995). The high concentrations of ammonia arising from the biochemical reactions of urease, as well as pH elevation result into important negative implications in medicine and agriculture (Zonia et al., 1995; Collins et al., 1993; Montecucco et al 2001; Zhengping et al). High concentrations of ureases cause significant environmental and economic problems by releasing abnormally large amounts of ammonia into the atmosphere during urea fertilization. It induces plant damage primarily by depriving plants of their essential nutrients and secondarily by ammonia toxicity which result in pH increase of the soil (Bremner et al., 1995). The urease inhibitors have played a vital role in the controlling the Helicobacter pylori which caused an imbalance of ammonia levels in cirrhotic patients (Zullo et al., 1998). The urease based liquid crystal-sensors are also used for the detection of heavy metals; these urease based sensors are useful in monitoring the inhibition of enzymatic activities that has been widely investigated for the detection of heavy metals (HMs) (Verma et al., 2005). In many cases, after inhibition of immobilized enzymes with HMs, chelating agents such as EDTA are usually used to regenerate the active site of the enzyme (Bracka et al., 2000). Urease-based sensors, which are inexpensive and sensitive to HMs, have been widely exploited in the scientific community (Volotovsky et al., 1997). Several analytical methods for monitoring HMs are based on the enzyme-catalyzed hydrolysis of urea into ammonia and carbon dioxide, as the consumption of urease inhibitors is vital for the production of grains over the long-term, meanwhile the coating of urea to reduce ammonia volatilization loss from urea fertilizer is gaining great attention of researchers in order to control the depletion of active ingredients in the soil. The use of fertilizer is very much common in the areas where the active ingredients do not occur in sufficient quantities. The nitrogen losses can be reduced in these situations and the use of urease inhibitor is applied to the fertilizer to maintain the optimum pH of the soil (Schwedt et aj., 1993; Saboury et al., 2010). Recently, the LCs has been doped with functional molecules that have been used to develop novel types of sensors. There are series of compounds with 4-pentyl-biphenyl-4-carboxylic acid (PBA), which contains pH-sensitive functional groups could be used in an LC-based pH sensor to monitor small amounts of H^+ released from enzymatic reactions, especially in solutions with a high buffer capacity (Bi et al., 2009). Literature is rich on the role *of H. pylori which* is the main cause for the gastric infections. Urease is a prominent antigen of the *H. Pylori* which has served as a powerful immunogen for this organism. There have been various reports on large number of patients who have shown gastritis significantly elevated response in immunoglobulins G and A along with urease in the blood serum samples with relative comparison of pre-infected levels. Range of enzymatic assays is reported to measure the immune responses and these are used for the diagnostic purposes in monitoring the antibiotic therapy of *H-Pylori* for the epidemiological studies (Amtul et al., 2002).

2. Inhibition mechanism

The inhibition mechanism is mainly based on the binding of the substrate and enzyme. The urease has a bi metallic centric structure which binds with range of molecules depending on the kinetics, dynamics, mechanism of action and the possible secondary interactions involved in this phenomenon. The mechanism of inhibition can be categorized on the basis of inhibitor, either as irreversible inhibitors or the reversible inhibitor where as the later one is further classified into two categories based on the binding with enzyme either in a competitive manner or in a non competitive manner. The process involved in case of reversible inhibitors mostly occur through non covalent interactions such as hydrogen bonding, hydrophobic interaction and orientation of inhibitor and enzyme in an organized fashion (Amtul et al., 2002). The irreversible inhibitors interact through its functional groups with amino acids in the active site, irreversibly e.g. the nerve gases and pesticides, which contain an organophosphorus that bind with serine residues in the enzyme acetylcholine esterase. The reversible inhibitors are of two types either as competitive or non-competitive. The competitive inhibitors compete with the substrate molecules for the active site while the inhibitor's action is proportional to its concentration that resembles with the substrate's structure closely. The non-competitive inhibitors are not influenced by the concentration of the substrate. It inhibits by binding irreversibly to the enzyme but not at the active site. The non-competitor inhibitor follows the allosteric sites in the enzyme for the binding (Figure 2.1). The reversible inhibitors can be washed out of the solution of enzyme by dialysis.

Fig. 1. Model representation of the binding of the enzyme active site and inhibitor.

In the competitive inhibition the inhibitor directly binds to the active site of the enzyme *via* a specific host and guest complex, while in the non-competitive mechanism the substrate binds somewhere else and not specifically on the active site in the enzyme this binding of substrate affect the structure of the enzyme and induce inhibition. Furthermore another mode of binding of inhibitor is the un-competitive mechanism, in this mechanism the inhibitor irreversibly binds with the enzyme and result an intermediate complex (E-I*) between the enzyme and the inhibitor. This portion of enzyme is known as allosteric site (Figure 1). The binding of enzyme with the substrate occurs in an organized fashion and the substrates have specific affinity toward the enzyme. Though the active site is a specific region in the enzyme and this specificity depends on the pro-non covalent interaction available for interaction in the inhibitors. These non covalent interactions between the enzyme and inhibitor are the main feature for the chemoselectivity of the substrate and enzymes during formation of the complex.

These interactions mainly provide a surface to regenerate the enzyme for the reaction cycle as a catalyst. The unique geometrical shapes and topology of the active site is complimentary for the substrate molecule, which fit like lock and key arrangement. The enzymes specifically react with only few compounds of the similar structural features, these structural feature in the inhibitor is known as pharmacohores, while the potency of the inhibitor is managed by appending various functional groups to the main skeleton of the inhibitor by making target oriented derivative containing these features and consequently resulted in the controlled target oriented synthesis of the potent and fascinating inhibitors. As there are so many theories which explain the mechanistic perspective of the enzymatic action, only the 'lock and key' and the 'induced fit' theories are well accepted (Donald et al., 2005). The mechanistic studies of the biological systems undergoes *via* a complex phenomenon, a large number of literature reports are available involving the structure-based design and testing of novel pharmacophores model for the recognition of urease inhibitors. The intimated details of the molecular geometry of urease enzyme as well as its mechanism of urea hydrolysis have been confirmed with XRD-analysis.

2.1 Kinetics of enzyme activity

The rate of reaction provide a clue about the affinity of the inhibitor towards the enzyme, while at the same time it also effect the potency of the inhibitor. The kinetic effect of rebeprazole has been investigated on the urease inhibition, it is reported that rebeprazole act as an irreversible noncompetitive inhibitor. Mainly the associated inhibitory potency of rabeprazole is dependent on the pH of reaction mixture and K_i value which varies at the progressive inactivation of urease by rabeprazole, initially it proceeds according to pseudo-first-order kinetics with respect to the remaining enzymatic activity at pH 7.0 and 37 °C, with a second-order rate constant of 0.0017 μM^{-1} s^{-1}. This inhibitor competes with substrate to bind to the enzyme and form as an E-I (enzyme and inhibitor) complex, that slowly transformed to an E-I* complex which is stable (Scheme 1) (Park et al., 1996).

The use of non-steady-state kinetic measurements is important to properly analyze the non competitive inhibition (Mobley et al., 1989; Rosenstein et al, 1984). A number of mechanistic studies have been carried out on various inhibitors. It is reported that hydroxamic acid which suggests that a bidentate complex form with one of the nickel ions, quite few scientists

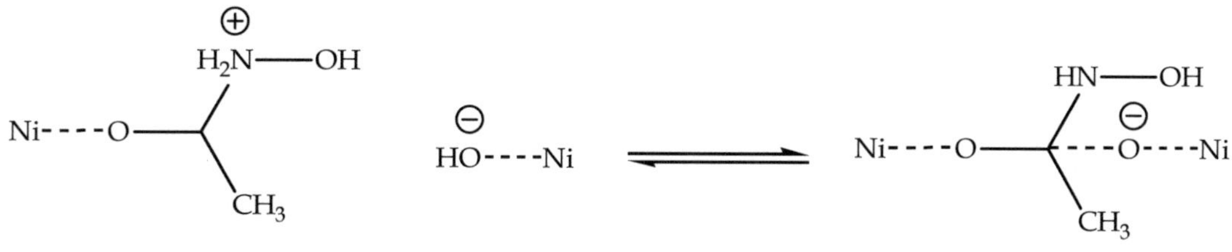

E.I E.I*

E.I = Enzyme-Inhibitor Complex
E.I* = Stable Enzyme-Inhibitor Complex

Scheme 1. Bi-dentate ligand complex of substrate and enzyme.

believe that it make a bridge between the two nickle ions. Similarly it is also suggested that the E-I and E-I* type of interactions occurs, forming an initial monodentate species that bridges the two nickel ions at the active site (Amtul et al., 2002). The non-steady-state studies have also been employed with other competitive inhibitors, the specific examples for the kinetic analysis is the interaction of phenylphosphoro diamidate with urease (*K. aerogenes*) with dissociation constant of 4.7×10^{-5} s^{-1}. The proposed studies suggests that the E-I state where an inhibitor bound in unidentate mode to one nickel ion while the EI* species involves the bridging of the two nickel ion with inhibitor where R is the aromatic ring *via* the formation of tetrahedral intermediate rather that the hypothetical trigonal bipyramidyl intermediate. Here the hydrolysis occurs with simple in-line displacement (Andrews et al., 1984).

2.2 Action of inhibition

The modes of action of enzyme have been investigated on the thiol based compounds. As the presence of an auxochrome sulphar (thio) moiety in the inhibitor help in characterization *via* UV-Visible spectroscopy for a variety of enzymes *e. g* jack bean and *K aerogenes* ureases, these spectroscopic perturbation is consistent with the development of thiolate. It is well documented that the Ni (II) charge-transfer complex is kinetically controlled process. Since a competition of thiol with urea for the active sites of the enzyme in binding the nickel provide a relative reference to further study the action of inhibitor over the enzyme.

A B

Scheme 2. Structural hypothesis for the competitive interaction of the urease active sites with thiourea (A) and hydroxyurea (B).

The mechanism of action related to hydrolysis of urease to ammonia and carbamate has been investigated in depth, the later compound spontaneously decomposes to the carbamic acid, accompanying the deprotonation generating equilibrium as a result of increase in pH. The research on urease has contributed extensively towards development of medicinal organic chemistry (Summer et al., 1926) and crystallization (Amtul et al., 2002). The urea binds into the pocked adjacent to Ni near Wat-502. The H2-H4 mobile flap may open to allow access into the active site of enzyme, once the active functional group binds to Ni-1 *via* coordination. The urea is then placed in a manner that allows the coordination of its oxygen with the N epsilon of His 219, by forming hydrogen bond. The urea also accepts a hydrogen bond from the interaction of its amide nitrogen with side chain of His 320. These non covalent interactions are oriented to urea in an organized fashion so that the carbonyl group of the urea undergoes polarization and attacked by hydroxide (Wat-502), that coordinates other metal atom of Ni-2 and forms tetrahedral intermediate. As one proton is transferred from His-320 to the nitrogen (leaving group), meanwhile the intermediate collapses into ammonia and carbamate. The carbamate undergoes hydrolysis to form carbamic acid and second molecule of ammonia (Amtul et al., 2002). The suboptimal interactions provide a significant source of substrate binding energy; this energy is commonly found in the enzymes with mobile active site loops which undergo induced fit. The side chain of His 320 in urease is in close proximity to the side chain of Asp 221 and Arg 336, whereas the geometry is not optimal for the hydrogen bonding between the residues. His 320 also interacts with Wat-170 *via* hydrogen bonding. The Wat-170 also forms hydrogen bond with two main chain carbonyls. Wat-170 does not have the ability to donate three hydrogen bonds, so one of the H-bond at any given time is unable to form, making the interactions of water suboptimal. There are also suboptimal interactions between the nickel metallo center and solvent. The active site water molecules (Wat-500, Wat, 501, Wat-502) have inter-oxygen distance that is too short to allow for simultaneous occupancy. Since the occupancy of the active site is too small to accommodate all three water molecules, there appears a competition for occupancy at the three positions. All these interactions must contribute to the flexibility of the mobile flap by the destabilizing the closed positions. These interactions are the main driving force for the catalytic mechanism of the urease in the various biological processes. The most of the biochemical reactions occurs by decreasing the free energy difference between the unbound enzyme with substrate and enzyme-substrate complex which accomplished in two ways with enzyme binding transition state which is highly favorable or the initial state which has higher free energy, possibly the addition of suboptimal interactions. Such interaction makes possible transition state which usually is not favorable, therefore most of the urease mechanisms involve a number of suboptimal interactions to derive the hydrolysis of urea in the systems (Mobley et al., 1989; Mulvaney et al., 1981).

3. Potent inhibitors

A vast range of naturally isolated and synthetic urease inhibitors has been reported in the literature though pharmacological importance and the developed resistance in the bacterial infections have attracted the attention of the scientific community to explore new versatile urease inhibitors.

Urease inhibitors are classified broadly into two major classes as metal-organic and organic compounds, the later examples are such as acetohydroxamic acid, humic acid and 1,4-

benzoquinone etc. The inhibitors have broad applications in combination with urea fertilizers that has been proposed for soil urease activity control (Bundy et al, 1973; Amtul et al., 2002). There is an example of kinetic studies on quinone-induced inhibition of urease has also been investigated, one among those is polyhalogenated benzoand naphthoquinones, in which the quinones were found to be non-competitive inhibitors of jack bean and bacterial ureases of different strengths (Zaborska et al., 2002; Pearson et al., 1997). Several other potent inhibitors are being used as first line of treatment for infections causes by urease-producing bacteria which are available in current market. The most effective inhibitors with safe great potency profile for considerable control of urease-related ailments are still needed. The available data provide a basis for urease inhibition properties of thiol-compounds, different cysteine derivatives (CysDs) with cysteine-like scaffold arylidene barbiturates have been studied as urease inhibitors such as *N*-Substituted Hydroxyureas (Amtul et al., 2002; Amtul et al., 2004; Todd et al., 1989). Acetohydroxamates are the bacterial urease inhibitor and used as therapeutic potential in hyperammonaemic states (Carlini et al., 2002). The urease inhibitors have also been employed for the coating of urea in the agricultural soils to reduce the ammonia mitigating. The nitrogen loss depends on the rate of biodegradation and constant volatilization in agricultural soils of urea fertilizer, such coating can improve the bioavailability of nitrogen and consequently increase dry matter yield and nitrogen uptake. These increases of urea contents in soil result from delayed urea hydrolysis by urease inhibitors and coating materials. The value of inhibitors in nitrogen mitigating is depending on the rate of biodegradation and persistence in soils (Amtul et al., 2007). The urease inhibitors have been investigated for the treatment of bacterial infections and also for the excessive urea break down in soil, therefore studies on the potent and specific inhibitors have been an active area of research. Taking in view the great potential of urease in medicine and soil sciences, several classes of urease inhibitor have been discovered in the recent years (Ara et al., 2007; Mobley et al., 1989; Mulvaney et al, 1981; Rosenstein et al., 1984). The urease inhibitors can also be classified into two categories either as metal complexes or organic compounds. The organic compounds are further classified into the three main classes such as hydroxamic acid analogs, phosphoramide compounds and thiourea derivatives. Hydroxamic acid (Scheme 3) analogs are the most commonly known urease inhibitors and first member of its series was hydroxamic acid which was discovered in 1962. The acetohydraxamic acid with other numerous derivatives have been synthesized and studied against ureases of plants and bacterial origin (Mobley et al., 1989). The other members of the series include *n*-aliphatic hydroxamic acid, ortho or para substituted benzohydroxamic acid, *trans*-cinnamoyl hydroxamic acid with other functional groups and substituents.

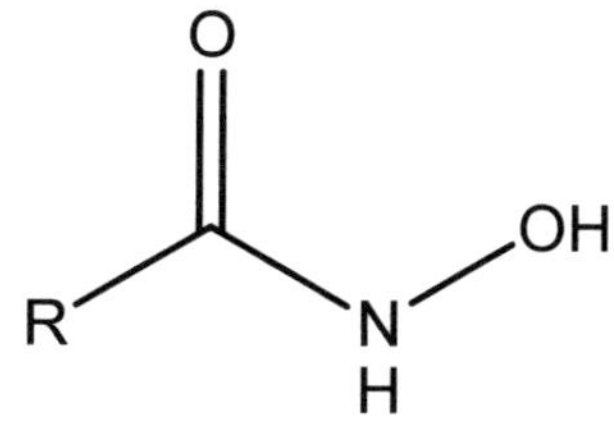

Scheme 3. General structure of hydroxamic acid.

Phosphoramide compounds are another class of potent inhibitors of the urease enzyme, this class of inhibitors comprise of a large number of simple and complex compounds. Mainly the less complex examples includes of phosphoramidates and diamidophos phate with substituted phenylphosphoradiamidates, while a range of N-acyl phosphoric triamidates (Humayun et al., 2010).

Thio substituted compounds are also strong inhibitors of the urease enzyme. This includes cysteamine containing a cationic β-amino cysteamine, which exhibit highest affinity for the urease, on the other hand the thiolates containing anionic carboxyl groups are uniformly poor inhibitors. The pH dependence studies demonstrate that the actual inhibitor is the thiolate anion (Ansari et al., 2005). A series of bezothiazepines have also been investigated bearing significant urease inhibitory activities (Kühler et al., 1998). The studies on structure activity relationship of the thio substituted benzimidazoles with *in-vitro* and *in-vivo* efficacy model, where the potency of the substrate is depend on the substituents positions in the pharmacophore (Creason et al., 1990). The inhibitory effects of *N-(n-*butyl) thiophosphora-triamides (NBPT) and its oxon analogs in the soil are also been reported (Zaborska et al., 2004). There are several other potent inhibitors readily available in the market the most common is NBPT (N-[n-butyl] thiophosphoratriamide) which is sold worldwide under the trade name of Agrotain. This fertilizer has amended the coats of urea preventing the urease enzyme from breaking down the urea for up to fourteen days of life. As this increases the probability that urea will be absorbed into the soil after a rain rather than volatilized into the atmosphere which reduce the nitrogen content in the soil. The rich content of urea in soils causes hydroxylation under the surface of soil and decreases atmospheric losses of the major ingredients of fertilizers. The use of potent inhibitors also decreases the localized zones of high pH which is common in untreated urea agricultural fields. These remedies are commonly employed to maintain the pH of the soil and controlled ammonia volatilization. The use of inhibitors increases the efficiency of urea in delivering nitrogen to soil.

A wide range of metal complexes with meager to good urease inhibitory activities are reported which includes heavy metal ions, such as Cu^{2+}, Zn^{2+}, Pd^{2+} and Cd^{2+} (Zaborska et al., 2001; Asato et al., 1997). There are reports on the urease inhibitory activities of organotin(IV), vanadium(IV), bismuth(III), copper(II), and cadmium(II) complexes (Khan et al., 2007; Ara et al., 2007; Hou et al., 2008; You et al., 2008; Tanaka et al., 2003; Cheng et al., 2007; You et al., 2010). A series of metal complexes of Schiff bases such as thiocyanato-coordinated manganese(III) complexes based on N-ethylethane-1,2-diamine and N-(2-hydroxyethyl)ethane-1,2-diamine have shown significant urease (Huang et al., 2011). There is library of compounds are available in various databases and the recent developments is continuously adding numerous class of of hetero-dinuclear CuII–ZnII complexes (You et al., 2011).

4. Structure Activity Relations (SAR)

The hydroxamic is the most commonly used urease inhibitor coordinating to nickel of urease through carboxylic acid and hydroxyl amine (Scheme 3). The structural studies on the urease from *Klebsiella aerogenes, Bacillus pasteurii,* and *Helicobacter pylori* have revealed that it contains a dinuclear Ni active site with a modified amino acid side chain-containing a carbamylated lysine residue that bridges the deeply buried metal atoms as shown in Scheme 3 (Benini et al., 1999; Jabri et al., 1995; Ha et al., 2001). The crystal structure has been resolved only for bacterial ureases; it has an active centre which contains two simple coordinated water molecules and a bridging OH group. The specificity of the enzyme is closely related to the shape of its active centre (Mobley et al., 1995).

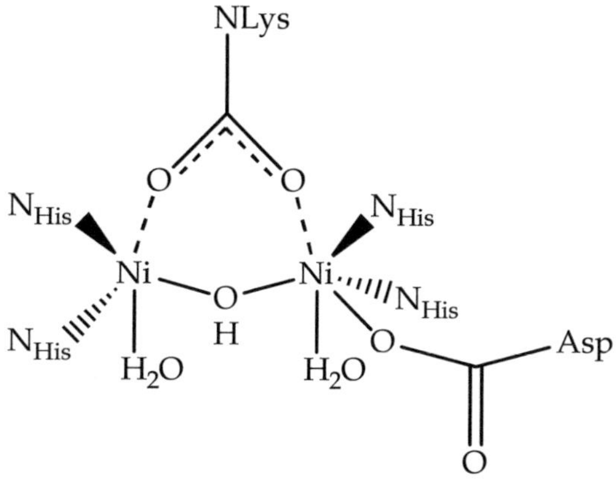

Scheme 4. Active center of urease.

There is vast reports on the structure of urease, the *Klebsiella areogenes,* the urease structure has a trimer of alpha, beta and gamma units in a triangular arrangement with three active sites per enzyme. These subunits include a 60.3 KD apha, a 11.7 KD beta and a 11.1 KD gamma subunit (Jabr et al., 1992). The interactions between the individual subunits of all three α subunits make extensive contacts with each other to build a trimer. Each alpha subunit packs between two beta and two gamma subunits to form the side of the triangle arrangement (Jabri et al., 1995). Each beta subunit packs between the two of the adjacent alpha subunits at each corners of the triangle. The remaining gamma subunit interacts with two gamma subunits at the crystallographic three fold axis. Because of the high degree of interaction, it is not obvious that which alpha, beta and gamma chains make up the primary units. High conservation of sequences and the extensive interactions of the trimer suggest a similar trimer structure in all the urease which is further subdivided into sub-units. The gamma subunit consists of an alpha-beta domain with four helices and two anti parallel stands. Two helices (b and c) and the two strands pack together with right-handed up-down-up-down topological order. One of the remaining helices (a) is lined up peripherally and interacts at the three fold axis with the other two gamma subunits, tightly packing against itself and to other (a) helices in an orthogonal manner. The last helix (d) experience single turn. The first fifteen residues of the beta subunits form two antiparallels beta sheets with the amino-terminal residues of the alpha subunits. The core of the beta subunit forms an imperfect six-stranded antiparallel beta sheet (Mobley et al., 1989; Hausinger et al., 1993). This subunit stabilizes the trimer by interacting with domain two of its own alpha subunits and domain one of a symmetry related alpha subunit. The alpha subunit consists of two domains, an (alpha, beta) 8 barrel domain and a primarily beta domain. The (alpha, beta) 8 barrel is the only domain that contributes to the active site. Its barrel is rather elliptical with the long axis connecting strands 1 and 7. Strands 9 and 10 extend the lower portion of strands 1 and 3. The active site is located at the carboxyl terminal of the strands. A flap is formed that covers the active site by a helical excursion between strand 7 and helix 7. Mixed four-strands and eight-stranded beta sheets form the wall of U-Shaped canyon which makes up the second beta domain. The walls are connected by the long strands 5 and 6 together with strands 10 and 11 which go down on one side, and up on the other side. The structure activity relationship of the various urease inhibitors are reported in literature, the studies on *in-vitro* with *in-vivo* efficacy of the effect benzimidazoles suggested that the orientation and position of the pyridyl moiety larger and more lipophillic chromophore of the substrate has affinity which affect the minimum bacterial concentration of the helicobacter pylori (Kühler et al., 1998).

5. Conclusion

The chapter is focused on the importance and future implications of the urease. The emphases have been laid on urease keeping in mind its role as virulence factor for the urinary tract infections and gastrointestinal infections while the inhibition of urease is also effective in the agricultural sector. Urease enzyme inhibition has become a growing area of research at the interface of the biomedical sciences. The broad range of potent inhibitors has been reported in literature from simple organic molecules to complex molecular building block with metal complexes. These molecular building blocks are expected to bring revolution in pharmaceutical, agricultural and environmental fields. Urease based sensors, adsorbent and other devices have also been discussed that are commercially used.

6. References

Amtul Z, Follmer, C., Mahboob, S., Atta, Ur. R., Mazhar, M., Khan K. M., Siddiqui, R. A., Muhammad, S., Kazmi, S.A., Choudhary, M. I. (2007) Germa-gamma-lactones as novel inhibitors of bacterial urease activity, *Biochem Biophys Res Commun.*, Vol 4, No. 356 (2) pp 457-63.

Amtul, Z., Kausar, N., Follmer, C., Rozmahel, R.F.,Atta, Ur, R., Kazmi, S. A., Shekhani, M., S., Jason L., Khan Khalid M., Choudhary, M. I. (2006) Cysteine based novel noncompetitive inhibitors of urease(s)-Distinctive inhibition susceptibility of microbial and plant ureases, *Bioorganic and Medicinal Chemistry*, Vol 14, pp 6737-6744.

Amtul, Z., Rasheed, M., Choudhary, M. I., Rosanna, S., Khan, K. M., Atta, Ur. R. (2004). Kinetics of novel competitive inhibitors of urease enzymes by a focused library of oxadiazoles/thiadiazoles and triazoles, *Biochem. Biophys. Res. Commun.*, Vol 319, pp 1053-1063.

Amtul, Z. Atta, Ur. R., Siddiqui, R.A., Choudhary, M. I. (2002). Chemistry and mechanism of urease inhibition. *Current Medicinal Chemistry*, Vol 9, pp 1323-1348.

Andrews, R. K., R. L. Blackeley, B. Zerner, (1984) Urea and urease, *Adv. Inorg. Biochem.* Vol 6, pp 245-283.

Ansari, F. L., Umbreen, S., Hussain, L., Makhmoor, T., Nawaz, S. A., Lodhi, M. A., Khan, S. H., Shaheen, F., Choudhary, M I. Atta, Ur. R. (2005) Bioassay models for antidiabetic activity of natural products, *Chemistry & Biodiversity*, Vol 2, No. (4), pp 487-496.

Ara, R., Ashiq, U., Tahir, M. M., Maqsood, Z.T., Khan, K.M., Lodhi, M.A., Choudhary, M.I. (2007) Chemistry, urease inhibition, and phytotoxic studies of binuclear vanadium(IV) complexes. *Chem. Biodivers.* Vol 4, pp 58-71.

Asato, E., Akamine, K. Y., Fukami, T., Nukada, R., Mikuriya, Deguchi, M. S., Yokota, Y. (1997) Bismuth(III) complexes of 2-mercaptoethanol: Preparation, structural and spectroscopic characterization, antibactericidal activity toward Helicobacter pylori , and inhibitory effect toward H. pylori -produced urease, *Bull. Chem. Soc. Jpn.* Vol 70, pp 639-648.

Benini, S., Rypniewski, W. R., Wilson, K. S., Miletti, S., Ciurli, S., Mangani, S. (1999) A new proposal for urease mechanism based on the crystal structures of the native and inhibited enzyme from Bacillus pasteurii: why urea hydrolysis costs two nickels", *Structure*, Vol 7, pp 205-216.

Bremner, J. M. *Fertilizer Research* (1995) Recent research on problems in the use of urea as a nitrogen fertilizer, Vol 42, pp 321-329.

Bracka, W., Paschke, A., Segner, H., Wennrich, R., Schuurmanna, G. (2000) Urease inhibition: a tool for toxicity identification in sediment elutriates, *Chemosphere*, Vol 40, pp 829-834.

Bi, X. Hartono, D., Yang, K.L. (2009) Real-time liquid crystal pH Sensor for monitoring enzymatic activities of penicillinase, *Advance Functional Material*, Vol 19, pp 3760-3765.

Bundy, L. G.; Bremner, J. M. (1973) Effects of substituted *p*-benzoquinones on urease activity in soils, *Soil Biol. Biochem.* Vol 5, No. 6, pp 847-853.

Carlini, C. R., Grossi-de-Sa, M. F. *Toxicon* (2002) Plant toxic proteins with insecticidal properties. A review on their potentialities as bioinsecticides, Vol 40, pp 1515-1539.

Creason, G.L., Schmitt, M.R., Douglass, E.A., Hendrickson , L.L. (1990) Urease inhibitory activity associated with *N*-(n-butyl)thiophosphoric triamide is due to formation of its oxon analog, Soil Biology and Biochemistry, Vol 22(2), pp 209-211.

Cheng, K., You, Z.-L., Zhu, H.-L. (2007) New method for the synthesis of a mononucleating cyclic peptide ligand, crystal structures of its Ni, Zn, Cu, and Co complexes, and their Inhibitory bioactivity against urease, *Aust. J. Chem.* Vol 60 pp 375-379.

Donald, V., and Judith, G. V. (2005)

Hagar, B. H, Magrath, T. B. (1925) The etiology of incrusted cystitis with alkaline urine, Journal of the American Medical Association, Vol 185, pp 1353-1358.

Ha, N. C., Oh, S. T., Sung, J. Y., Cha, K. A., Lee, M. H., Oh, B. H. (2001) The crystal structure of urease from Klebsiella aerogenes, *Nat. Struct. Biol*, Vol 8, pp 505-509.

Hausinger, R. P. (1993) Biochemistry of Nickel, *Plenum Press, New York*, pp 23-57.

Hou, P., You, Z.-L., Zhang, L., Ma, X.-L., Ni, L.-L. (2008) Synthesis, characterization, and DNA-binding properties of copper(II), cobalt(II), and nickel(II) complexes with salicylaldehyde 2-phenylquinoline-4-carboylhydrazone, *Transition Met. Chem.* Vol 33, pp 267-273.

Huang, C. Y., Shi, D. H., You, Z. L. (2011) Unprecedented preparation of *bis*-Schiff bases and their manganese(III) complexes with urease inhibitory activity, *Inorganic Chemistry Communications*, Vol 14, No. 10, pp 1636-1639.

Humayun, P., Nazia, M., Muhammad, Y., Ajmal, K, Khan, K. M., Faiz-ul-Hassan, N., Choudhary, M. I. (2010) Letters in Drug Design & Discovery, Vol 7, No. 2, pp 102-108.

Jabri, E., Carr, M. B., Hausinger, R. P., Karplus, P. A. (1995) The crystal structure of urease from Klebsiella aerogenes , *Science*, Vol 268, pp 998-1004.

Jabr, E., M. H. Lee, R. P. Haisomger., P.A. Karplus. (1992) Preliminary crystallographic studies of urease from jack bean and from Klebsiella aerogenes, *J. Mol. Biol*, Vol 227, pp934-937.

Khan, M.I., Baloch, M.K., Ashfaq, M. (2007) Spectral analysis and in vitro cytotoxicity profiles of novel organotin(IV) esters of 2-maleimidopropanoic acid, *J. Enzym. Inhib. Med. Chem.*, Vol 22, pp343-350

Kühler, T. C., Swanson, M., Shcherbuchin, V., Larsson, Håkan., Mellgård, B., Sjöström, JE. (1998) Structure-Activity Relationship of 2-[[(2-Pyridyl)methyl]thio]-1H-benzimidazoles as anti-helicobacter pylori agents in vitro and evaluation of their in vivo efficacy, *J. Med. Chem.*, Vol 41, No. 11, pp 1777-1788.

Mobley, H. L., Island, M. D., Hausinger, R. P. (1995) Molecular biology of microbial urease, *Microbiol. Rev.*, Vol 59, pp 451-480.

Mobley, H. L. T., Hausinger, R. P., (1989) Microbial urease: Significance, regulation andmolecular characterization *Microbiol Rev*, Vol 53, pp 85-108.

Montecucco, C., Rappuoli, R., (2001) Living dangerously: how Helicobacter pylori survives in the human stomach, *Mol. Microbiol. Nat. Rev. Mol. Cell Biol.* No. 2, No. 6, 457-466.

Mulvaney, R. L., Bremner, J. M. (1981) Control of urea transformations in soils, *Soil Biochem.* Vol 5, pp 153-196.

Park, J.B., Imamura, L., Kobashi, K. (1996) *Biological Pharmaceutical Bulletin,* Vol 19, No. 2, pp 182-187.

Pearson, M.A., Michel, L.O., Hausinger, R.P., Karplus, P.A. (1997) Structures of Cys319 variants and acetohydroxamate-inhibited *Klebsiella aerogenes* urease, *Biochemistry,* Vol 36, pp 8164-8172.

Rosenstein, I. J. M., Hamilton-Miller, J. M. T. (1984) Inhibitors of urease as chemotherapeutic agents, *Crit. Rev., Microbiol,* Vol 11, pp 1-12.

Saboury, A.A., Poorakbar-Esfahani, E., Rezaei-Behbehani, G. (2010) *J. Sci. I. R. Iran,* A thermodynamic study of the interaction between urease and copper ions, Vol 21, No. (1), pp 15-20.

Schwedt, G., Waldheim, D.O., Neumann, K.D., Stein, K. *J. Fresenius, Analytical Chemistry* (1993) Trace analysis and speciation of copper by application of an urease reactor, Vol 346, No. 6-9, pp 659-662.

Summer, J. B. (1926) The isolation and crystallization of the enzyme of the urease, *J. Biol. Chem.,* Vol 69, pp 435-441.

Tanaka, T.,Kawase, M., Tani, S. (2003) Urease inhibitory activity of simple α,β-unsaturated ketones, *Life Sci.* Vol 73, No. 23, pp 2985-2990.

Todd, M. J., Hausinger, R. P. (1989) Competitive Inhibitors of Klebsiella aerogenes Urease,*J. Biol. Chem.,* Vol 264, No. 27, No. 15835-15842.

Verma, N., Singh, M. (2005) Biosensors for heavy metals, *BioMetals,* Vol 18, pp 121-129.

Volotovsky, V., Nam, Y.J., Kim, N., (1997) Urease-based biosensor for mercuric ion determination. *Sensors 1. Actuators B,* Vol 42, pp 233-237.

You, Z. L.,Lu, Y., Zhang, N., Ding, B. W., Sun, H., Hou, P., Wang C. (2011) Preparation and structural characterization of hetero-dinuclear Schiff base copper(II)–zinc(II) complexes and their inhibition studies on Helicobacter pylori urease, *Polyhedron,* Vol 30, pp 2186-2194.

You, Z.-L., Ni, L.-L., Shi, D. H., Bai, S. (2010) Synthesis, structures, and urease inhibitory activities of three copper(II) and zinc(II) complexes with 2-{[2-(2-hydroxyethylamino) ethylimino]methyl}-4-nitrophenol, *Eur. J. Med. Chem.* Vol 45, pp 3196-3199.

You, Z.-L., Han, X., Zhang, G.-N. Z. (2008) Synthesis, crystal structures, and urease inhibitory activities of three novel thiocyanato-bridged polynuclear schiff base cadmium(II) complexes, *Anorg. Allg. Chem.* Vol 634, pp 142-146.

Zaborska, W., Krajewska, B., Olech, Z., (2004) Heavy metal ions inhibition of jack bean urease: potential for rapid contaminant probing, *J. Enzym. Inhib. Med. Chem.* Vol 19, No. 1, pp 65-69.

Zaborska, W., Kot, M., Superata, K. (2002) Evaluation of the *inhibition* mechanism, *J. Enzym. Inhib. Med. Chem.* Vol 17, pp 247-253.

Zaborska, W., Krajewska, B., Leszko, M., Olech, Z. (2001) Inhibition of urease by Ni2+ ions: Analysis of reaction progress curves, *J. Mol. Catal. B: Enzym.,* Vol 13, pp 103-108.

Zonia, L.E., Stebbins, N. E., Polacco, J. C. (1995) Essential Role of Urease in Germination of Nitrogen-Limited Arabidopsis thaliana Seeds, *Plant Physiology,* Vol 107, 1097-1103.

Zhengping, W., Cleemput, O. Van, Demeyer, P., Baert, L. (1991) Effect of urease inhibitors on urea hydrolysis and ammonia volatilization, *Biological. Fertilizer. Soils,* Vol 11, No. 1, pp 43-47.

Zullo, A., Rinaldi, V., Folino, S., Diana, F., Attili, A. F. (1998) Helicobacter pylori Urease Inhibition and Ammonia Levels in Cirrhotic Patients, *The American Journal of Gastroenterology,* Vol 93, pp 851-852.